M.G. Minty F. Zimmermann

Measurement and Control of Charged Particle Beams

With 172 Figures

Dr. Michiko G. Minty
DESY - MDE
Notkestrasse 85
22607 Hamburg
Germany
E-mail: michiko.minty@desy.de

Dr. Frank Zimmermann
CERN, AB Division, ABP Group
1211 Geneva 23
Switzerland
E-mail: frank.zimmermann@cern.ch

A catalog record for this book is available from the Library of Congress.

Bibliographic information published by Die Deutsche Bibliothek.
Die Deutsche Bibliothek lists this publication in the Deutsche Nationalbibliografie; detailed bibliographic data is available in the Internet at http://dnb.ddb.de.

The original version of this book was revised.
The correction to this book can be found at https://doi.org/10.1007/978-3-662-08581-3_13

We dedicate this book to the memory of Prof. Dr. Bjorn Wiik,
whose charismatic and visionary leadership continues to guide us
towards new directions in accelerator physics.

Preface

The intent of this book is to bridge the link between experimental observations and theoretical principles in accelerator physics. The methods and concepts, taken primarily from high energy accelerators, have for the most part already been presented in internal reports and proceedings of accelerator conferences, a portion of which has appeared in refereed journals. In this book we have tried to coherently organize this material so as to be useful to designers and operators in the commissioning and operation of particle accelerators.

A point of emphasis has been to provide, wherever possible, experimental data to illustrate the particular concept under discussion. Of the data presented, most are collected from presently existing or past accelerators and we regret the problem of providing original data some of which appear in less accessible publications – for possible omissions we apologize. Regarding the uniformity of the text, particularly with respect to symbol definitions, we have taken the liberty to edit certain representations of the data while trying to maintain the essence of the presented observations. Throughout the text we have attempted to provide references which are readily available for the reader.

In this monologue we describe practical methods for measuring and manipulating various beam properties, and illustrate these concepts with many examples, which are taken from our working experience at CERN, DESY, SLAC, IUCF, KEK, LBNL, FNAL, and other laboratories. In Chaps. 2, 3, 4, 7 and 8 we discuss a present various techniques which can be employed to verify or correct the transverse and longitudinal optics, to optimize the beam orbit, and to measure or vary the beam emittances. Other chapters are devoted to special topics, such as transverse manipulations in photoinjectors (Chap. 5), beam collimation (Chap. 6), polarization (Chap. 9), injection and extraction (Chap. 10), and beam cooling (Chap. 11). Some basic knowledge of accelerator physics is a necessary prerequisite for following the material presented.

This monologue results from many years of practice in accelerator physics and from teaching at various particle accelerator schools. We are grateful to our many students for their enthusiasm and especially for their interesting ideas and questions. We express our gratitude to Prof. S.Y. Lee, former organizer of the United States Particle Accelerator Schools, for suggesting and encouraging this work.

We thank most gratefully our mentors and colleagues with whom we had the pleasure to work or who have supported our professional carreers, including Chris Adolphsen, Ron Akre, Gianluigi Arduini, Ralph Assmann, Karl Bane, Desmond Barber, Walter Barry, Martin Breidenbach, Reinhard Brinkmann, Karl Brown, David Burke, John Byrd, Yunhai Cai, John Cameron, Alex Chao, Ernest Courant, Martin Donald, Frank-Josef Decker, Martin Donald, Jonathan Dorfan, Don Edwards, Helen Edwards, Paul Emma, Alan Fischer, Etienne Forest, John Fox, Joseph Frisch, Alexander Gamp, Hitoshi Hayano, Sam Heifets, Linda Hendrickson, Thomas Himel, Georg Hoffstätter, Albert Hofmann, John Irwin, Keith Jobe, Witold Kozanecki, Wilhelm Kriens, Alan D. Krisch, Kiyoshi Kubo, S.Y. Lee, Gregory Loew, Douglas McCormick, Lia Merminga, Phil Morton, Steve Myers, Yuri Nosochkov, Katsunobu Oide, Toshiyuki Okugi, Ewan Paterson, Nan Phinney, Robert Pollock, Pantaleo Raimondi, Ina Reichel, Tor Raubenheimer, Burton Richter, Robert Rimmer, Thomas Roser, Marc Ross, Francesco Ruggiero, Giovanni Rumolo, Ron Ruth, Shogo Sakanaka, Matthew Sands, Frank Schmidt, Peter Schmüser, John Seeman, Mike Seidel, Robert Siemann, William Spence, Christoph Steier, Gennady Stupakov, Mike Sullivan, Nobu Terunuma, Dieter Trines, James Turner, Junji Urakawa, Albrecht Wagner, Nick Walker, David Whittum, Helmut Wiedemann, Uli Wienands, Bjorn Wiik, Ferdinand Willeke, Perry Wilson, Mark Woodley, Yiton Yan, and Michael Zisman.

We would especially like to thank our colleagues who have gratuitously contributed to the examples and figures presented in this book. Last but not least, we also thank our editor Dr. Christian Caron and his team from Springer Verlag including Gabriele Hakuba, Sandra Thoms, and Peggy Glauch for their patience, continuous encouragement, and valuable help.

Hamburg and Geneva,
April 2003

Michiko G. Minty
Frank Zimmermann

Contents

Symbols

Constants

a	0.0011596	anomalous part of the electron magnetic moment
c	2.9979×10^{8} m/s	speed of light in vacuum
C_d	2.1×10^{3} m^2 GeV^{-3} s^{-1}	
C_q	3.84×10^{-1}	
C_Q	$\approx 2\times10^{-11}$ m^2 GeV^{-5}	
e	1.6×10^{-19} C	electric charge
G	1.79285	anomalous part of the proton magnetic moment
h	6.626075×10^{-34} J s	Planck's constant
m_e	511 keV/c	electron mass
m_p	928.28 MeV/c	protron mass
N_A	6.0221×10^{23} mol^{-1}	Avogadro's number
r_e	2.817940×10^{-15} m	classical electron radius
μ_B	5.78838×10^{-11} MeV T^{-1}	Bohr magneton

Frequent Abbreviations

BNS	damping named after Balakin, Novokhatsky, and Smirnov
BPM	Beam Position Monitor
CCS	Chromatic Correction Section
DF	Dispersion-Free
DFS	Dispersion-Free Steering
Drift	Drift space (a field-free region)
FEL	Free Electron Laser
FF	Final Focus
IP	Interaction Point
Linac	Linear accelerator
OTM	One Turn Map
Quad	Quadrupole magnet (QF focusing, QD defocusing)
SASE	Self-Amplified Spontaneous Emission
SVD	Singular Value Decomposition

Acronyms of Accelerator Facilities and Projects

AGS	Alternating Gradient Synchrotron at BNL
ALS	Advanced Light Source at LBNL
ANL	Argonne National Laboratory in Chicago
APS	Advanced Photon Source at ANL
ASSET	Accelerator Structure Setup Facility at SLAC
ATF (BNL)	Accelerator Test Facility at BNL
ATF (KEK)	Accelerator Test Facility at KEK
BEPC	Beijing Positron Electron Collider
BNL	Brookhaven National Laboratory on Long Island
CERN	European Organization for Nuclear Research in Geneva
CESR	Cornell Electron Storage Ring
CLIC	Compact Linear Collider
CTF	CLIC Test Facility
DESY	Deutsches Elektronen-Synchrotron in Hamburg
ESRF	European Synchrotron Radiation Facility in Grenoble
FFTB	Final Focus Test Beam at SLAC
FNAL	Fermi National Accelerator Laboratory near Chicago
HERA	Hadron-Elektron Ring-Anlage at DESY
IUCF	Indiana University Cyclotron Facility
ISR	Intersecting Storage Rings at CERN
JLC	Japanese or Joint Linear Collider
KEK	High Energy Accelerator Research Organizatioin Tsukuba
KEKB	KEK B factory
LBNL	Lawrence Berkeley National Laboratory in Berkeley
LEP	Large Electron Positron Collider at CERN
LHC	Large Hadron Collider under construction at CERN
NLC	Next Linear Collider
NLCTA	NLC Test Accelerator
PEP	Proton-Electron-Positron Project at SLAC
PEP-II	SLAC B factory
PETRA	Positron-Elektron Tandem Ring-Anlage
PS	Proton Synchrotron at CERN
Recycler	permanent magnet antiproton ring at FNAL
RHIC	Relativistic Heave Ion Collider
SLAC	Stanford Linear Accelerator Center near San Francisco
SLC	SLAC Linear Collider
SPEAR	Stanford Positron Electron Accelerating Ring
SPring-8	third generation synchrotron radiation facility in Japan
SPS	Super Proton Synchrotron at CERN
TESLA	Tera Electron Volt Energy Superconducting Linear Accelerator
Tevatron	TeV proton collider at FNAL
TRISTAN	former electron-positron collider at KEK
TTF	TESLA Test Facility
ZGS	Zero Gradient Synchrotron at ANL

Alphanumeric Symbols

A	atomic mass in units of the proton mass $[m_p]$
$B_{x,y,z}$	transverse and longitudinal magnetic fields [T]
B_r, B_ϕ	radial and angular components of magnetic field [T]
$B_\perp, B_\parallel$	components of magnetic fields perpendicular and parallel to the particle velocity [T]
$\mathrm{B}_{\mathrm{mag}}$	mismatch parameter [1]
C	circumference of a circular accelerator [m]
D(s)	dispersion function [m]
$\mathcal{D}_{x,y}$	Sands' number for total guide field configuration [1]
E	particle energy [GeV]
$E_{x,y,z}$	transverse and longitudinal electric fields [V/m]
E_r, E_ϕ	radial and angular components of electric field [V/m]
f	quadrupole focal length [m]
f_{coll}	average bunch collision frequency in a collider [Hz]
f_{rev}	revolution frequency in a circular accelerator [kHz]
f_{rf}	accelerating rf frequency [MHz]
$f_{x,y}$	transverse betatron frequencies [kHz]
f_s	synchrotron oscillation frequency [Hz]
$F_{x,y,z}$	transverse and longitudinal Lorentz force [N]
F_r, F_ϕ	radial and angular components of Lorentz force [N]
$F(q)$	longitudinal aperture function [1]
g	Lande g-factor [1]
G	curvature function of the design orbit [1/m]
h	harmonic number, $h = \frac{f_{\mathrm{rf}}}{f_{\mathrm{rev}}}$ [1]
H	Hamiltonian [m]
$H_{x,y}$	horizontal/vertical dispersion invariant [m]
$i(t)$	beam current in time domain [A]
$I(\omega)$	beam current in frequency domain [As]
i_{dc}	dc component of beam current [A]
i_b	component of beam current at rf frequency (=$2i_{dc}$) [A]
$I_{x,y}$	action variables [m]
$J_{x,y}$	transverse damping partition numbers [1]
J_ϵ	longitudinal damping partition number [1]
k	normalized quadrupole strength [m^{-2}]
k_h	ratio of voltages of harmonic cavities and accelerating rf [1]
k_l	loss factor [V/C]
$k_{\gamma,b}$	ratio of energies of emitted photons and beam energy [1]

Alphanumeric Symbols, continued

K	integrated quadrupole strength [m^{-1}]
L	superperiod length in a periodic lattice [m]
$\mathcal{L}$	luminosity of a collider [$cm^{-2}s^{-1}$]
m	normalized sextupole strength [m^{-3}]
M	integrated sextupole strength [m^{-2}]
m_x	mass of particle x [GeV/c^2]
$\boldsymbol{n_s}$	stable spin direction [1]
N_{ppb}	number of particles per bunch [1]
N_t	number of turns [1]
p_x, p_y, p_z	components of the particle momentum vector [GeV/c]
q	overvoltage factor [1]
Q	quality factor [1]
Q_x, Q_y	transverse betatron tunes, also called $\nu_{x,y}$ [1]
Q_s	synchrotron tune [1]
$Q_{I,II}$	eigenmodes of betatron oscillations (for coupled systems) [1]
Q'_x, Q'_y	horizontal and vertical chromaticity [1]
R	cavity impedance [Ω]
R_l	loaded cavity impedance [Ω]
$R_{i,j}$	point-to-point transfer matrix from i to j [m, 1, m^{-1}]
s	longitudinal coordinate along beamline [m]
S_x, S_y, S_z	components of the beam polarization [1]
t	time measured in laboratory rest frame [s]
T_{rev}	revolution period [s]
T_{rf}	period of rf acceleration [s]
U_0	energy loss per turn due to synchrotron radiation [eV]
U_{hom}	energy loss per turn due to higher order modes [eV]
V_c	cavity voltage [MV]
$W_\perp, W_\parallel$	transverse and longitudinal components of the wakefields, [m^{-2}] and [m^{-1}]
x, y	horizontal and vertical position coordinates [m]
x', y'	horizontal and vertical angle coordinates [1]
x_{co}, y_{co}	transverse coordinates representing central trajectory offset [m]
x_β, y_β	tranverse coordinates representing offset due to betatron motion [m]
x_δ, y_δ	transverse coordinates representing offset due to energy deviation [m]
x_d, y_d	position offset due to quadrupole misalignment [m]
x_b, y_b	position offset due to BPM electronic offset [m]
$x_m y_m$	measured position offset seen by a BPM [m]
X_0	radiation length, [m] or [m^4/g]
z	longitudinal coordinate (relative to bunch center) [m]
Z	atomic number [1]

Greek and Latin Symbols

α_c	momentum compaction factor [1]
α_p	rate of spin precession [s^{-1}]
α_x, α_y	Twiss parameter, $\alpha = -\frac{1}{2}\frac{d\beta}{ds}$ [1]
β	relativistic velocity factor, $\beta = \frac{v}{c}$ [1]
β_c	cavity coupling parameter [1]
β_x, β_y	Twiss parameter, beta function [m]
$\beta^*_{x,y}$	beta function at a collider interaction point [m]
χ^2	chi-squared parameter used in minimization algorithms [1]
δ	relative momentum deviation from ideal particle [1]
ϵ	strength of depolarizing resonances [1]
$\epsilon_{x,y}$	transverse beam emittance [m rad]
$\epsilon_{x,y,N}$	normalized transverse beam emittance [m rad]
ϵ_z	longitudinal beam emittance, [m rad] or [eV s]
γ	Lorentz factor, $\gamma = \frac{E}{mc^2}$ [1]
γ_x, γ_y	Twiss parameter $(\beta_{x,y}\gamma_{x,y} = (1 + \alpha_{x,y}{}^2))$ [1/m]
γ_t	transition energy [1]
$\kappa_\pm$	coupling parameter [1]
λ	an eigenvalue
λ_{rf}	rf wavelength, $\lambda_{\mathrm{rf}} = 2\pi f_{\mathrm{rf}}$ [m]
μ	nonlinear tune shift with amplitude parameter [1/m]
$\boldsymbol{\mu}$	particle magnetic moment [MeV/T]
$\mu_{x,y}$	phase advance argument, $\mu_{x,y} = 2\pi\phi_{x,y}$ [1]
$\nu_{x,y}$	transverse betatron tunes, also called $Q_{x,y}$ [1]
ν_s	spin tune [1]
ω_r	angular revolution frequency [s^{-1}]
Ω	solid angle [steradian]
$\Omega_{x,y}$	transverse angular betatron frequencies [s^{-1}]
Ω_s	angular synchrotron frequency [s^{-1}]
ϕ_b	phase of beam relative to rf crest [1]
ϕ_l	loading angle [1]
$\phi_{x,y}$	horizontal and vertical betatron phase [1]
ϕ_x	synchronous phase angle [1]
ϕ_z	tuning angle [1]
Ψ	spin wave function [1]

Greek and Latin Symbols, continued

ρ	local bending radius [m]
ρ_w	bending radius in a wiggler magnet [m]
σ	cross section for scattering processes, [m^2] or [barn]
σ_{ij}	ij-th element of the beam matrix, Σ_{beam}, [m^2, m, 1]
σ_δ	rms relative momentum spread [1]
$\sigma_{x,y}$	rms transverse beam sizes [m]
σ_z	rms bunch length [m]
Σ_{beam}	beam matrix
θ	kick angle induced by a corrector magnet [1]
τ	beam lifetime [s]
τ_f	fill time of a structure or cavity [s]
$\tau_{x,y}$	transverse damping times [s]
τ_δ	longitudinal damping time [s]
τ_q	quantum lifetime [s]

1 Introduction

Particle accelerators were originally developed for research in nuclear and high-energy physics for probing the structure of matter. Over the years advances in technology have allowed higher and higher particle energies to be attained thus providing an ever more microscopic probe for understanding elementary particles and their interactions. To achieve maximum benefit from such accelerators, measuring and controlling the parameters of the accelerated particles is essential. This is the subject of this book.

In these applications, an ensemble of charged particles (a 'beam') is accelerated to high energy, and is then either sent onto a fixed target, or collided with another particle beam, usually of opposite charge and moving in the opposite direction. In comparison with the fixed-target experiments, the center-of-mass energy is much higher when colliding two counter-propagating beams. This has motivated the construction of various 'storage-ring' colliders, where particle beams circulate in a ring and collide with each other at one or more dedicated interaction points repeatedly on successive turns. A large number of particles, or a high beam current, is desired in almost all applications. The colliders often require a small spot size at the interaction point to maximize the number of interesting reactions or 'events'.

The charged particles being accelerated are typically electrons, positrons, protons, or antiprotons, but, depending on the application, they can also be ions in different states of charge, or even unstable isotopes. Often the beams consist of several longitudinally separated packages of particles, so-called 'bunches', with empty regions in between. These bunches are formed under the influence of a longitudinal focusing force, usually provided by the high-voltage rf field, which also serves for acceleration.

If the trajectory of a high-energy electron or positron is bent by a magnetic field, it emits energy in the form of synchrotron radiation. The energy loss per turn due to synchrotron radiation increases with the fourth power of the beam energy and decreases only with the inverse of the bending radius. This limits the energy attainable in a ring collider. The maximum energy ever obtained in a circular electron-positron collider – more than 104 GeV per beam – was achieved in the Large Electron Positron Collider (LEP) at the European laboratory CERN in Geneva, Switzerland, with a ring circumference of almost 27 km.

M. G. Minty et al., *Measurement and Control of Charged Particle Beams*,
https://doi.org/10.1007/978-3-662-08581-3_1

The most promising option for accomplishing electron-positron collisions at even higher energy are linear colliders, where the two beams are rapidly accelerated in two linear accelerators ('linacs') and collide only once. In order to obtain a reasonable number of interesting events, the spot sizes at the collision point must be much smaller than those obtained in all previous colliders. Design values for the root-mean-square (rms) vertical spot size at the collision point are in the range 1–6 nm, for center-of-mass energies between 500 GeV and 3 TeV. The one and only high energy linear collider to date is the Stanford Linear Collider (SLC), which was operated from 1988–1998 at Stanford University in California. The SLC collided electrons and positron beams with an energy of about 47 GeV each, and the vertical rms beam size at the collision point varied between 500 nm and 2 μm. In order to be able to achieve the small spot size, the beam must have a high density; e.g., a small emittance.

A positive feature of the synchrotron radiation is that at lower energy it leads to a shrinkage of the beam volume in a storage ring via radiation damping. The beam volume is usually characterized in terms of three emittances, which are proportional to the area in the phase space occupied by the beam for each degree of motion. The radiation damping acts with a typical exponential time constant of a few ms. This damping property is exploited in the linear-collider concept by first producing a high-quality dense beam in a damping ring, at a few GeV energy, prior to its acceleration in a linear accelerator (which consists essentially of a long series of accelerating rf cavities with intermediate transverse focusing by quadrupole magnets of alternating polarity) and subsequent collision.

Synchrotron radiation itself is also used directly for numerous applications in biology, material science, X-ray lithography, e.g., for microchip fabrication, and medicine, to mention a few. Many synchrotron radiation centers have been established all over the world. In these facilities, the photon beam quality depends on the properties of the electron or positron beam stored in the ring, thus placing high demands on the beam quality and trajectory control, similar to those required by the colliders.

Recent developments have demonstrated the possibility to produce substantially (6–7 orders of magnitude) brighter light at even shorter wavelength. These are based on the coherent amplification of photons spontaneously emitted as an extremely dense beam traverses a series of alternating bending magnets with short period (an 'undulator') in a single pass. This concept of a free-electron laser (FEL) based on self-amplified spontaneous emission (SASE) presently draws much attention around the world. While in a conventional light source, the light power increases in proportion to the number of particles, in a SASE FEL it increases in proportion to its square.

There are many other types of accelerators and their uses, not all of which can be covered in detail in this book. Noteworthy are perhaps the ion or pion accelerators which are used for cancer therapy, and of which there are several in operation, e.g., in Canada and Japan.

We also note that, unlike the collider operation, in the preparation of high-intensity proton beams for a fixed target, the emittance is often intentionally diluted, so that the beam fills the entire available aperture. This 'painting' stabilizes the beam and reduces the effect of the beam space-charge forces. Also in this case performance may further be improved by optics corrections and by a more precise knowledge of the beam properties.

In Tables 1.1, 1.2, and 1.3 we list a selection of typical parameters for a few ring colliders, linear colliders, and light sources, respectively.

Table 1.1. Parameters of Storage Ring Colliders

Variable	Symbol	Tristan	PEP-II	KEKB	HERA	LEP	LHC
Species		e^+e^-	e^+e^-	e^+e^-	$pe^\pm$	e^+e^-	pp
Beam energy [GeV]	E_b	30	9, 3.1	8, 3.5	920, 27.5	104	7000
No. of bunches	n_b	2	1658	5000	174	4	2800
Bunch population [10^{10}]	N_b	20	2.7, 5.9	1.4, 3.3	10, 4	40 40	11 11
Rms IP beam size [µm]	$\sigma^*_{x,y}$	300, 8	157, 4.7	90, 1.9	112, 30	250, 3	16
Normalized rms emittance [µm]	$\gamma\epsilon_{x,y}$	6000, 90	400, 15 (e^+)	125, 2.5 (e^+)	5, 1000	8000, 40	3.75
Circumference [km]	C	3.02	2.20	3.02	6.34	26.66	26.66

The reaction rate in a collider, R, is given by the product of the cross section of the reaction σ and the luminosity L:

$$R = \sigma L \,. \tag{1.1}$$

Considering two beams with Gaussian transverse profiles of rms size σ_x (in the horizontal direction) and σ_y (in the vertical direction), with $N_{b,1}$ and $N_{b,2}$ the number of particles per bunch per beam respectively, the luminosity for head-on collisions is expressed by

$$L = \frac{f_{\text{coll}} N_{b,1} N_{b,2}}{4\pi\sigma_x\sigma_y} \,, \tag{1.2}$$

with f_{coll} the average bunch collision frequency. In a storage ring, the number of particles per bunch N_b is related to the total stored current I by

Table 1.2. Parameters of (Planned) Linear Colliders

Variable	Symbol	FFTB	SLC	NLC	TESLA	CLIC
Beam energy [GeV]	E_b	47	47	250	250	1500
No. of bunches / train	n_b	1	1	190	2820	154
Rep. rate [Hz]	n_b	10	120	120	5	100
Bunch population [10^{10}]	N_b	0.5	4	0.75		0.4
Rms IP beam size [nm]	$\sigma^*_{x,y}$	60 (y)	1400, 500	245, 2.7	553, 5	43, 1
Normalized rms emittance [μm]	$\gamma\epsilon_{x,y}$	2 (y)	50, 8	3.6, 0.04	10, 0.03	0.58, 0.02
Luminosity [10^{34} cm^{-2}s^{-1}]	L	—	2×10^{-4}	2	3.4	10

Table 1.3. Parameters of Light Sources and SASE FELs

Variable	Symbol	ALS	ESRF	SPring-8	TTF FEL	TESLA FEL
Beam energy [GeV]	E_b	1.5	6	8	1	15–50
No. of bunches	n_b	300	662	1760	800	11500/ pulse
Bunch population [10^{10}]	N_b	0.5	0.5	0.2	0.6	0.6
Rms beam size [μm]	$\sigma_{x,y}$	200, 31	400, 20	150, 20	50	27
Norm. transv. emittance [μm]	$\gamma\epsilon_{x,y}$	10, 0.7	47, 0.35	94, 0.04	2	1.6
Bunch length [mm]	σ_z	4.0	4.0	4.0	0.05	0.025

$$I = n_t n_b N_b e \frac{f_{\mathrm{rf}}}{h} , \tag{1.3}$$

where n_t is the number of trains, n_b is the number of bunches per train, e is the electric charge, f_{rf} is the accelerating frequency, and h is the harmonic number. The rms beam sizes $\sigma_{x,y}$ are related to the beam volume, or to the emittance, and to a focusing parameter $\beta_{x,y}$, via $\sigma_{x,y} = \sqrt{\epsilon_{x,y}\beta_{x,y}}$. Hence, for a linear collider smaller emittances $\epsilon_{x,y}$ translate into higher luminosity[1]. In (1.2), we have omitted a number of correction factors, which are sometimes

[1] for storage ring colliders this is not necessarily true since $N_b/\epsilon_{x,y}$ may be limited by the beam-beam interaction

important. For example, if the beta functions at the collision point are comparable or smaller than the bunch length, the luminosity is lower than that predicted by (1.2). This is referred to as the 'hourglass effect'. In addition, at high current the focusing force of the opposing beam may significantly change the single-particle optics. As a result, the beta functions at the interaction point either increase or decrease ('dynamic beta function'), and the luminosity changes accordingly.

The parameter describing the photon-beam quality of a synchrotron-radiation light source is the spectral brightness B, which refers to the photon flux in the six-dimensional phase space. Again considering a Gaussian beam, and assuming that the beam sizes are above the photon diffraction limit ($\epsilon_{x,y} > \lambda_\gamma/4\pi$, where λ_γ is the photon wavelength), the average spectral brightness at frequency ω is

$$B(\omega) = \frac{C_\psi E I S(\omega/\omega_c)}{4\pi^2 \epsilon_x \epsilon_y} , \tag{1.4}$$

where E is the beam energy, I the beam current, $C_\psi = 4\alpha/(9em_ec^2) \approx 3.967\times10^{19}$ photons / (sec rad A GeV), where α is the fine structure constant,

$$S(\omega/\omega_c) = \frac{9\sqrt{3}}{8\pi}\frac{\omega}{\omega_c}\int_{\omega/\omega_c}^{\infty} K_{5/3}(\bar{x})\,\mathrm{d}\bar{x} , \tag{1.5}$$

and $\omega_c \equiv (3/2)c\gamma^3/\rho$ the critical photon frequency (where ρ is the bending radius, and γ the electron beam energy divided by the rest energy m_ec^2). The important point is that the average spectral brightness depends strongly on the beam emittance and on the beam current.

In this book we will describe commonly used strategies for the control of charged particle beams and the manipulation of their properties. These are strategies aimed towards improving the accelerator performance and meeting the ever more demanding requirements. Emphasis is placed on relativistic beams in storage rings and linear accelerators. Only one chapter is devoted to problems associated with low energy beams. We assume that the reader is familiar with fundamental accelerator optics as described, for example, in [1, 2, 3, 4]. In the remainder of this introduction we nonetheless review some fundamentals of accelerator optics thereby also introducing the notations to be used in this text. In the following chapters, we discuss basic and advanced methods for measuring and controlling fundamental beam properties, such as transverse and longitudinal lattice diagnostics and matching procedures, orbit correction and steering, beam-based alignment, and linac emittance preservation. Also to be presented are techniques for the manipulation of particle beam properties, including emittance measurement and control, bunch length and energy compression, bunch rotation, changes to the damping partition number, and beam collimation issues. Finally, we discuss a few special topics, such as injection and extraction methods, beam cooling, spin transport, and polarization.

The different techniques are illustrated by examples from various existing or past accelerators, for example, the large electron-positron collider LEP [5] at CERN, the SLAC PEP-II B factory [6], the linac of the KEK B factory [7], the Stanford Linear Collider (SLC) [8, 9], TRISTAN at KEK [10], the synchrotron light sources SPEAR at SLAC [11] and the ALS at Berkeley [12], the CERN Low Energy Antiproton Ring (LEAR) [13], the Accelerator Test Facility (ATF) at KEK [14], the electron-proton collider HERA at DESY [15], the final-focus test beam at SLAC [16], the CERN $p\bar{p}$ collider SPS [17], the ASSET experiment at SLAC [18], the TESLA Test Facility at DESY [19], the FNAL recycler ring [20], RHIC [21], and the ISR at CERN [22]. At various places, we also refer to planned or proposed future accelerators, such as the Large Hadron Collider [23], the Next Linear Collider [24], the TESLA Linear Collider [25], and the Muon Collider [26].

1.1 Review of Transverse Linear Optics

We can distinguish two types of accelerator systems: rings and transport lines both with and without acceleration. In a storage ring the optical functions, such as the dispersion D or the beta function β, are well defined by the periodic boundary conditions. For a transport line, on the other hand, there is no such boundary condition, and here it is convention to determine the initial values of the optical functions from the initial beam size and the correlations contained in the initial beam distribution (see (1.17–1.19)).

Often a 3-dimensional coordinate system (x, s, y) is employed to describe the particle motion, where the local tangent to s points in the direction of the beam line, x is directed in the radial outward direction, and y in the vertical upward direction. These coordinates are illustrated in Fig. 1.1. In a beam line without any bending magnets, or if there is bending in more than one plane, some ambiguity exists in the definition of the x and y. While s gives the location around the ring, the particle coordinates x and y measure the transverse distance from an ideal reference particle, e.g., a particle passing through the center of perfectly aligned quadrupole magnets. Further, it is customary to introduce a second longitudinal coordinate $z = s - v_0 t$ where v_0 denotes the velocity of the ideal particle and t the time. The coordinate z thus measures the longitudinal distance to the ideal reference, which may be taken to be the center of the bunch. For example, if $z > 0$ the particle is moving ahead of the bunch center and arrives earlier in time than the bunch center at an arbitrary reference position.

In a linear approximation, the transverse motion of a single particle in an accelerator can be described as the sum of three components [4, 27]

$$u(s) = u_{\text{c.o.}}(s) + u_\beta(s) + D_u(s)\delta\,, \tag{1.6}$$

where $u(s) = x(s)$ or $y(s)$ is the horizontal or vertical coordinate at the (azimuthal) location s. Here $u_{\text{c.o.}}$ denotes the closed equilibrium orbit (or,

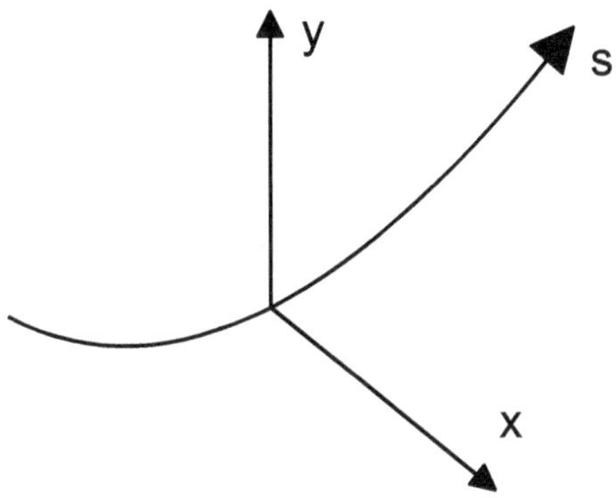

Fig. 1.1. Schematic of the reference trajectory and the transverse coordinate system

in a transport line, some reference trajectory), u_β the orbit variation due to betatron motion (transverse oscillations), and $D_u\delta$ the orbit change resulting from an energy offset; D_u is the dispersion function, and $\delta = \Delta p/p$ is the relative deviation from the design momentum given by the difference of the particle momentum from the design momentum both divided by the design momentum.

The beam is transversely focused by quadrupole magnets usually of alternating polarity. The linear equation of motion for the horizontal motion is

$$\frac{\mathrm{d}^2 x}{\mathrm{d}s^2} = -k(s)x\,, \tag{1.7}$$

where x is the offset from the quadrupole center. The focusing coefficient $k(s)$ is given in units of m^{-2} and is nonzero only in a quadrupole field, in which it is given by

$$k = \frac{B_T}{(B\rho)a}\,, \tag{1.8}$$

where B_T denotes the quadrupole pole-tip field, a the pole-tip radius, and

$$B\rho\ [\mathrm{T-m}] \approx 3.356\ p\ [\mathrm{GeV/c}] \tag{1.9}$$

is the magnetic rigidity in units of Tesla-meters. Often, especially in large acclerators, one can employ a 'thin-lens' or 'kick' approximation, and express the effect of the quadrupole simply by a change in the trajectory slope $x' \equiv dx/ds$ according to

$$\Delta x' = -Kx\,, \tag{1.10}$$

where

$$K \equiv k l_{\mathrm{quad}} \tag{1.11}$$

is the integrated strength of the quadrupole in units of m^{-1} and l_{quad} is the quadrupole length. Here and in the following, we use the prime to signify a derivative with respect to the longitudinal position s.

Note that the strength of other magnets can be normalized to the beam momentum in a similar way as for quadrupoles. As an example, we consider a sextupole magnet. This is a nonlinear element which is often installed in storage rings at a location with nonzero dispersion and is used for correcting the variation of the quadrupole focusing strength with the particle momentum, i.e., the chromaticity. The local sextupole field in units of m^{-3} is expressed as $m = 2B_T/(a^2(B\rho))$, where B_T now denotes the sextupole pole-tip field, and the integrated sextupole strength becomes $M = ml_{\text{sext}}$. In a kick approximation, the effect of the sextupole on the horizontal trajectory slope is $\Delta x' = Mx^2/2$.

In the remainder of this section, however, we ignore the effect of nonlinear elements and restrict the discussion to particle trajectories with small amplitudes, which evolve according to the linear optics. Then, for constant beam energy, the horizontal or vertical betatron motion, i.e., the solution to (1.7), can be parametrized by a pseudo-harmonic oscillation of the form [4]

$$u_{\beta,x,y}(s) = \sqrt{2I_{x,y}\beta_{x,y}(s)}\ \cos(\phi_{x,y}(s) + \phi_0)\,, \tag{1.12}$$

where $\beta_{x,y}(s)$ is called the beta function, $\phi_{x,y}(s)$ the betatron phase ('angle variable'), and $I_{x,y}$ is an 'action variable'. Action variables are known from classical mechanics. There the Hamiltonian of a harmonic oscillator is expressed by $H(I, \phi, \theta) = QI$, where θ is a time-like variable and Q is a constant equal to the number of oscillations per revolution (in accelerator physics this is called the 'tune'). In the absence of nonlinear perturbations and using a proper choice of coordinates, the betatron motion in an accelerator can be described by exactly the same Hamiltonian as the harmonic oscillator.

In addition to the beta function β, two closely related functions are often introduced to characterize the betatron motion. These are

$$\alpha(s) = -\frac{1}{2}\beta'(s) \qquad \text{and} \qquad \gamma(s) = \frac{1+\alpha^2(s)}{\beta(s)}\,, \tag{1.13}$$

where as before the prime indicates a derivative with respect to the longitudinal position s, and we have dropped the subindex 'x, y'. Together β, α, and γ are referred to as the Twiss parameters. Henceforth, we will use x instead of u, but, here and in the following, the same equations apply in the horizontal and in the vertical plane. The main difference is that quadrupoles which are focusing in one plane are defocusing in the other.

The functions $\phi_{x,y}(s)$ and $\beta_{x,y}(s)$ in (1.12) vary with the azimuthal location s, while the action $I_{x,y}$ and initial phase ϕ_0 are constants of motion. The beam is matched to the lattice if the betatron phases are distributed randomly. In this case, the value of $I_{x,y}$ averaged over all particles of the beam is equal to the rms beam emittance. For example, in the horizontal plane, we then have

$$\epsilon_x^{rms} \equiv \sigma_x^2/\beta_x = \langle I_{x,y}\rangle \tag{1.14}$$

and

$$I_x = \frac{x_\beta^2 + (\beta_x x'_\beta + \alpha_x x)^2}{\beta_x}\,. \tag{1.15}$$

The 'betatron oscillation' described by (1.12) refers to a particle at a fixed design energy. Later, we will discuss how the motion is modified if the energy is not constant, introducing the two concepts of dispersion and chromaticity.

If the beam is accelerated, as in a linac, the beam energy is not constant and the right-hand side of (1.12) must be multiplied by $\sqrt{\gamma(0)/\gamma(s)}$, since the increase in longitudinal momentum p_s reduces the transverse beam size by effectively introducing a damping force $d^2x/ds^2 \approx -p_x/p_s^2\, dp_s/ds$. Inside an accelerating structure with energy gradient E', the formula for α becomes

$$\alpha(s) = -\frac{\beta'(s)}{2} + \frac{\beta E'(s)}{2E(s)}\,, \tag{1.16}$$

$E(s)$ is the beam energy at location s.

The three optical functions $\beta(s)$, $\alpha(s)$ and $\gamma(s)$ are proportional to the three second moments of the beam distribution, with the rms beam emittance as the constant of proportionality:

$$\langle x^2\rangle_s = \beta(s)\;\epsilon, \tag{1.17}$$
$$\langle xx'\rangle_s = -\alpha(s)\;\epsilon, \tag{1.18}$$
$$\langle x'^2\rangle_s = \gamma(s)\;\epsilon\,, \tag{1.19}$$

where $\langle\ldots\rangle_s$ denotes an average over the beam distribution at the location s. Thus, the actual values of β, α and γ can be deduced from the measured beam distribution. It is a challenge to the accelerator physicist to make them coincide with their design values.

In a storage ring, the optical functions α, β and γ are periodic: $\beta(s) = \beta(s+C)$, $\alpha(s) = \alpha(s+C)$, and $\gamma(s) = \gamma(s+C)$, where C is the ring circumference. For a transport line, or linac, no such periodic boundary condition exists; and the values of the optical functions depend on the incoming beam distribution, e.g., via (1.17), (1.18), and (1.19).

Another description, alternative to the 'Twiss parameters' (α, β and γ) represents the motion of a single particle in terms of a transport matrix [28, 29]. Here, a trajectory is denoted by a point in the phase space (x, x') which is transformed from the initial location i to a new (final) location f through a linear transformation

$$\begin{pmatrix} x \\ x' \end{pmatrix}_f = \begin{pmatrix} R_{11} & R_{12} \\ R_{21} & R_{22} \end{pmatrix}_{fi} \begin{pmatrix} x \\ x' \end{pmatrix}_i . \tag{1.20}$$

This can also be generalized to a (6×6) transport matrix for motion with coupling between the horizontal, vertical and longitudinal planes. In the 6-dimensional case, the vector (x, x') is replaced by $(x, x', y, y', z, \delta)$, where δ is the relative energy error and z is the longitudinal distance to a co-moving reference particle, are the coordinates in the longitudinal phase space.

Let us review a few examples. In lowest-order approximation, for a drift space of length l, the 2-dimensional transport matrix is

$$\mathbf{R}_{\text{drift}} = \begin{pmatrix} 1 & l \\ 0 & 1 \end{pmatrix} . \tag{1.21}$$

The matrix for a focusing quadrupole of gradient $k = (\partial B/\partial x)/(B\rho)$ and of length l_q is

$$\mathbf{R}_{\text{quad}} = \begin{pmatrix} \cos\phi & \sin\phi/\sqrt{|k|} \\ -\sqrt{|k|}\sin\phi & \cos\phi \end{pmatrix} , \tag{1.22}$$

where $\phi = l_q\sqrt{|k|}$. If we take the limit of vanishing quadrupole length $l_{\text{quad}} \to 0$ while holding the integrated gradient $K = |k|l_q$ constant, we arrive at the matrix for an idealized 'thin-lens' quadrupole

$$\mathbf{R}_{\text{thin-lens}} = \begin{pmatrix} 1 & 0 \\ -K & 1 \end{pmatrix} . \tag{1.23}$$

Thus, the focal length of the thin quadrupole is given by $1/K$. The R matrix for a sequence of quadrupoles and drift spaces is simply the product of the R matrices for the individual elements.

It is important to note that the description in terms of optical functions and the R matrix formalism are equivalent and complementary. We can transform the optical functions from one location to another using the elements of the R matrix:

$$\begin{pmatrix} \beta \\ \alpha \\ \gamma \end{pmatrix}_f = \begin{pmatrix} R_{11}^2 & -2R_{11}R_{12} & R_{12}^2 \\ -R_{11}R_{21} & 1+2R_{12}R_{21} & -R_{12}R_{22} \\ R_{21}^2 & -2R_{21}R_{22} & R_{22}^2 \end{pmatrix}_{fi} \begin{pmatrix} \beta \\ \alpha \\ \gamma \end{pmatrix}_i . \tag{1.24}$$

Alternatively, we can express the elements of the R matrix from i to f in terms of the optical functions at the initial and final locations,

$$\mathbf{R}_{fi} = \begin{pmatrix} \sqrt{\frac{\beta_f}{\beta_i}}(\cos\phi_{fi} + \alpha_i\ \sin\phi_{fi}) & \sqrt{\beta_f\beta_i}\ \sin\phi_{fi} \\ -\frac{1+\alpha_f\alpha_i}{\sqrt{\beta_f\beta_i}}\ \sin\phi_{fi} + \frac{\alpha_i-\alpha_f}{\sqrt{\beta_f\beta_i}}\ \cos\phi_{fi} & \sqrt{\frac{\beta_i}{\beta_f}}\ (\cos\phi_{fi} - \alpha_f\ \sin\phi_{fi}) \end{pmatrix} , \tag{1.25}$$

where $\phi_{fi} = (\phi_f - \phi_i)$ is the betatron phase advance between the two locations.

In a storage ring, one can compute the 1-turn matrix by demanding that the final location f is equal to the initial one i, after a full revolution. The matrix (1.25), or one-turn-map (OTM), then simplifies to

$$\mathbf{R}_{\text{otm}} = \begin{pmatrix} \cos\mu + \alpha\ \sin\mu & \beta\ \sin\mu \\ -\gamma\sin\mu & \cos\mu - \alpha\ \sin\mu \end{pmatrix} , \tag{1.26}$$

where μ denotes the 1-turn phase advance. The betatron tune Q or ν, is defined as the number of betatron oscillations per turn. It is related to the 1-turn phase advance μ by $Q \equiv \nu = \mu/(2\pi)$.

We note explicitly that the tranport matrices for a single particle and for the beam centroid are identical at low beam currents and without nonlinearities.

1.2 Beam Matrix

The beam distribution can be characterized by its first and second moments. The first moments give the centroid motion. The second moments are combined in a 'beam matrix'. For example, the beam matrix Σ_{beam} for the horizontal plane is defined as

$$\begin{aligned}\Sigma^x_{\text{beam}} &= \epsilon_x \begin{pmatrix} \beta & -\alpha \\ -\alpha & \gamma \end{pmatrix} \\ &= \begin{pmatrix} \langle x^2 \rangle - \langle x \rangle^2 & \langle x x' \rangle - \langle x \rangle \langle x' \rangle \\ \langle x' x \rangle - \langle x' \rangle \langle x \rangle & \langle x'^2 \rangle - \langle x' \rangle^2 \end{pmatrix} . \end{aligned} \tag{1.27}$$

Here α, β, and γ are the ellipse (e.g., Twiss) parameters (compare (1.17–1.19), and Fig. 1.2), ϵ is the beam emittance, and the bracketed terms are various moments of the beam distribution, i.e., $\langle x \rangle$ is the first moment, or mean, of the distribution in position, $\langle x' \rangle$ is the first moment, or mean, of the distribution in angle, and $\langle x^2 \rangle$, $\langle x'^2 \rangle$ are the second moments of the beam distribution. Specifically, for a beam intensity distribution $f(x)$,

$$\langle x \rangle = \frac{\int_0^\infty x f(x)\, \mathrm{d}x}{\int_0^\infty f(x)\, \mathrm{d}x} , \tag{1.28}$$

and

$$\langle x^2 \rangle = \frac{\int_0^\infty x^2 f(x)\, \mathrm{d}x}{\int_0^\infty f(x)\, \mathrm{d}x} . \tag{1.29}$$

The root-mean-square (rms) of the distribution σ_x is (usually) the physical quantity of interest:

$$\sigma_x = \sqrt{\langle x^2 \rangle - \langle x \rangle^2} . \tag{1.30}$$

If the mean of the distribution is neglected (i.e., either disregarding the static position offset of the core of the beam, or defining the coordinates with respect to this offset), (1.27) reduces to

$$\Sigma^x_{\text{beam}} = \begin{pmatrix} \langle x^2 \rangle & \langle x x' \rangle \\ \langle x x' \rangle & \langle x'^2 \rangle \end{pmatrix} \tag{1.31}$$

and the rms of the distribution is simply $\sigma_x = \langle x^2 \rangle^{\frac{1}{2}}$.

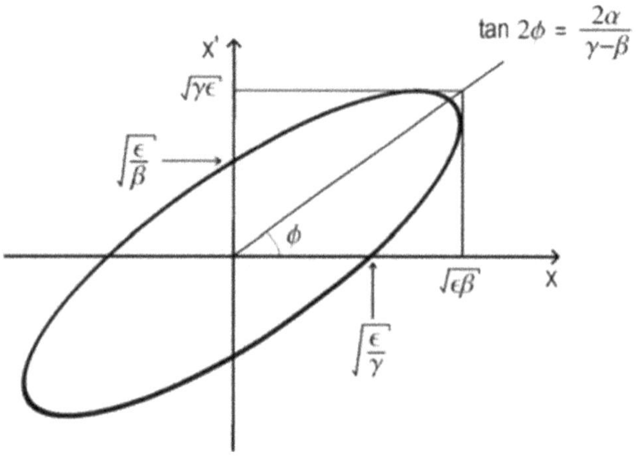

Fig. 1.2. Ellipse parameters for the beam matrix [28, 29]

The transformation between an initial beam matrix $\Sigma_{\text{beam},0}$ to the beam matrix Σ_{beam} at a desired observation point is

$$\Sigma_{\text{beam}} = R\Sigma_{\text{beam},0}R^t \, , \tag{1.32}$$

where R is the transfer matrix and R^t is the transpose of R. Here, depending on how many degrees of freedom are considered, Σ_{beam} and R can be 2×2, 4×4 or 6×6 matrices. For an uncoupled system, the 4×4 matrix $\Sigma^{xy}_{\text{beam}}$, characterizing the transverse beam distribution in the horizontal and vertical phase space, is of block-diagonal form:

$$\Sigma^{xy}_{\text{beam}} = \begin{pmatrix} \Sigma_{11} & \Sigma_{12} & 0 & 0 \\ \Sigma_{21} & \Sigma_{22} & 0 & 0 \\ 0 & 0 & \Sigma_{33} & \Sigma_{34} \\ 0 & 0 & \Sigma_{43} & \Sigma_{44} \end{pmatrix} , \tag{1.33}$$

and

$$R = \begin{pmatrix} R_{11} & R_{12} & 0 & 0 \\ R_{21} & R_{22} & 0 & 0 \\ 0 & 0 & R_{33} & R_{34} \\ 0 & 0 & R_{43} & R_{44} \end{pmatrix} . \tag{1.34}$$

Note that the $\Sigma^{xy}_{\text{beam}}$-matrix is symmetric with $\Sigma^{xy}_{12} = \Sigma^{xy}_{21}$ (cf. (1.27)) but that, in general, $R_{12} \neq R_{21}$.

1.3 Review of Longitudinal Dynamics

If the energy of the beam, or of a particle in the beam, differs from the design energy its trajectory may deviate from the trajectory of an ideal particle which has the desired energy. In first order, this deviation is linear in the momentum deviation $\delta = \Delta p/p$ (p here is the nominal momentum). For a transport line we can write the horizontal displacement resulting from the energy error as

$$\Delta x(s) = R_{16}\delta\,, \tag{1.35}$$

where R_{16} is the (1,6) transport matrix element from the location where the energy error δ was induced to the location s. In a (planar) storage ring, the orbit deviation due to an energy offset is given by the periodic dispersion function $D_x(s)$ as

$$\Delta x(s) = D_x(s)\delta\,. \tag{1.36}$$

Also, in a transport line, the R_{16} matrix element of (1.35) is often called dispersion, but it should be kept in mind that this term is not uniquely defined as the measured values depend on the location where an energy change is introduced (and may be different from the real energy-position correlation within the bunch). As a result, the correction of dispersion in a transport line or a linac can become conceptually quite complicated.

If the beam (or particle) energy is varied, the radius of curvature and, thus, the path length in the bending magnet changes. The first order path length change is characterized by the momentum compaction factor α_c:

$$\alpha_c = \frac{\Delta C/C}{\delta} = \frac{1}{C}\oint \frac{D(s)}{\rho(s)}\,\mathrm{d}s\,. \tag{1.37}$$

If $\gamma > 1/\sqrt{\alpha_c}$, a ring is said to operate 'above transition'; this is the case for most electron and high-energy proton rings. For a transport line, one may replace α_c in (1.37) by the R_{56} matrix element, and D_x by R_{16}, and consider

$$R_{56}(s) = \int_{s_0}^{s} \frac{R_{16}(s')}{\rho(s')}\,\mathrm{d}s'\,. \tag{1.38}$$

Just as the beam, or an individual particle in the beam, executes betatron oscillations, it also performs oscillations in the longitudinal phase space, e.g., in a storage ring with nonzero rf voltage. The frequency of the synchrotron motion is usually much lower than the betatron-oscillation frequencies (one synchrotron period typically corresponds to 100s of turns). It can be expressed in terms of a synchrotron tune Q_s (which is the synchrotron frequency f_s in units of the revolution frequency f_{rev}):

$$Q_s = \frac{f_s}{f_{\mathrm{rev}}} = \sqrt{\frac{(\alpha_c - \gamma^{-2})he\hat{V}\sin\phi_s}{2\pi\beta c p_0}}\,, \tag{1.39}$$

where α_c again is the momentum compaction factor, $\hat{V}$ the amplitude of the rf voltage (assumed as simply cosine-like: $V_{\mathrm{rf}} = \hat{V}\cos(\omega_{\mathrm{rf}}t + \psi)$), h the rf harmonic number ($f_{\mathrm{rf}} = hf_{\mathrm{rev}}$), e the particle charge, p_0 the equilibrium momentum, c the speed of light, and ϕ_s the synchronous phase angle measured with respect to the crest of the rf. The latter is determined by the equality $e\hat{V}\cos\phi_s = U_0$, where U_0 is the average energy loss per turn, and by the condition for phase stability is $0 < \psi_s < \pi/2$ above transition. The transition energy corresponds to $Q_s = 0$.

1.4 Transverse and Longitudinal Equations of Motion

The smooth approximation to the beam motion considers only the average focusing force and ignores the discrete locations of rf cavities or quadrupole magnets. In the longitudinal plane, we assume further that the oscillations are small compared with the amplitude of the focusing rf wave so that only the linear part of the sinusoidal rf force is sampled by the beam. With these approximations both longitudinal and transverse oscillations are then described by the equation of a harmonic oscillator.

In the transverse plane

$$x'' + (\omega_\beta/c)^2 x = 0\,, \tag{1.40}$$

$$p_x'' + (\omega_\beta/c)^2 p_x = 0\,, \tag{1.41}$$

with $p_x \equiv (\alpha x + \beta x')$, where ω_β/c^2 is the appropriate average of $k(s)$ in (1.7) and $\omega_\beta = 2\pi Q$. In the longitudinal plane

$$\phi'' + \omega_s^2 \phi = 0\,, \tag{1.42}$$

$$\delta'' + \omega_s^2 \delta = 0\,, \tag{1.43}$$

with $\phi' = (\alpha_c - (1/\gamma)^2)\omega_{\rm rf}\delta$ and $\omega_s = 2\pi f_s$.

The second moments of the beam distribution have an immediate physical significance. For example, $\sigma_x = \langle x^2\rangle^{\frac{1}{2}}$ is the horizontal rms beam size, $\sigma_{x'} = \langle x'^2\rangle^{\frac{1}{2}}$ the horizontal rms beam divergence, $\sigma_y = \langle y^2\rangle^{\frac{1}{2}}$ the vertical rms beam size, $\sigma_{y'} = \langle y'^2\rangle^{\frac{1}{2}}$ the vertical rms beam divergence, $\sigma_z = \langle z^2\rangle^{\frac{1}{2}}$ the rms bunch length (or, in terms of rf phase, $\sigma_\phi = \omega_{\rm rf}\sigma_z/c$), and $\sigma_\delta = \langle \delta^2\rangle^{\frac{1}{2}}$ the rms momentum spread.

For completeness we list the beam matrix $\Sigma_{\rm beam}$ for a transverse (e.g., horizontal) plane,

$$\Sigma^x_{\rm beam} = \begin{pmatrix} \langle x^2\rangle & \langle xx'\rangle \\ \langle xx'\rangle & \langle x'^2\rangle \end{pmatrix}, \tag{1.44}$$

and its analogue for the longitudinal one,

$$\Sigma^z_{\rm beam} = \begin{pmatrix} \langle \phi^2\rangle & \langle \phi\delta\rangle \\ \langle \delta\phi\rangle & \langle \delta^2\rangle \end{pmatrix}. \tag{1.45}$$

The transverse and longitudinal emittances are obtained from

$$\epsilon_x = \sqrt{\det \Sigma^x_{\rm beam}} = \sqrt{\langle x^2\rangle\langle x'^2\rangle - \langle xx'\rangle^2}\,, \tag{1.46}$$

$$\epsilon_y = \sqrt{\det \Sigma^y_{\rm beam}} = \sqrt{\langle y^2\rangle\langle y'^2\rangle - \langle yy'\rangle^2}\,, \tag{1.47}$$

and

$$\epsilon_z = \sqrt{\det \Sigma^z_{\rm beam}} = \sqrt{\langle \phi^2\rangle\langle \delta^2\rangle - \langle \phi\delta\rangle^2}\,, \tag{1.48}$$

where det denotes the determinant of the corresponding matrix. It is common to consider the normalized emittances $\epsilon_{x,y,N} \equiv \gamma\beta\epsilon_{x,y}$ and $\gamma\beta\epsilon_z$ which are constant under acceleration (here β and $\gamma = 1/\sqrt{1-\beta^2}$ are the relativistic factors).

Exercises

1.1 Beam Emittance in Terms of Action Angle Variables

Individual particles within a bunch (index i) perform betatron oscillations, which can be described in terms of action-angle variables by $x_i(s) = \sqrt{2I_x\beta_x(s)}\cos\phi_x(s)$. Consider a beam whose distribution function depends only on the action variable; i.e., $\rho(I_x, \phi_x) = \rho(I_x)/(2\pi)$. Define the horizontal emittance as $\epsilon_x = \langle x^2(s)\rangle/\beta_x(s)$. Show that $\epsilon_x = \langle I_x\rangle$, where the square brackets denote an average over the beam distribution at location s.

1.2 Projected Beam Emittances

a) Consider now a 2-dimensional particle distribution which is Gaussian and uncorrelated in the 4 variables x_0, x_0', y_0 and y_0' with emittances ϵ_{x0} and ϵ_{y0}. Suppose the beam passes a skew quadrupole of strength K_s with beta functions at the quadrupole equal to β_x and β_y. Afterwards the coordinates of the new distribution are x, x', y and y'. They are correlated as

$$x' = x_0' + K_s y_0 \,, \tag{1.49}$$

$$y' = y_0' + K_s x_0 \,, \tag{1.50}$$

$$x = x_0 \,, \tag{1.51}$$

$$y = y_0 \,. \tag{1.52}$$

Calculate the beam matrix in terms of the Twiss parameters and the initial uncoupled emittances.

b) The projected horizontal and vertical emittances are given by the square root of the determinant of the 2×2 submatrices. Calculate the projected emittances and express them in terms of the initial uncoupled emittances, and the skew quadrupole strength K_s.

2 Transverse Optics Measurement and Correction

In order to preserve the beam quality, accurate knowledge of the transverse optics and its correction is most often mandatory. For example, if the distribution of a beam injected into a storage ring is not matched to the ring optics, the emittance will grow due to filamentation. Or, if there is a significant optics error, e.g., induced by a strength error in a quadrupole magnet, the beam envelope may vary strongly. The resulting reduction in dynamic aperture may then lead to enhanced beam loss.

There are several approaches for verification of the beam optics. Some of these are based on exciting the beam using a transverse kicker magnet, others vary the strength of individual quadrupoles. In either case, the optics can be evaluated using measurements of the beam response, which is measured on one or more transverse electrodes ("pick ups").

In this chapter, we discuss several techniques used to measure the betatron tune, betatron phase advances, and beta functions. We then describe how the source of an optics error may be localized, and, alternatively, how an imperfect optics can be matched to the design values by the use of orthogonal 'multiknobs'. We next present general and powerful data processing techniques. The beam response to (large) transverse deflections is discussed next which includes a combination of effects including nonlinear detuning, radiation damping, head-tail damping, and chromaticity. The last section is devoted to measurements and correction techniques for betatron coupling.

2.1 Betatron Tune

2.1.1 Introduction

In a storage ring the betatron tune, or Q value, is defined as the number of betatron oscillation periods per revolution[1]:

$$Q = \frac{\phi(C)}{2\pi} = \frac{1}{2\pi} \oint_C \frac{ds}{\beta(s)} , \tag{2.1}$$

where the integral is taken around the ring of circumference C. A schematic of a betatron oscillation is shown in Fig. 2.1.

[1] often, especially in the American literature, it is also denoted by ν

M. G. Minty et al., *Measurement and Control of Charged Particle Beams*,
https://doi.org/10.1007/978-3-662-08581-3_2

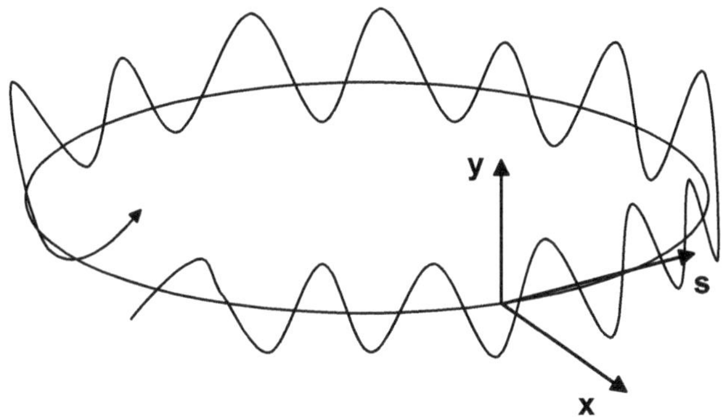

Fig. 2.1. Schematic of a betatron oscillation around a storage ring

The integer part of the tune is easily inferred from the orbit distortion induced by exciting a single steering corrector. The resulting distortion produces a coherent betatron oscillation and is easily measured by taking the difference in the orbits as measured using beam position monitors (BPMs) both with and without the kick; counting the number of oscillation periods around the ring circumference determines the integer value of the tune. A more intricate method, discussed in Sect. 2.2, involves performing a harmonic analysis of the betatron oscillations recorded by multi-turn BPMs. In this case the betatron phase advance between adjacent BPMs can be determined. The total phase advance around the ring gives the tune. If the integer part of the tune agrees with model predictions, large optics errors can be excluded. As important as the integer value of the tune is its fractional part, since the latter can have a strong effect on the beam lifetime or emittance.

Tune measurements are useful for quite a variety of applications: the tune shift with quadrupole strength gives the local beta function, the tune shift with rf frequency the chromaticity, the tune shift with current the effective transverse impedance, and the tune shift with betatron amplitude

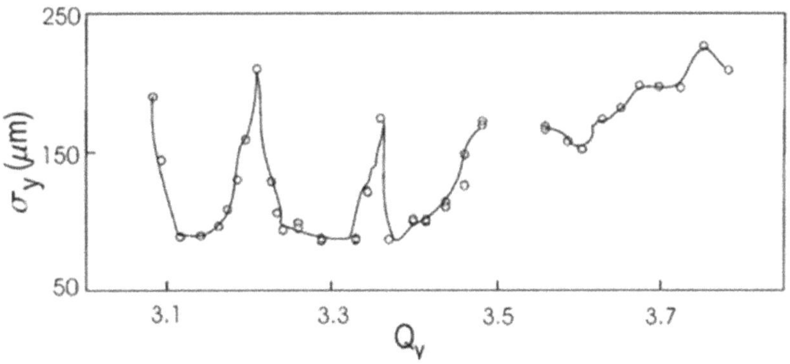

Fig. 2.2. Rms vertical size of the electron beam extracted from the SLC damping ring as a function of the vertical betatron tune. This measurement was performed under unusually poor vacuum conditions [1]

the strength of nonlinear fields. Furthermore, optimizing and controlling the tunes improves the beam lifetime and the dynamic aperture, and can reduce beam loss or emittance growth during acceleration. For example, Fig. 2.2 shows the variation of the extracted vertical beam size as a function of the vertical betatron tune which was measured at the SLC electron damping ring. Other effects including space charge, ionized gas molecules, the beam-beam interaction, and radiation damping can affect the tune signal, which may be seen for example in the shape of the beam response to a swept-frequency excitation. An example showing the dramatic effect of the nonlinear beam-beam force is shown in Fig. 2.3.

Fast decoherence and filamentation, head-tail damping or instabilities may make it difficult to extract a clean and reproducible tune signal. Conversely, the strong influence of various phenomena on the tune signal also implies that all these processes can be studied by means of tune measurements.

In the following we will describe three methods of measuring the fractional part of the tune. These approaches fall into two different categories: (1) precision tune measurements and (2) tune tracking (the latter aims at monitoring and controlling fast changes, e.g., during acceleration). For simplicity, the fractional part of the tune will also be denoted by Q.

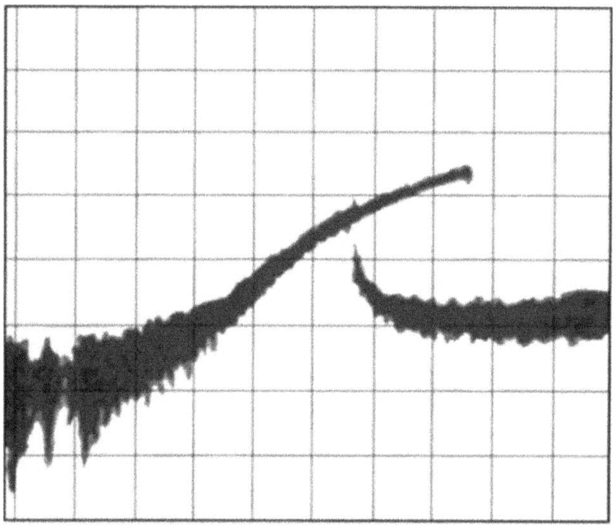

Fig. 2.3. Transverse tune measurement (swept-frequency excitation) with 2 colliding bunches at Tristan. Vertical axis: 10 dB/div., horizontal axis: 1 kHz/div. [2] (Courtesy K. Hirata and T. Ieiri, 1998)

2.1.2 Fast Fourier Transform (FFT) and Related Techniques

A common method to measure the fractional part of the betatron tune involves exciting transverse beam motion and detecting the transverse beam position over a number of successive turns N. The excitation may consist either of white noise or of a single kick. Beam oscillations after injection are also often measured as these are often naturally present so the measurement does not interfere with accelerator operation. The power density of the detected signal is computed via a Fourier transformation and the betatron tunes are identified as the frequencies with the highest amplitude peak (this is not always the case as sometimes the beam is strongly excited at other frequencies). Figure 2.4 shows typical multi-turn BPM measurements. Alternatively, a spectrum analyzer can be used to frequency analyze the signal detected by a pick up.

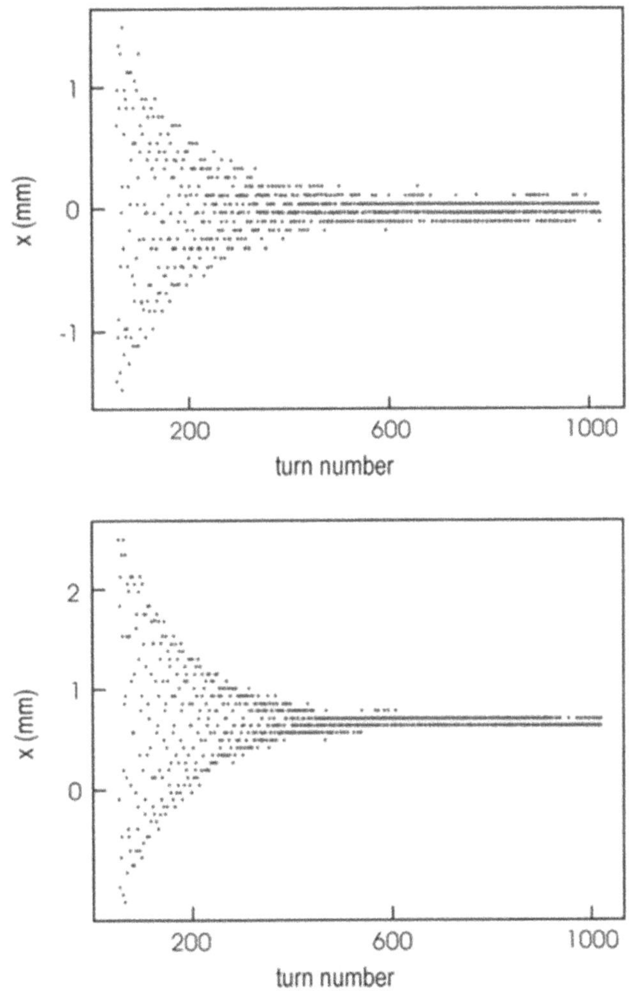

Fig. 2.4. Multi-turn orbit measurement for the motion of a single bunch in a train of 3 bunches at LEP-I. Shown are horizontal BPM orbit readings as a function of turn number: (*top*) BPM in a dispersive arc region; (*bottom*) BPM in a non-dispersive straight section. At the start of the measurement the bunch was deflected by a kicker. The corresponding FFT spectra are displayed in Fig. 2.5 (Courtesy A.-S. Müller, 2001)

A Fourier analysis uses a time series $x(1),\ x(2), ..., x(N)$ of N orbit measurements for consecutive turns as input. This time series is expanded as a linear combination of N orthonormal functions,

$$x(n) = \sum_{j=1}^{N} \psi(Q_j)\ \exp(2\pi i n Q_j)\,. \tag{2.2}$$

where the coefficients $\psi(Q_j)$ are calculated from the inverse formula

$$\psi(Q_j) = \frac{1}{N} \sum_{n=1}^{N} x(n) \exp(-2\pi i n Q_j)\,. \tag{2.3}$$

The expansion can be done efficiently with a Fast Fourier Transform algorithm. The frequency corresponding to the largest value of ψ is taken as the approximate tune (see Fig. 2.5). The frequency error is given by the finite sample size, N.

$$|\delta Q| \leq \frac{1}{2N}\,. \tag{2.4}$$

Thus, to obtain a tune value with a resolution of 0.001 or better requires orbit data for about $2^{10} \approx 1000$ turns. As an illustration, Fig. 2.5 displays FFT spectra of the orbit motion measured at the two BPMs of Fig. 2.4. The FFT confirms that a large part of the orbit motion in the dispersive region is due to synchrotron oscillations.

The resolution can be improved by an interpolated FFT. If we use a simple Fourier analysis based on the peak amplitude of ψ in (2.2), typically we need about 1000 turns of orbit data to obtain an adequate tune resolution. During this time the beam could filament or the oscillation amplitude could decrease significantly, giving rise to spurious results. Fortunately, interpolating the shape of the Fourier spectrum around the main peak improves the resolution quite dramatically [3]. Thereby the same resolution can be achieved by processing data over considerably fewer turns.

The underlying assumption is that the shape of the Fourier spectrum is known and corresponds to that of a pure sinusoidal oscillation with tune Q_{int},

$$|\psi(Q_j)| = \left| \frac{\sin N\pi(Q_{\text{int}} - Q_j)}{N \sin \pi(Q_{\text{int}} - Q_j)} \right|\,. \tag{2.5}$$

The formula for the interpolated tune Q_{int} reads [4]:

$$Q_{\text{int}} = \frac{k}{N} + \frac{1}{\pi} \arctan\left(\frac{|\psi(Q_{k+1})| \sin\left(\frac{\pi}{N}\right)}{|\psi(Q_k)| + |\psi(Q_{k+1})| \cos\left(\frac{\pi}{N}\right)} \right)\,, \tag{2.6}$$

where $|\psi(Q_k)|$ is the peak of the Fourier spectrum in (2.2), and $|\psi(Q_{k+1})|$ its highest neighbor. In other words, instead of using only the peak value of the

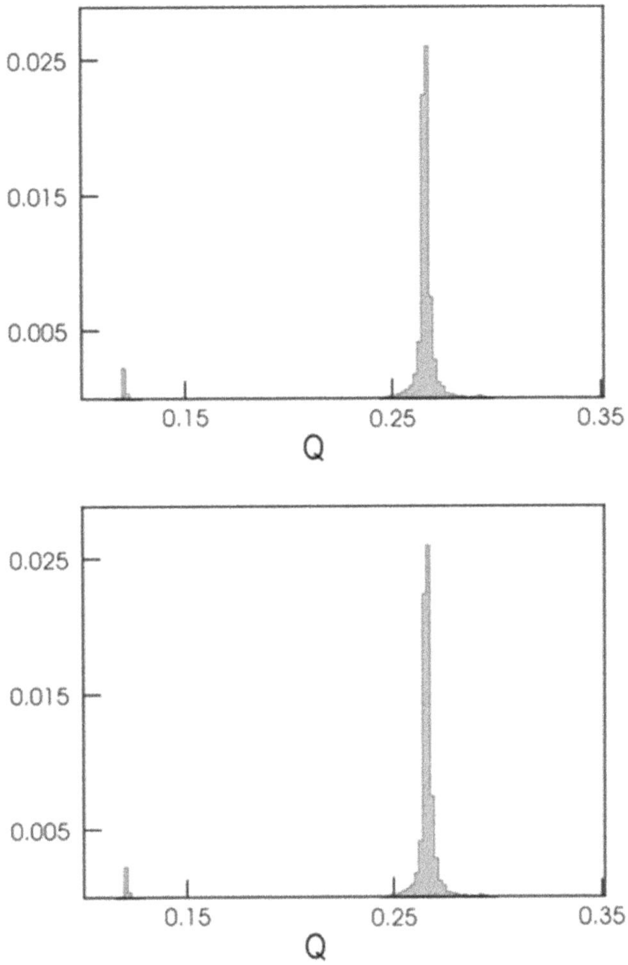

Fig. 2.5. FFT power spectra for the two BPM measurements of Fig. 2.4 in a dispersive (*top*) and in a nondispersive region (*bottom*). The horizontal scale is in tune units, the vertical scale in arbitrary units (Courtesy A.-S. Müller, 2001)

FFT, one interpolates between the two highest points. For large N the error decreases as

$$|\delta Q| \propto \frac{1}{N^2} \, . \tag{2.7}$$

So, the resolution improves quadratically with the number of turns and already from a beam signal recorded over 30–60 turns, fairly accurate tune values can be obtained. For $N \gg 1$, (2.6) may be approximated by the simpler form [3]

$$Q_{\text{int}} \approx \frac{k}{N} + \frac{1}{N} \arctan\left(\frac{|\psi(Q_{k+1})|}{|\psi(Q_k)| + |\psi(Q_{k+1})|}\right) \, . \tag{2.8}$$

A further refinement is an interpolated FFT with data windowing, which further increases the accuracy of the Fourier analysis [4, 5]. Here, the data

$x(n)$ are weighted with filter functions $\chi(n)$ before the interpolated FFT is applied. The Fourier coefficients of the filtered signal are

$$\psi(Q_j) = \frac{1}{N} \sum_{n=1}^{N} x(n)\chi(n) \exp(-2\pi i n Q_j) . \tag{2.9}$$

Note that the regular FFT, (2.3), corresponds to (2.9) without filter, or to $\chi(n) = 1$. Applying instead a Hanning-like filter of order l, $\chi_l(n) = A_l \sin^l(\pi n/N)$, where A_l is a normalization constant, in the limit $N \gg 1$ the interpolated tune reads

$$Q_{\mathrm{Han}} = \frac{k}{N} + \frac{1}{N}\left(\frac{(l+1)\psi(Q_{k+1})}{\psi(Q_k)+\psi(Q_{k+1})} - \frac{l}{2}\right) . \tag{2.10}$$

The resolution improves with the $(l+2)$th power of the number of samples N:

$$|\delta Q| \propto \frac{1}{N^{l+2}} . \tag{2.11}$$

An example comparing the precision of different FFT procedures is shown in Fig. 2.6 [4], which clearly demonstrates the superiority of the interpolated FFT with data windowing (Hanning filter). Unfortunately, the beneficial effect of the Hanning filter disappears when the signal contains a small noise component [4], in which case the resolution decreases as $\sim N^{-2}$ only, just as with the simple interpolated FFT. Another potential problem may arise if the signal measured in the control room is far from a purely sinusoidal shape, e.g., due to a nonzero chromaticity, filamentation, or collective effects.

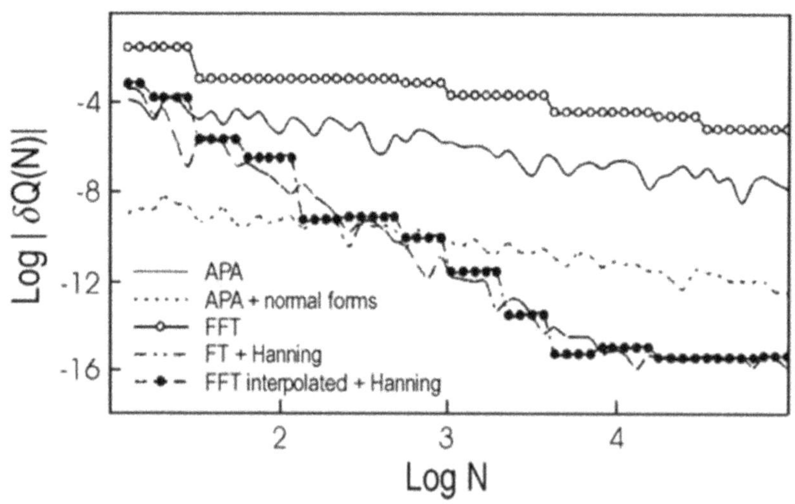

Fig. 2.6. Tune precision vs. number of turns, considering different FFT techniques applied to tracking data for the 4-D Hénon map [4]. The abbreviation 'APA' refers to a calculation of the average phase advance, which can be computed either in the original phase-space coordinates or in so-called normal-form coordinates. See [4] for more details on these alternative methods (Courtesy M. Giovannozzi, 1998)

Accurate computations of fundamental oscillation frequencies, using techniques similar to those described above, allow for a global view of the underlying beam dynamics, and in particular the identification of chaotic region of phase space. Such 'frequency map analysis' [6] has been applied to both simulation data [7, 8, 9] and measurements [10].

Another approach to obtain higher accuracy than an FFT is Lomb's method [11], which was applied at LEP with great success [12]. This method can be applied to an arbitrary data sample without any constraints on the number of data points or on the time interval between two successive points. It weights the data on a 'per point basis' and not 'per time interval' as an FFT [12].

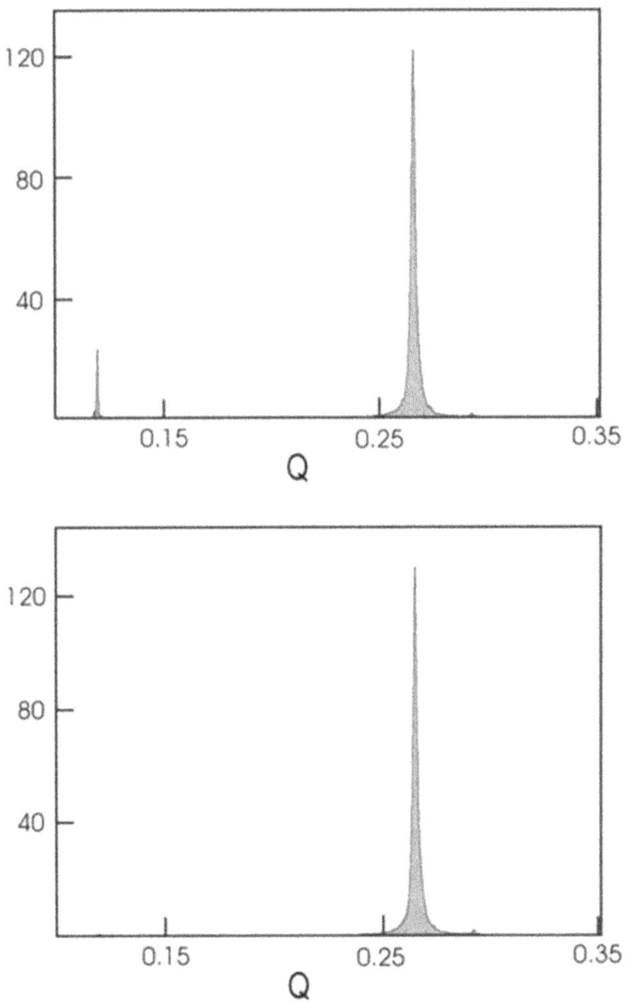

Fig. 2.7. Lomb normalized periodogram [12] for the two BPM measurements of Fig. 2.4 in a dispersive (*top*) and in a nondispersive region (*bottom*). The horizontal scale is in tune units, the vertical scale in arbitrary units (Courtesy A.-S. Müller, 2001)

For N data points h_i measured at times t_i, the so-called 'Lomb normalized periodogram' $P_N(Q)$ is defined by

$$P_N(Q) = \frac{1}{2\sigma^2}\left\{\frac{\left[\sum_n \cos(2\pi Q(n-n_0))\right]^2}{\sum_j \cos^2(2\pi Q(n-n_0))} + \frac{\left[\sum_n \sin(2\pi Q(n-n_0))\right]^2}{\sum_n \sin^2(2\pi Q(n-n_0))}\right\} \tag{2.12}$$

with

$$d_n = h_n - \frac{1}{N}\sum_{n=1}^{N} h_n \;, \tag{2.13}$$

$$\sigma^2 = \frac{1}{N-1}\sum_{n=1}^{N} d_n^2 \;, \tag{2.14}$$

$$\tan(4\pi Q n_0) = \frac{\sum_n \sin(4\pi Q n)}{\sum_n \cos(4\pi Q n)} \;. \tag{2.15}$$

For any test frequency Q the Lomb method determines the contents of that frequency in the data set. The constant n_0, which in general is not an integer, is computed so as to eliminate the phase of the original harmonic in the expression for the power spectrum $P_N(Q)$. Since the phase dependence is removed, the Lomb periodogram can be more accurate than the FFT.

As an example, Fig. 2.7 displays the Lomb normalized periodograms for the two LEP data sets of Fig. 2.4. Comparison with the corresponding FFT power spectra in Fig. 2.5 reveals the higher quality of the Lomb transformation.

2.1.3 Swept-Frequency Excitation

A different method to measure the tune involves exciting the beam with a steady sinusoidal wave and detecting the amplitude and phase of the beam response. The excitation frequency is increased in steps, or at a constant rate. The strength of the harmonic excitation is adjusted so as to produce beam oscillations of adequate amplitude at the resonant frequency.

The result of this measurement is a 'transverse beam-transfer function', which is the (complex) response of the beam to a harmonic excitation as a function of frequency. The beam-transfer function contains important information, for example, about the transverse impedance or about radiation damping [13]. It is easy to see from (2.2) that, in frequency domain, the tune signal from a single bunch repeats itself in frequency intervals corresponding to multiples of the revolution frequency $f_{\rm rev}$ (i.e., a spectrum analysis of the signal from one pick up contains no information about the integer part of the tune). If n_b equidistant bunches are stored in a ring and the combined signal of all bunches is detected, the periodicity of the FFT signal is $n_b f_{\mathrm{rev}}$. In addition, the tune spectrum from 0 to $n_b f_{\mathrm{rev}}/2$ and that from $n_b f_{\mathrm{rev}}/2$ to $n_b f_{\mathrm{rev}}$ are mirror images of each other. Therefore, for the study of multibunch instabilities, it is sufficient to measure the beam transfer function around each revolution harmonic between zero and $n_b f_{\mathrm{rev}}/2$.

The concept of the beam-transfer function can be extended to higher-order beam excitations. At the CERN Antiproton Accumulator a quadrupole pick-up was used to measure the quadrupole mode beam-transfer function of an antiproton beam [14].

The frequency-sweep method as discussed so far requires a relatively long time in order to measure the response at each frequency with sufficient accuracy. However, there exists a fast version of this method, called a chirp excitation. Here the frequency of the excitation is ramped rapidly across the tune resonance, while the beam response is observed [15]. This is useful for monitoring fast tune changes as is done for example during acceleration in the SPS [16, 17].

2.1.4 Phase Locked Loop

Exactly at the betatron tune the amplitude of the beam-transfer function has zero slope as a function of excitation frequency, whereas the phase of the beam-response exhibits maximum slope. The phase difference between excitation and beam motion changes from 0 degree to 180 degree when the excitation frequency is ramped through the resonance. Directly at the betatron tune, the phase difference is 90 degrees. The phase difference can be monitored continually by a phase locked loop (PLL) circuit (see, for example, [15, 18]) and may be used in feedbacks which regulate the betatron tune [19].

The signal flow diagram of a phase locked loop is sketched in Fig. 2.8. The phase detector compares the frequency of a beam position signal, e.g., from a BPM, with the frequency of a local oscillator. The phase-detector output voltage is a measure of the frequency difference of its two input signals. After low-pass filtering and amplification, this signal is used to adjust the frequency

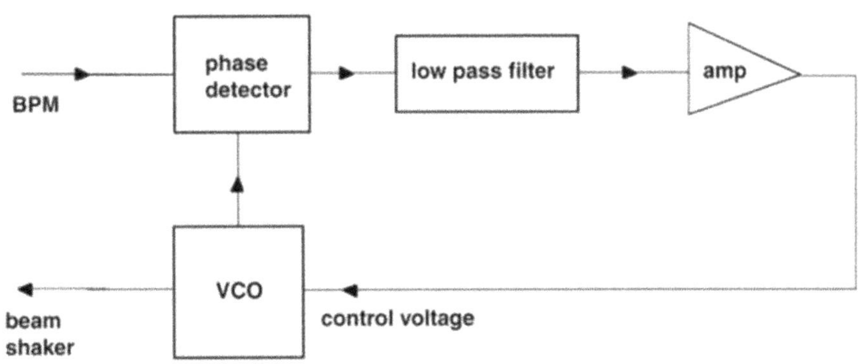

Fig. 2.8. Schematic of phase locked loop for continuous tune control

of the local oscillator (VCO), such that the oscillator 'locks' to the frequency of the input beam signal. The oscillator frequency serves as the betatron tune signal which is displayed or processed by the accelerator control system. Sometimes the oscillator signal is also used to excite the beam, in which case the phase locked loop becomes part of a 'lock-in amplifier'. PLL circuits allow a continuous tracking of the time evolution of the betatron tune.

2.1.5 Schottky Monitor

All the techniques reported so far measure the coherent betatron tune, i.e., the oscillation frequency of the beam centroid. In the case of unbunched proton, antiproton or ion beams it is also possible to measure the incoherent betatron tune, i.e., the oscillation frequency of individual particles in the beam (in the absence of centroid motion). The incoherent signal is proportional to $\sqrt{\epsilon N \Delta f}$, where ϵ is the beam emittance, N the number of particles in the beam, and Δf the frequency bandwidth. Though this signal is small, it can be detected with sensitive 'Schottky monitors' [20].

2.1.6 Multi-Bunch Spectrum

Spectrum analysis of the beam position signal reveals differences between single bunch and multi-bunch operation. For a single bunch, the spectrum analyzer displays two betatron side bands (upper and lower) at frequencies $nf_{\mathrm{rev}} \pm f_\beta$ around each revolution harmonic. There are two sidebands because negative frequencies are not displayed, and instead appear at positive value (as mirrored about the zero-frequency axis). All sidebands have equal amplitudes. Only at very high frequencies, comparable to the bunch frequency $f \sim c/(2\pi\sigma_z)$ does the amplitude decrease.

For n_b bunches uniformly distributed around the ring, the different betatron sidebands may have different amplitudes, which are determined by the strength of different possible multibunch modes. There are a total of n_b independent multibunch modes, which can all be found in the frequency range extending from 0 to $n_b f_{\mathrm{rev}}/2$. The amplitudes of the sidebands are in general unequal.

In the simplest multi-bunch case, we have two bunches diametrically opposed to each other. The position as a function of time seen by a beam position monitor is

$$x(t) \propto \sum_{n=-\infty}^{\infty} \delta(t - nT_{\mathrm{rev}})\, \mathrm{e}^{\mathrm{i}\omega_x t} + r \sum_{n=-\infty}^{\infty} \delta(t - nT_{\mathrm{rev}} - T_{\mathrm{rev}}/2)\, \mathrm{e}^{\mathrm{i}(\omega_x t + \phi)} , \quad (2.16)$$

where r is a coefficient which relates the oscillation amplitudes of the two bunches. We assume that the beam position $x(t)$ is obtained from two pick ups, which are located on the horizontally inward and outward sides of the beam pipe, respectively, by taking the difference signal $(A - B)$ and normalizing it by the sum signal $(A + B)$, so that the measurement $x(t)$ does not vary with the beam current. Without such normalization, the coefficient r will also depend on the intensity ratio of the bunches.

The Fourier transform of (2.16) is

$$\tilde{x}(\omega) \propto \sum_{k=-\infty}^{\infty} \delta(\omega - \omega_x - k\omega_{\text{rev}}) \left(1 + re^{\mathrm{i}(k\pi+\phi)}\right) . \tag{2.17}$$

If $r = 1$ and $\phi = 0$, the two bunches oscillate in phase. This is the so-called 0 mode or σ mode. If this mode is excited, only betatron sidebands around the even revolution harmonics are visible. On the other hand if $r = 1$ and $\phi = \pi/2$, the bunches oscillate out of phase. This is the so-called π mode, related to the odd-harmonic betatron sidebands.

A multi-bunch oscillation mode can be excited and its amplitude grow, due to an interaction with the accelerator impedance at the mode frequency. The accelerator impedance entails, e.g., the resistive chamber wall, or cavity-like objects in the beam pipe. Usually, the dominant impedance has a narrow frequency band width, and, therefore, in multi-bunch operation with n_b bunches only a few of the n_b multi-bunch modes can be driven by the impedance.

2.2 Betatron Phase

2.2.1 Harmonic Analysis of Orbit Oscillations

By exciting transverse oscillations, sampling the beam position over N turns, and performing a harmonic analysis, one can determine the betatron phase at the location of the pick up as follows [21].

The oscillation detected by the BPM is a harmonic function

$$x_{km} = A_k \cos(2\pi Q_x m + \phi_k) , \tag{2.18}$$

where the index k specifies the BPM, m is the turn number, and A_k the measured amplitude, which depends on the local beta function, on the magnitude of the oscillation, and on the BPM calibration. Here ϕ_k is the measured phase of the oscillation, which may be aliased (using a single BPM the integer part of the tune is undetermined).

In the limit of large N, the two Fourier sums

$$C_k = \sum_{m=1}^{N} x_{km} \cos(2\pi m Q_x) , \qquad S_k = \sum_{m=1}^{N} x_{km} \sin(2\pi m Q_x) , \tag{2.19}$$

approach the asymptotic values

$$C_k \approx \frac{A_k N}{2} \cos \phi_k \,, \qquad S_k \approx \frac{A_k N}{2} \sin \phi_k \,. \tag{2.20}$$

The betatron phase at the kth monitor can therefore be expressed as

$$\phi_k \approx \tan^{-1}\left(\frac{S_k}{C_k}\right) \,. \tag{2.21}$$

The amplitude is given by $A_k \approx 2\sqrt{C_k^2 + S_k^2}/N$. Figure 2.9 shows 5 consecutive measurements of the betatron phase advance around the PEP-II HER using (2.21). The phase advance predicted by the model was subtracted from the measured phase. The figure demonstrates that the measurement is highly reproducible, and that, for this example, it is in good agreement with the model. The offset of about 40° is due to different reference points in model and measurement.

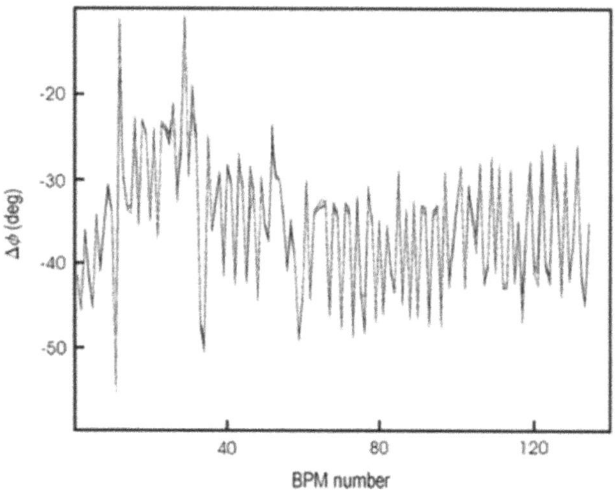

Fig. 2.9. Difference between measured and predicted betatron phase advance (in degrees) as a function of position around the PEP-II HER (BPM number) for 5 consecutive measurements; the 5 curves are superimposed. The total phase advance around the ring is about 9000° (Courtesy M. Donald, 1998)

Application: Transverse Impedance Measurement. Measuring the betatron phase advance for different bunch currents provides information about the effective transverse impedance, a quantity which describes the electromagnetic coupling of the beam to its environment [22]. A measurement of the current-dependent phase advance around the LEP ring is shown in Fig. 2.10. Clearly visible as step changes are the locations of the rf cavities in the straight sections.

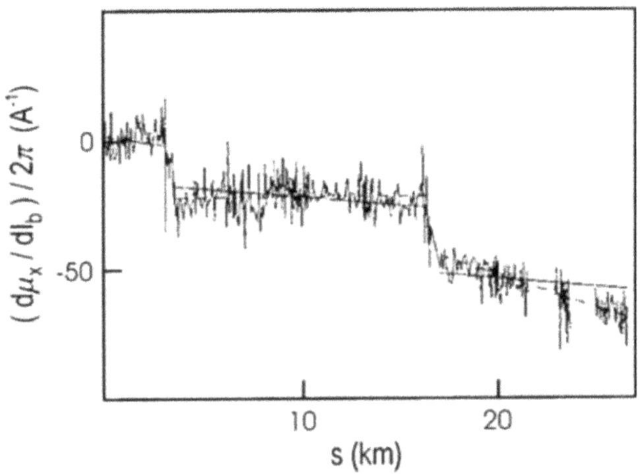

Fig. 2.10. Dependence of the horizontal betatron phase advance on the bunch current, $d\phi/dI_b/(2\pi)$ in units of A^{-1}, measured at LEP [23] (Courtesy A. Hofmann, 1998)

2.3 Beta Function

2.3.1 Tune Shift Induced by Quadrupole Excitation

Presumably the simplest beta-function measurement is to detect the shift in the betatron tune as the strength of an individual quadrupole magnet is varied. This shift can be computed using the 'thin-lens' approximation of (1.10). The tune shift induced by a gradient change for a long quadrupole can then be obtained by linear superposition.

A focusing quadrupole exerts a restoring force on the particle trajectory. For example, as discussed in Chap. 1, the linear equation for the horizontal particle motion inside a quadrupole of strength k is

$$\frac{\mathrm{d}^2 x}{\mathrm{d}s^2} = -kx \ , \tag{2.22}$$

where

$$k = \frac{B_T}{(B\rho)a} \tag{2.23}$$

is measured in units of m^{-2}, and B_T denotes the pole-tip field, a the pole-tip radius, and $(B\rho)$ the magnetic rigidity of the beam: $B\rho[\mathrm{Tm}] \approx 3.356p[\mathrm{GeV/c}]$. We already mentioned in Chap. 1 that the effect of the quadrupole can often be integrated over its length, and represented as a deflection or 'kick':

$$\Delta x' = -Kx \ . \tag{2.24}$$

The integrated quadrupole strength K, in units of m^{-1}, is obtained by multiplying k with the length l_Q of the quadrupole.

With a quadrupole excitation of ΔK, the 2×2 transport matrix for the entire ring is the product of the original transport matrix, (1.12), with $f = i$,

$$\begin{pmatrix} \cos(2\pi Q_{x,y}) + \alpha_{x,y}\sin(2\pi Q_{x,y}) & \beta_{x,y}\sin(2\pi Q_{x,y}) \\ -\sin(2\pi Q_{x,y}) & \cos(2\pi Q_{x,y}) - \alpha_{x,y}\sin(2\pi Q_{x,y}) \end{pmatrix}, \quad (2.25)$$

and a perturbation matrix representing the effect of the change in gradient,

$$\begin{pmatrix} 1 & 0 \\ -(\pm\Delta K) & 1 \end{pmatrix}, \quad (2.26)$$

where Q_x is the original tune, $\beta_{x,y}$ and $\alpha_{x,y}$ the optical functions at the quadrupole, and the plus or minus sign refers to the horizontal and vertical plane, respectively. The function $\beta_{x,y}$ is to be determined.

The trace of the product matrix must be equal to $2\cos(2\pi\bar{Q}_{x,y})$, where $\bar{Q}_{x,y} = (Q_{x,y} + \Delta Q_{x,y})$ is the new tune, and $\Delta Q_{x,y}$ is the tune shift induced by a quadrupole excitation of ΔK. Explicit evaluation of the trace gives the equation

$$2\cos(2\pi(Q_{x,y} + \Delta Q_{x,y})) = 2\cos(2\pi Q_{x,y}) - \beta_{x,y}(\pm\Delta K)\ \sin(2\pi Q_{x,y})\,. \quad (2.27)$$

Solving for $\beta_{x,y}$ we find [24, 25]:

$$\beta_{x,y} = \pm\frac{2}{\Delta K}\ [\cot(2\pi Q_{x,y})\{1 - \cos(2\pi\Delta Q_{x,y})\} + \sin(2\pi\Delta Q_{x,y})]\,, \quad (2.28)$$

where the $\pm$ sign refers to the horizontal and vertical planes, respectively. For a small tune change, (i.e., $2\pi\Delta Q_{x,y} \ll 1$), and far from the integer or half integer resonance (i.e., $\cot(2\pi Q_{x,y}) \leq 1$), we can further simplify and obtain

$$\beta_{x,y} \approx \pm 4\pi\,\frac{\Delta Q_{x,y}}{\Delta K}, \quad (2.29)$$

which is commonly applied.

Figure 2.11 illustrates the error involved in approximating (2.28) by (2.29). The difference between the two expressions becomes important if $Q_{x,y}$ is close to an integer or half integer resonance, and for large changes ΔK [24].

A recent beta function measurement at the Fermilab Recycler is depicted in Fig. 2.12. The nonlinear dependence is well described by the complete (2.28).

Care has to be taken that the applied change in quadrupole strength does not alter the beam orbit, which happens if the beam is off-center in the quadrupole whose strength is varied. If the orbit changes, part of the measured tune shift could then be caused by the closed-orbit variation at the sextupole magnets elsewhere in the accelerator. If a strong effect on the orbit is observed, the orbit should first be corrected with the help of steering correctors before the new (shifted) tune value is measured. Sometimes, several magnets are connected to the same power supply, and then the strengths K_i $(i = 1, ..., m)$ of m quadrupoles must be changed simultaneously, all by the same amount ΔK. The above result is easily generalized to this case: the induced tune change is related to the average beta function at the m quadrupoles via $\langle\beta_{x,y}\rangle_m \approx \pm 4\pi\ \Delta Q_{x,y}/(m\Delta K)$. However, the disadvantage of averaging over several quadrupoles is that beta beating may be less evident.

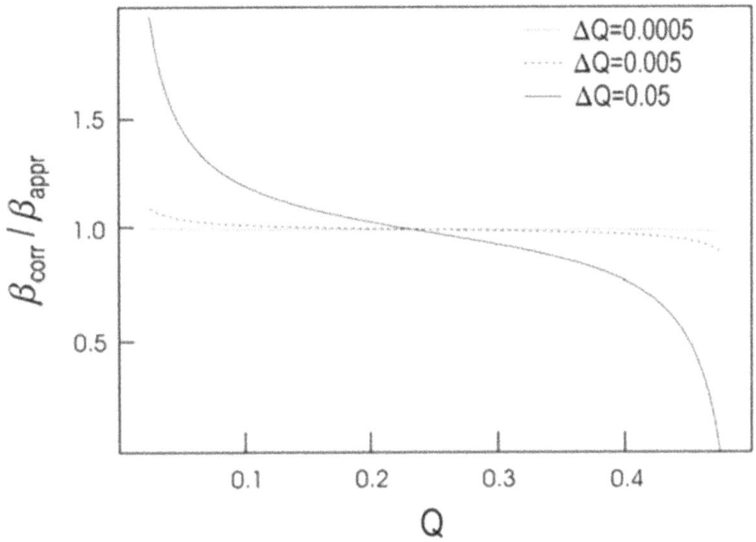

Fig. 2.11. The ratio $\beta_{\mathrm{corr}}/\beta_{\mathrm{appr}}$ of the correct beta function β_{corr}, inferred from (2.28), to the approximation β_{appr} given in (2.29) as a function of the nominal tune Q. The three curves correspond to different magnitudes of ΔQ

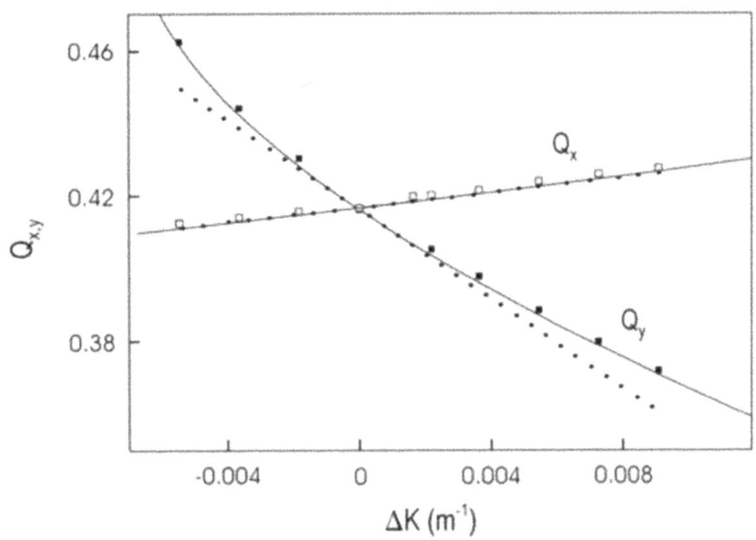

Fig. 2.12. Betatron tunes in the Fermilab Recycler Ring are plotted versus the strength of quadrupole QT601 [26]. The measurements are compared with theoretical predictions using the exact nonlinear dependence of (2.28) [*solid lines*], or its linear approximation (2.29) [*dotted lines*], and taking the beta functions at quadrupole QT601 to be equal to their design values

2.3.2 Betatron Phase Advance

A different method determines the beta function from betatron oscillations measured by multi-turn beam position monitors (BPMs). The underlying idea is to calculate the beta function from the betatron phase advance between three adjacent BPMs. The betatron phase at each BPM can be obtained with a high precision using (2.19) and (2.21) [27]. Since the oscillation amplitude may be subject to calibration errors, it is not taken as an input to this calculation. Instead, the computed beta functions can be used to check and correct the BPM calibration.

The first row of the matrix $\mathbf{R}_{fi}$ in (1.25) can be rewritten as

$$\tan \phi_{fi} = \frac{R_{12}}{R_{11}\beta(s_i) - R_{12}\alpha(s_i)}, \tag{2.30}$$

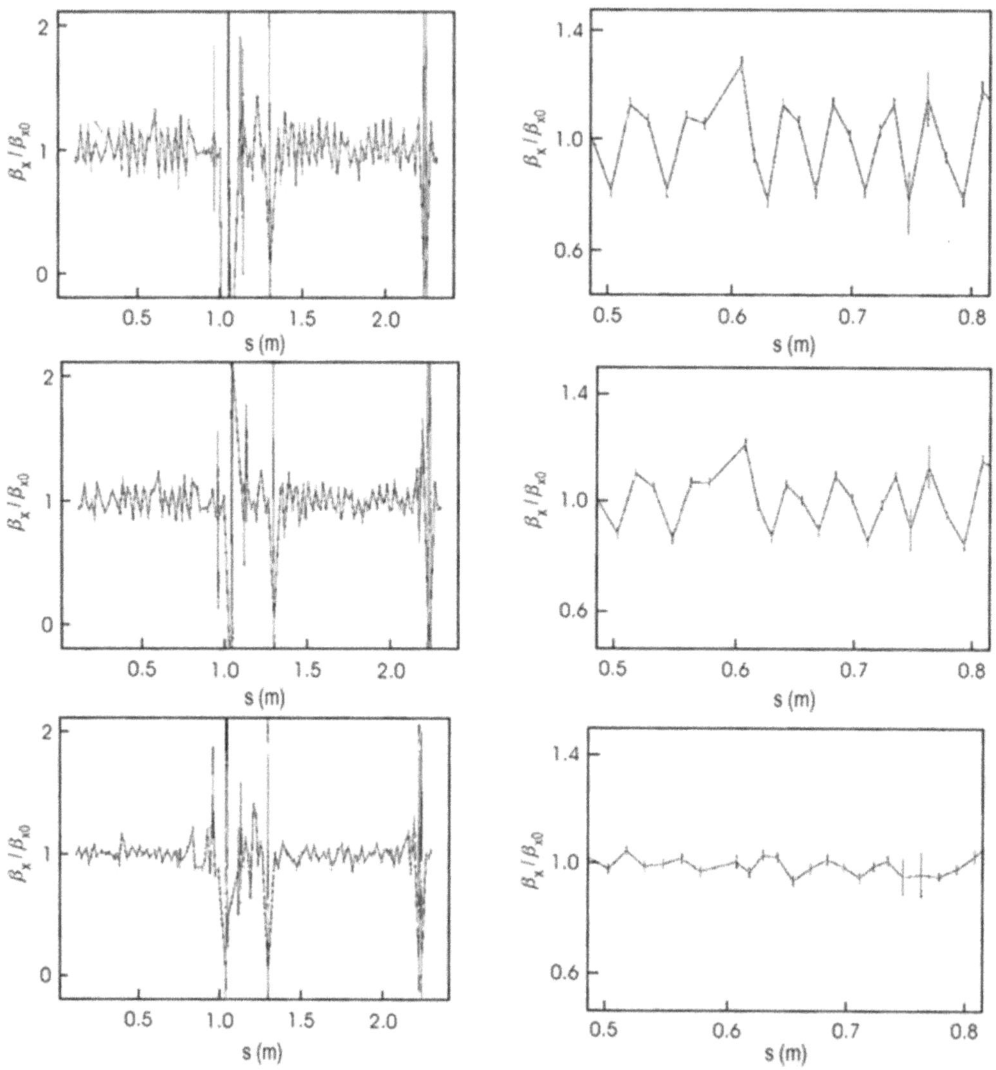

Fig. 2.13. Ratio of the horizontal beta function inferred from phase advance measurements to the model beta function: (*left column*) for the entire PEP-II High Energy Ring (HER); (*right column*) for a limited region only; (*top row*) with all magnets at nominal strength; (*middle row*) for an increased strength of a single quadrupole pair (called QF5) by 0.1%; (*bottom row*) for a strength increase of 0.15%. The "fliers" evidencing large error bars correspond to either malfunctioning BPMs or to a phase advance between successive BPMs equal to 0, $\pi/2$ or π (Courtesy M. Donald, 1998)

where ϕ_{fi} is the phase advance from monitor i to monitor f, and the R_{kl} are transport matrix elements between the same two locations. These matrix elements can be calculated from the geometry of the beam line assuming that the quadrupole magnets located between the BPMs are at their nominal strength. For a set of three BPMs, there are two independent equations of the form (2.30), which we can solve for the two unknowns α and β [27].

To be more explicit, let us denote the transport matrix from BPM 1 to 2 by $\mathbf{M}$ and the matrix from BPM 1 to 3 by $\mathbf{N}$:

$$\mathbf{M}(1 \to 2) = \begin{pmatrix} m_{11} & m_{12} \\ m_{21} & m_{22} \end{pmatrix} \qquad \mathbf{N}(1 \to 3) = \begin{pmatrix} n_{11} & n_{12} \\ n_{21} & n_{22} \end{pmatrix} \tag{2.31}$$

and denote the phase advance from BPM 1 to 2 and that from 1 to 3 by ϕ_{21} and ϕ_{31}, respectively. Applying (2.30) twice, we arrive at two expressions for the values of α and β at the first BPM [27]:

$$\beta(s_1) = \left(\frac{1}{\tan\phi_{21}} - \frac{1}{\tan\phi_{31}}\right) \Big/ \left(\frac{m_{11}}{m_{12}} - \frac{n_{11}}{n_{12}}\right) \tag{2.32}$$

$$\alpha(s_1) = \left(\frac{n_{11}}{n_{12}\tan\phi_{21}} - \frac{m_{11}}{m_{12}\tan\phi_{31}}\right) \Big/ \left(\frac{m_{11}}{m_{12}} - \frac{n_{11}}{n_{12}}\right) . \tag{2.33}$$

Figure 2.13 gives an example of beta functions obtained by this method, and it also shows the optics correction achieved by changing the strength of a quadrupole pair. The method encounters numerical degeneracies if the phase advance between adjacent BPMs is $\frac{\pi}{2}$, since then $\phi_{31} = \pi$ and $\tan\phi_{31} = 0$.

2.3.3 Orbit Change at a Steering Corrector

A simple method to measure the local beta function at a steering corrector magnet next to a BPM consists of exciting the corrector and detecting the orbit change at that BPM [28].

The formula for the closed-orbit distortion $\Delta x_{\rm co}$ of a relativistic beam induced by a single dipole kick (measured by measuring the difference in beam orbits obtained with and without the kick) is

$$\begin{aligned} \Delta x_{\rm co}(s) = \Delta\theta \, & \frac{\sqrt{\beta(s)\beta(s_0)}\ \cos(|\phi(s) - \phi(s_0)| - \pi Q)}{2\ \sin\pi Q} \\ & + \Delta\theta \, \frac{D_x(s) D_x(s_0)}{(\alpha_c - 1/\gamma^2)C} \end{aligned} \tag{2.34}$$

where s is the location of the BPM, C the ring circumference, α_c the momentum compaction factor defined as the relative change in circumference per relative momentum deviation or $\alpha_c \equiv (\Delta C/C)/(\Delta p/p)$, and s_0 the location where the kick ($\Delta\theta$) is applied. The last term is a small correction reflecting the change in beam energy induced by a kick at a dispersive location, for a bunched beam and constant rf frequency. If the locations s and s_0 are the same, and if we ignore the small correction due to the energy change, the formula simplifies, and the beta function at the BPM-corrector pair can be obtained from

$$\beta_{\mathrm{BPM/cor}} \approx 2 \ \tan \pi Q \ \frac{\Delta x_{\mathrm{co}}}{\Delta \theta} \ . \tag{2.35}$$

2.3.4 Global Orbit Distortions

One can also infer the beta functions at all BPMs simultaneously by measuring the average (static) orbit response to two, or more, steering correctors [29]. We call x_{ia} and x_{ib} the orbit change measured at the ith BPM when deflections θ_a or θ_b are applied at the correctors a or b. The beta function is computed from the relation [29]

$$\beta_i = \frac{4 \sin^2 \pi Q}{\sin^2 \Delta} \left(\frac{x_{ia}^2}{\beta_a \theta_a^2} + \frac{x_{ib}^2}{\beta_b \theta_b^2} - \frac{2 x_{ia} x_{ib} \cos \Delta}{\sqrt{\beta_a \beta_b} \theta_a \theta_b} \right) , \tag{2.36}$$

where $\Delta = |\phi_a - \phi_b|$ is the phase advance between the two correctors, which should not be a multiple of π. Prior to applying this equation, the three quantities Δ, θ_a and θ_b are determined by fitting a few BPM readings $x_{ja,b}$ in the vicinity of the correctors to the model optics. The computed beta functions can be verified by exciting other corrector pairs in different sections of the ring and comparing the results. Figure 2.14 shows an example from the KEK B factory, where this method was developed.

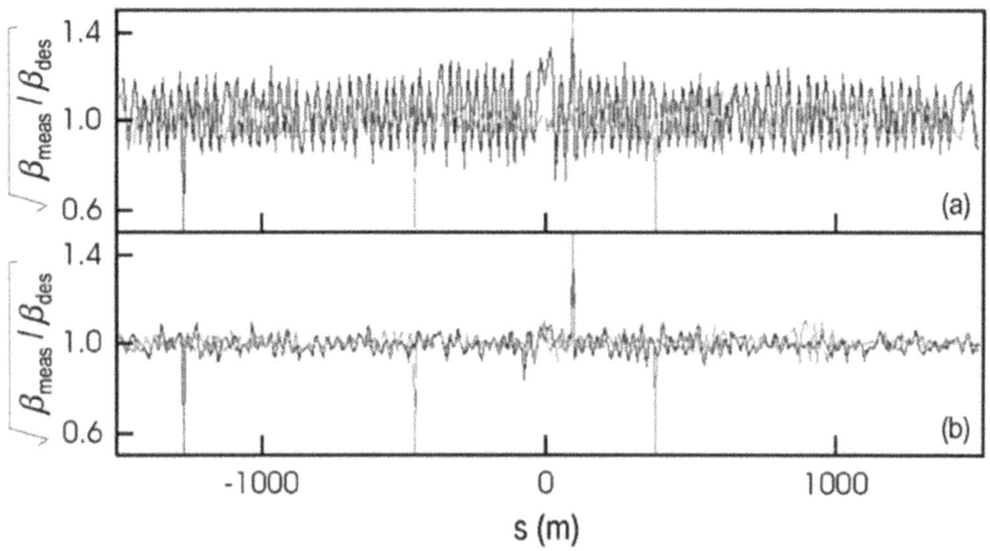

Fig. 2.14. Beta function measurement at KEKB, based on (2.36). Ratio of measured β function to the design value is shown before (*top*) and after optics correction (*bottom*) (Courtesy H. Koiso, 2000)

Fitting difference orbits to a betatron oscillation is a simpler variant, which has been used at many accelerators. An example from the heavy-ion collider RHIC is shown in Fig. 2.15.

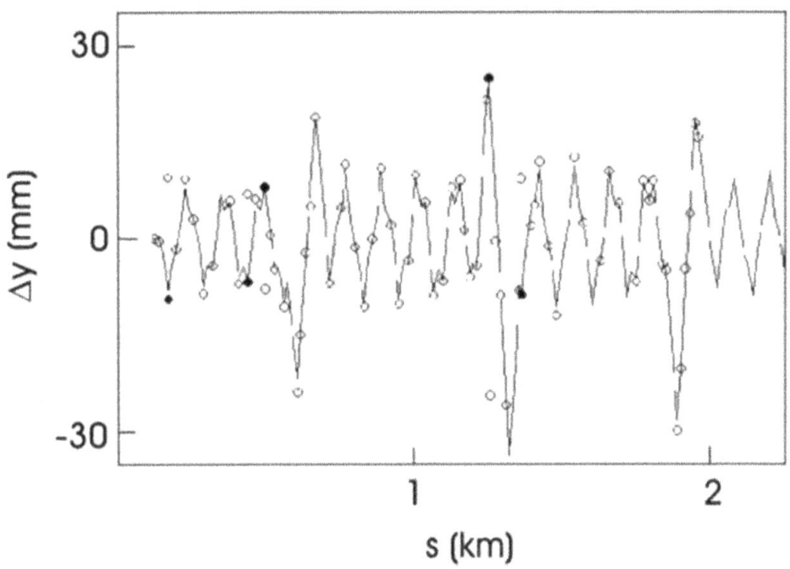

Fig. 2.15. Measured difference orbit in RHIC (*circles*) together with a fit using the online model (*dashed line*) [30] (Courtesy S. Peggs, 2000)

2.3.5 β^* at Interaction or Symmetry Point

To determine the beta function at the interaction point of a collider ring, or at any other symmetry point, e.g., in a light source), one can excite a pair of symmetrically placed quadrupoles, by an amount $\pm\Delta K$ (asymmetric excitation), where K is the integrated quadrupole gradient in units of m^{-1}. From (2.29), the total tune shift is given by

$$\Delta Q_{\mathrm{tot}} = \Delta Q_+ - \Delta Q_- \approx \frac{\Delta K}{4\pi} \left[\langle\beta_+\rangle - \langle\beta_-\rangle\right] , \tag{2.37}$$

where $\langle ... \rangle$ indicates the average over the effective length of the quadrupole and the $\pm$ sign refers to the left or right quadrupole. The advantage of the asymmetric excitation of two quadrupoles is that if the phase advance between the two quadrupoles is about 180 degrees, which is usually the case, almost no beta-beating is induced around the accelerator. In addition, if the optics is perfect and the beam waist is centered at the collision (or symmetry) point, the beta functions at the two quadrupoles are the same and, to first order, there is no net tune change: $\Delta Q_+ - \Delta Q_- = 0$.

The beta function at the collision (symmetry) point β^* is a quadratic function of the ratio

$$\eta = \langle\beta_+\rangle - \langle\beta_-\rangle = 4\pi \, \frac{\Delta Q_+ - \Delta Q_-}{\Delta K} , \tag{2.38}$$

which takes the form [31]

$$\beta^* = \beta^*_{\text{design}} \left(1 + a_{\text{optics}} \eta^2\right) , \tag{2.39}$$

where β^*_{design} is the nominal interaction-point beta function. The coefficient a_{optics} depends on the optics between the quadrupoles which are being varied and the interaction point and can be calculated with using an optics program, e.g., MAD [32]. For the LEP low-β insertions, $a_{\text{optics}} \approx 1/15$ [31]. The optics is optimally adjusted when $\Delta Q_{\text{tot}} = 0$.

2.3.6 R Matrix from Trajectory Fit

Consider a set of three BPMs, which are not a multiple of π apart in betatron phase and with nonzero dispersion for at least one. The horizontal orbit readings at these three BPMs, which we denote by $x^{(1)}$, $x^{(2)}$, and $x^{(3)}$, then contain complete information about the betatron motion (x and x') and the energy offset (δ) of each trajectory. This means we can express the orbit at every other BPM as a linear combination of the orbit reading for these three BPMs:

$$x(s) = B(s)x^{(1)} + C(s)x^{(2)} + D(s)x^{(3)} \tag{2.40}$$

If the three BPMs are adjacent and the optics between them is known, (2.40) is equivalent to the more familiar form given in terms of the transport matrix elements as

$$x(s) = R_{11}^{s_0 \to s} x(s_0) + R_{12}^{s_0 \to s} x'(s_0) + R_{16}^{s_0 \to s} \delta, \tag{2.41}$$

because then the three variables $x(s_0)$, $x'(s_0)$ and δ are known linear combinations of $x^{(1)}$, $x^{(2)}$ and $x^{(3)}$.

If multiple data sets for many different turns (in a ring) or for many bunch passages (in a transport line) are acquired, the coefficients $B(s)$, $C(s)$, and $D(s)$, or equivalently the R matrix elements R_{11}, R_{12} and R_{16} may be obtained by a fitting procedure. However, care has to be taken: a simple least squares fit may produce spurious results if the BPM readings on the left and right side of (2.41) have noise components.

The effect of the noise in the horizontal coordinates can be illustrated by a simple example, taken from [33]. Consider a linear fit of the form $y = px + q$, where p and q are to be determined and both x and y are smeared stochastically. Figure 2.16 shows the reconstructed slope normalized to the true slope as a function of the signal-to-noise ratio in the horizontal coordinate, R_x. Even for a signal-to-noise ratio of 3 the fitted slope still has a 10% error. This result is independent of the noise in the y coordinate.

A better approach, which takes into account the noise in the horizontal coordinates, is to 'find the principal axes of the set of data points and then turn the parameter vector parallel to the principal axis along which the data points fluctuate the least' [33]. The general problem and its solution are as

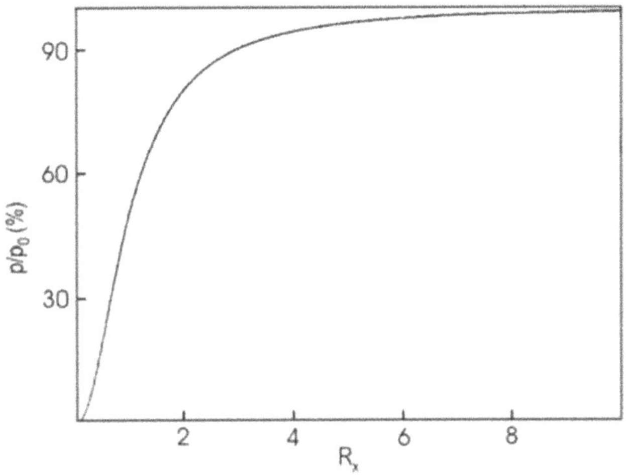

Fig. 2.16. Systematic slope error introduced in a linear χ^2 fit of the form $y = px+q$, neglecting the noise smearing in the x measurement [33]. Shown is the reconstructed slope normalized to the true slope as function of signal-to-noise ratio R_x in the horizontal coordinate (Courtesy P. Emma, 1998)

follows [33]. Let x_n be a measured variable which is linearly correlated with $(n-1)$ other measurements $x_1, ..., x_{n-1}$, and suppose there are a total of N data sets. Introducing normalized coordinates,

$$z_i = \frac{x_i - \langle x_i \rangle}{\sigma_i}, \tag{2.42}$$

the fit equation (2.40) or (2.41) may be reexpressed as

$$\mathbf{u}^T \cdot \mathbf{z} = 0\,. \tag{2.43}$$

Introducing the symmetric covariance matrix

$$C_{ij} = \sum_{l=1}^{N} z_{li} z_{lj}, \tag{2.44}$$

Equation (2.43) may then be solved in a least squares sense by

$$\mathbf{C} \cdot \mathbf{u} = \lambda u\,, \tag{2.45}$$
$$|\mathbf{u}|^2 = 1\,, \tag{2.46}$$
$$\chi^2 = \lambda\,, \tag{2.47}$$

where the solution $\mathbf{u}$ is simply the normalized eigenvector corresponding to the smallest eigenvalue λ of $\mathbf{C}$, where λ gives also the χ^2 of the fit.

If we assume that $\lambda = \lambda_1$ (the smallest eigenvalue) is not degenerate, and consider a scalar function $f(\mathbf{u})$ of the fit parameters $\mathbf{u}$, the rms fit error in f is given by

$$\sigma(f)^2 = (\nabla_{\mathbf{u}} f)^T \cdot \mathbf{T} \cdot (\nabla_{\mathbf{u}} f)\,, \tag{2.48}$$

where $\mathbf{T}$ is a symmetric $n \times n$ matrix defined by

$$T_{ij} = \sum_{r=2}^{n} \frac{\lambda_r + \lambda}{(\lambda_r - \lambda)^2} (u_r)_i (u_r)_j \,. \tag{2.49}$$

In particular, the rms error of the coefficient u_i in the normalized equation (2.43) is $\sigma(u_i) = \sqrt{T_{ii}}$.

The reconstruction of lattice parameters from orbit and energy fluctuations can be studied by computer simulations. Figure 2.17 presents simulation results for the SLC final focus [33] with an assumed BPM resolution of 20 μm, employing both a standard χ^2 fit and a principal axes transformation. The results of the former differ strongly from the underlying model parameters despite the good fits and small error bars. The improvement using the principal axes method shows that the optics are reconstructed almost perfectly.

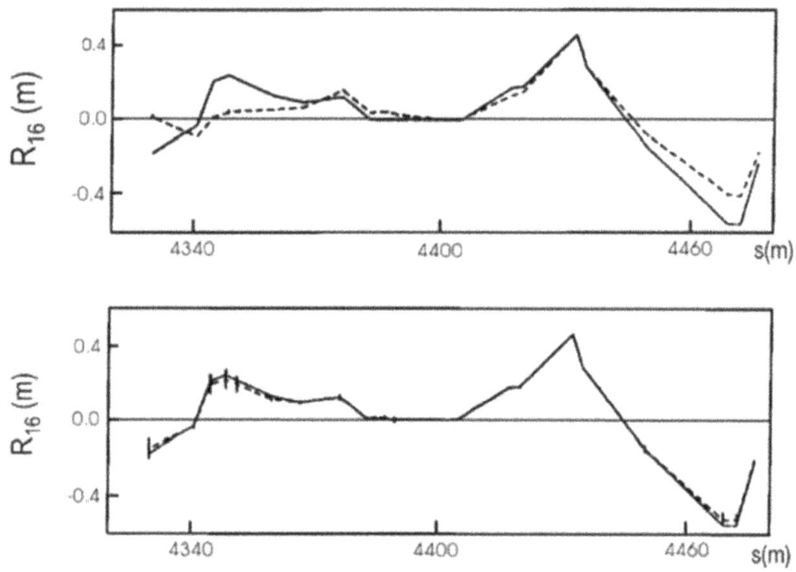

Fig. 2.17. Reconstructed R matrix element R_{16} in the SLC final focus from a sample of 100 simulated trajectories with fluctuations in betatron orbit and energy assuming a 20 μm BPM resolution [33]: (*top*) standard χ^2 fit; (*bottom*) principal axes transformation. The fit results (*dashed*) are compared to the model used for the trajectory generation (*solid*) (Courtesy P. Emma, 1998)

A Fitting Example. As a much simpler illustration, let us consider the single linear equation

$$y = px \,, \tag{2.50}$$

and assume that N pairs of data (x_i, y_i) are taken. The slope p is to be determined from a fit. A standard fit minimizes the χ^2 defined as

$$\chi^2 = \frac{1}{N} \sum_{i=1}^{N} (y_i - px_i)^2 \,. \tag{2.51}$$

Setting the derivative $\mathrm{d}\chi^2/\mathrm{d}p$ equal to zero, one obtains

$$\sum_i x_i y_i = \sum_i x_i^2 p = 0 \tag{2.52}$$

or

$$p = \frac{\sum_i x_i y_i}{\sum_i x_i^2} \,. \tag{2.53}$$

In the alternative 'eigenfit' method [33] we instead minimize

$$\chi^2 = \frac{1}{N}\sum_{i=1}^{N}(u_1 y_i - u_2 x_i)^2 + \lambda(1 - u_1^2 - u_2^2) \,, \tag{2.54}$$

where λ is a Lagrange multiplier. Now, three derivatives are set to zero: $\mathrm{d}\chi^2/\mathrm{d}u_1 = \mathrm{d}\chi^2/\mathrm{d}u_2 = \mathrm{d}\chi^2/\mathrm{d}\lambda = 0$. The last equation constrains the magnitude of the vector $\boldsymbol{u} = (u_1, u_2)$: $u_1^2 + u_2^2 = 1$. The others can be written as a system of two linear equations:

$$\begin{pmatrix} \sum_i y_i^2 & \sum x_i y_i \\ \sum_i x_i y_i & \sum x_i^2 \end{pmatrix} \begin{pmatrix} u_1 \\ u_2 \end{pmatrix} = \lambda \begin{pmatrix} u_1 \\ u_2 \end{pmatrix} \,. \tag{2.55}$$

Suppose now that the real slope is $p = 1$, that the real x assumes the values $+1$ and -1 with equal probability, and that the data points for x and y have a random measurement error of ± 1. As an example, this is fulfilled for an 8-component real vector $\boldsymbol{x}$ equal to

$$\boldsymbol{x} = (-1, -1, -1, -1, +1, +1, +1, +1) \,, \tag{2.56}$$

a measured vector $\boldsymbol{x}_{\mathrm{meas}}$

$$\boldsymbol{x}_{\mathrm{meas}} = (-2, -2, 0, 0, 0, 0, +2, +2) \,, \tag{2.57}$$

and a measured vector $\boldsymbol{y}_{\mathrm{meas}}$

$$\boldsymbol{y}_{\mathrm{meas}} = (-2, 0, -2, 0, 0, +2, 0, +2) \,. \tag{2.58}$$

The relevant sums over the measured values are $\sum_i x_i^2 = 16$, $\sum_i y_i^2 = 16$, and $\sum_i x_i y_i = 8$. Hence, the standard fit (2.53) yields a slope $p = 1/2$, which is off by a factor of two. The eigenfit yields the characteristic equation $(16 - \lambda)^2 - 64 = 0$. Choosing the smallest solution $\lambda = 8$, one finds $u_2 = -u_1$, from which we can deduce that the slope is 1. This is the correct solution.

2.4 Detection of Quadrupole Gradient Errors

Once the beta functions have been measured and a significant difference from the model has been found, the source of the discrepancy must be determined. In most cases, the difference from the model beta function will be a beta beat (an oscillation of the measured beta function around the design beta function at twice the betatron frequency) and the source will be a gradient error in one (or more) of the quadrupole magnets.

A gradient error ΔK (in units of m^{-1}) at location s_0 will result in a beta beat of the form

$$\Delta\beta(s) = \frac{\beta(s)\beta(s_0)}{2\sin(2\pi Q)}\,\Delta K(s_0)\,\cos(2|\phi(s)-\phi(s_0)|-2\pi Q)\,. \qquad (2.59)$$

2.4.1 First Turn Trajectories

In a storage ring, a first attempt to find the error may consist of exciting steering correctors (or changing the amplitude of the injection kicker) and fitting first-turn difference trajectories to an on-line or off-line optics model. The difference of two trajectories measured for different injection amplitudes should match a betatron oscillation as predicted by the model. Using corrector excitations, this method applies also to gradient error detection in linear accelerators and transport lines.

The initial conditions can be determined by fitting the difference orbit, using only a few BPMs, to the model. The oscillation so obtained is then propagated along the beam line. It will agree with the measured difference trajectory until it passes the location of the gradient error, at which point the propagated betatron oscillation and the measurement will start to disagree. The location of the gradient error thus identified can be confirmed by back-propagating the solution obtained with an independent set of BPMS further downstream. If the previous hypothesis is correct, the backward fit should begin to deviate from the model at the same point as the forward fit.

In principle, by analyzing first-turn orbits gross optics errors are easily identified. In practice, it is not always so simple, as beam loss, BPM spray (from lost particles), or kicker noise may corrupt the BPM measurements on the first few turns.

2.4.2 Closed-Orbit Distortion

A variant of this method is to make use of the fact that, except for the location of the corrector, a closed-orbit distortion for a stored beam has exactly the same pattern as a betatron oscillation. Thus, in much the same manner as for the first turn, the model can be used to fit the change in the closed orbit (with and without corrector excitation) to a betatron oscillation and then to propagate this oscillation around the ring. Again, the location where a noticable disagreement starts identifies the magnet with a gradient error. The excitation of this magnet can be changed, and the measurement repeated, until the agreement with the model is satisfactory. Figure 2.18 shows an example of this method from the PEP-II commissioning. A gradient error close to the interaction point was clearly identified.

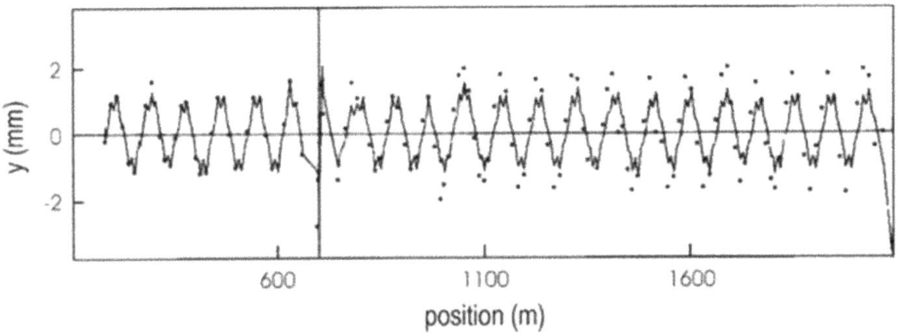

Fig. 2.18. Finding quadrupole gradient errors by fitting betatron oscillations to closed-orbit distortions: an example from the PEP-II HER commissioning using the codes LEGO [34] and RESOLVE [35]. The induced orbit change is fitted to a betatron oscillation over a small number of BPMs (*towards the left*). The betatron oscillation so obtained was propagated along the beam line using the model optics (*solid line*) and compared with the actual orbit variation (*symbols*). In this example, the measurements and fit agree well up to a region close to the interaction point, near $s = 700$ m on the horizontal axis. It was later verified that two quadrupole pairs in this region had gradient errors on the order of 0.1% (Courtesy Y. Cai, 1998)

It is possible to considerably extend this simple closed-orbit distortion scheme. For example, the response of all BPMs to every single steering corrector may be combined into a big matrix, which can be used as an input to a sophisticated statistical fitting program, such as LOCO [36, 37]. LOCO then varies the individual gradients of the quadrupoles in a computer model, e.g., using MAD [32], to find the modified quadrupole gradients that best reproduce the measured orbit response data. The advantage of using multiple data sets and multiple fits lies therein that the numerical solutions are overconstrained (which reduces the influence of systematic errors) and that multiple error sources, if present, can be more accurately ascertained.

2.4.3 Phase Advance

Instead of fitting trajectories, one can also use (2.32) to compute the beta functions from the measured phase advance around the ring. Then either the quadrupoles fields may be adjusted in the model or the actual magnet settings of the accelerator may be changed to improve the agreement between the measured and predicted phase advance and so identify the source of the discrepancy.

An example from PEP-II has been presented in Fig. 2.13 [38]. From top to bottom the improved agreement of model and measurement is clear as the strength of a quadrupole pair in the interaction region was changed by a total of 0.15%. For each quadrupole value, the left column shows the entire ring while the right column shown an expanded view of a particular section. As can be seen, the final quadrupole strength (bottom) yields a satisfactory agreement with the model.

2.4.4 π Bump Method

Another method which can be used to identify local gradient errors is the π-bump technique applied at Tristan and at the ATF [39, 40, 41]. Here, local orbit bumps are induced, one by one, across each quadrupole magnet (or across small groups of quadrupole magnets), and the orbit is measured. A non-closure of a theoretically closed bump is indicative of an optics error within the region of the bump. Of course, an error in the calibration of the bump dipole magnets may also result in non-closure of the bump, but the effects of dipole errors and optics errors can be separated by their betatron phase. In particular, for an ideal π-bump, the bump "leakage", or non-closure, due to a gradient error and that due to a dipole error are perpendicular to each other. In more complex situations, computer programs can be used to process multiple measurements involving overlapping bumps to determine both the dipole and the focusing errors.

2.5 Multiknobs, Optics Tuning, and Monitoring

A problem frequently encountered in practice is to correct an aberration or to match one optical function without degrading others. For example, we may want to cancel a residual dispersion without affecting the beta functions. This can be done very efficiently by using a "multiknob", which is a combination of quadrupoles and skew quadrupoles (or possibly dipole correctors, sextupoles, even octupoles,...) that are simultaneously changed with the proper ratio and relative sign in such a way that only the aberration of interest is generated, i.e., corrected.

Such knobs are very powerful. For example, by scanning a dispersion knob, one can minimize the beam size on a downstream profile monitor or wire scanner, thus eliminating any residual dispersion. In a ring, similar knobs may be used to correct skew coupling and dispersion: by minimizing the beam size on a synchrotron light monitor with these knobs, the vertical equilibrium emittance can be reduced.

Multiknobs were used to minimize the spot sizes at the SLC interaction point (IP) and so produce the maximum luminosity. The spot sizes at the IP, which could be estimated from scans of the beam-beam deflection angle versus the offset between the two beams [42, 43, 44, 45], were routinely optimized by correcting the most important low-order aberrations including the waist shift, dispersion and skew coupling of either beam using the multiknobs which consisted of orthogonal linear combinations of quadrupoles and/or skew quadrupoles. For many years, the fine-tuning with colliding beams was performed on an hourly basis by measuring the spot size for different (typically 5–7 knob values) and adjusting each knob to the best value, as determined by fitting the square of the measured spot size as a function of the knob setting to a parabola.

A typical set of optimization scans for both beams is illustrated in Fig. 2.19. Small waist shifts were obtained by changing the strength of the last three quadrupoles (the 'final triplet') and of one quadrupole further upstream. Part of the linear coupling was corrected by means of a skew quadrupole at the same betatron phase as the final triplet. The residual dispersion at the IP was corrected using two normal and two skew $-I$ quadrupole pairs in the chromatic correction system (CCS), excited equally with opposite sign. Chromaticity was compensated by the two pairs of CCS-sextupoles. The strength of these sextupoles could also be varied to minimize the vertical spot size and, thus, to minimize the change in focusing with particle energy. As it turned out, the IP spot sizes proved to be relatively insensitive to the exact

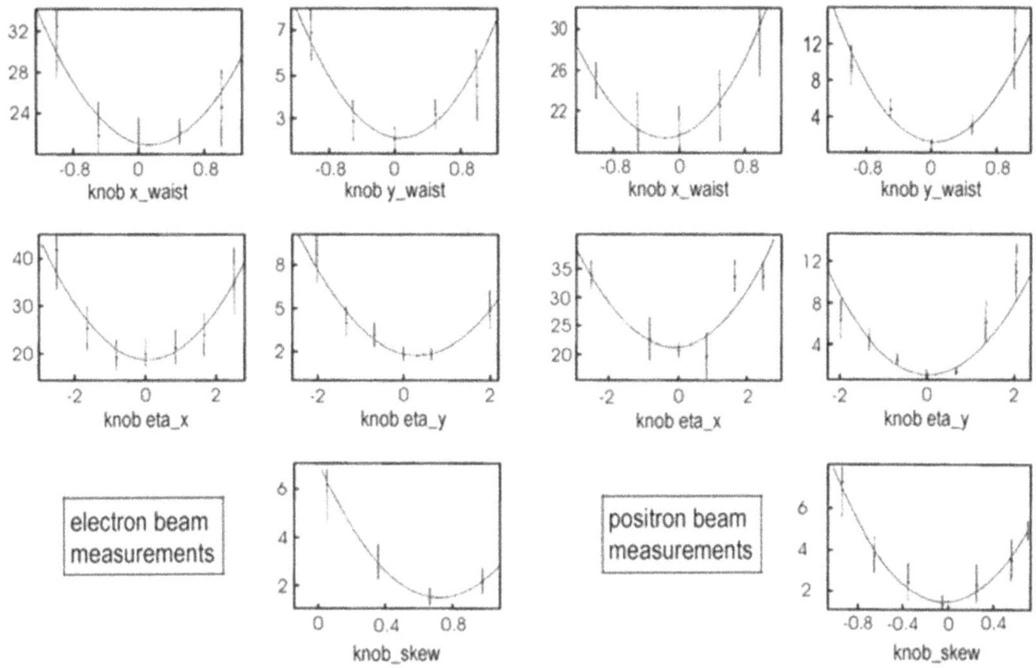

Fig. 2.19. Aberration scans at the SLC interaction point. Shown is the square of the measured horizontal or vertical convoluted spot size of the colliding electron and positron beams in units of μm^2 as a function of the multiknob setting in arbitrary units. In total 10 aberrations were controlled: the waist position, the horizontal and vertical dispersion, and the coupling from the horizontal into the vertical plane (for both beams). All these aberrations were corrected using different combinations of quadrupole and skew quadrupole magnets, which formed orthogonal multiknobs. Each multiknob was adjusted so as to optimize the IP spot size, which was determined using a parabolic fit to the beam sizes (squared). Until 1997, the beam size was inferred from the beam-beam deflection scans, with results as shown in the figure. The errors were significant [46]. In the last years of SLC operation, the aberrations were controlled and the luminosity was optimized using a new automatic 'dither' feedback, which made up/down changes to the multiknobs, and determined their effect on the luminosity from the correlated variation of the beamstrahlung-induced energy loss or other luminosity-related signals, such as the number of radiative Bhabha events [47]

settings of these chromatic sextupoles. In addition, several normal and skew sextupoles at non-dispersive locations in the final transformer were used for the correction of second-order geometric aberrations.

In general, the multiknob coefficients may be calculated in a variety of ways. A straightforward approach is to use the matching functions of the MAD program [32] to determine the relative changes in quadrupole strengths, ΔK_i, required to vary the parameter of interest (such as the beta function or the dispersion at some position).

A second approach to generate multiknobs involves using a singular value decomposition singular valued decomposition (SVD), which is a procedure commonly used for solving systems of linear equations with either too many or too few variables. The problem can be cast into a matrix equation of the form

$$\begin{pmatrix} \Delta\beta_x \\ \Delta\alpha_x \\ \Delta\mu_x \\ \Delta\eta_x \\ \Delta\eta'_x \\ \Delta\beta_y \\ \Delta\alpha_y \\ \Delta\mu_y \\ \Delta\eta_y \\ \Delta\eta'_y \\ \dots \end{pmatrix}_{s=s_0} = \begin{pmatrix} B_{11} & B_{12} & \dots & B_{1N} \\ B_{21} & B_{22} & \dots & B_{2N} \\ & \dots & \dots & \\ B_{M1} & B_{M2} & \dots & B_{MN} \end{pmatrix} \begin{pmatrix} \Delta K_1 \\ \Delta K_2 \\ \dots \\ \Delta K_N \end{pmatrix} = BK\,. \tag{2.60}$$

The sensitivity matrix **B** may be obtained using an optics model or it may be determined empirically by measurement. To optimally constrain the solution, the number of adjustable parameters N should be larger or equal to the number of constraints M. So unless there is an optical symmetry that can be exploited, one should require $N \geq M$.

We postpone the detailed discussion of the SVD algorithm to Chap. 3. Here, we only note that an explicit solution using SVD will be presented for (3.28). Since (2.60) is of exactly the same form, it can be solved accordingly. If, for example, we want to create a dispersion knob which changes ΔD_x by 1 mm at location $s = s_0$ while keeping all other parameters constant, we can solve this problem by SVD. The latter will determine a set of changes in the quadrupole strengths $(\Delta K_1, \dots, \Delta K_N)$, which fulfills the objective and which simultaneously minimizes the overall magnitude of the changes, i.e., the sum $\sum_i (\Delta K_i)^2$. This scheme can be generalized to include higher-order optics in an obvious way.

The knob coefficients are calculated only for small changes to the intermediate optics. If many quadrupoles are part of the knob, the R matrices between them will change as the knob is being varied. This means the coefficients determining the knob are not constant, but change depending on the knob setting. If needed, this problem can be overcome by using nonlinear knobs, where the differential change in quadrupole strengths is recalculated in many small steps, effectively performing an integration of (2.60), in order to determine the final quadrupole values at the desired knob setting. Such nonlinear knobs were developed for the SLC final focus [48].

In many cases, the beam optics varies diurnally, or is sensitive to outside temperature, air pressure, etc., with associated variations in the accelerator performance. A continuous optics monitoring on relevant time intervals can yield a better understanding of the source of the perturbation and may also serve as a first step towards a stabilizing feedback. As an example, in the SLC linac a diagnostic pulse was used to monitor, quasi-continuously, the stability of the linac optics [49]. Every few minutes fast kickers induced betatron oscillations for individual selected bunches, prior to their injection into the linac. The propagation of the measured betatron oscillations through the linac was recorded, and then decomposed into an amplitude and a phase component. The betatron phase advance along the linac, thus inferred, proved a sensitive indicator of optics changes, which were caused, e.g., by variations in the linac energy profile due to drifts in the rf phase reference system, or by variations in beam current, bunch length, and bunch distribution.

2.6 Model-Independent Diagnostics

Some accelerators may be so complex and their optics so variable that it is not possible to establish a model which reproduces reality sufficiently well over longer periods of time as to be useful for diagnostics and monitoring purposes. A good example of such a situation is the SLAC linac. There, for example, the beam energy profile along the linac, which changed due to time-dependent drifts in the klystron (power source) phases, was impossible to determine precisely as the beam energy was measurable only in a few select locations. Therefore, the required (energy-matched) quadrupole gradients were not fully determined. Also transverse and longitudinal wakefields contributed to changes in the beam optics. These depended on the bunch charge, the bunch distribution, and the beam orbit.

A method employed to evaluate relative changes in the optics (determined in this case not only by lattice parameters but also by the beam parameters which were influenced strongly by hard to quantify beam-environment effects) entailed model-independent analysis of the orbit data. In such applications, no attempt is made to accurately determine the parameter set for an optics model, but the beam information and the beam response to certain perturbations are used directly to monitor the accelerator stability, to determine misalignments of accelerator structures and so on.

An interesting approach has been developed at SLAC by J. Irwin, C.-X. Wang, Y. Yan *et al.* [50, 51]. The primary quantity on which the analysis is based is a matrix of BPM readings $\mathbf{B}$, where

$$\mathbf{B} = \begin{pmatrix} b_{11}\ b_{12} \dots b_{1m} \\ \dots \dots \\ b_{p1}\ b_{p2} \dots b_{pm} \end{pmatrix} \tag{2.61}$$

has m columns, representing m different BPMs, and p rows, for p different beam pulses. In an actual application m may be of the order of 100 and p can be several 1000. There are many contributions to $\mathbf{B}$, for example changes in the initial conditions, changes in the beamline components, ground motion, BPM noise etc.

One can assume a linear (or quadratic) expansion of the form

$$\mathbf{b_i} = \mathbf{b_0} + \sum_{s=1}^{S} \Delta\hat{q}_i^s \left[\sigma_s \frac{\partial \mathbf{b}}{\partial q^s}\right] + \text{noise}\,, \tag{2.62}$$

where $\mathbf{b}$ denotes a row vector of the matrix $\mathbf{B}$, $\hat{q}^s$ represents the sth variable affecting the BPM readings (such as an incoming betatron oscillation, or klystron amplitude), and σ_s the rms variation of this perturbation. The variable $\hat{q}_i^s$ is normalized so that the rms value over time is one, or $\langle \hat{q}^s \rangle_{\text{rms}} = 1$, and $\langle \Delta q_s \rangle_{\text{rms}} = \sigma_s \langle \hat{q}^s \rangle_{\text{rms}} = \sigma_s$.

The above equation can be rewritten in matrix form as

$$\mathbf{B} = \mathbf{Q} \cdot \mathbf{F}^t + \eta\,, \tag{2.63}$$

where now η represents the statistical variations in the BPM readings, and the rows of the matrix $\mathbf{F}^t$ represent the sensitivity vectors $\mathbf{f} = [\sigma_s \partial \mathbf{b}/\partial q^s]$, weighted by the amount of variation (σ_s) detected in each variable.

The relative contribution of different variables q^s to the observed orbit variation can be identified fairly easily, provided certain tagging signals $\mathbf{t_i}$ (e.g., the bunch length, energy, or the incoming betatron oscillation inferred from the first few BPMs) are available and correlated to one (or more) of the source perturbations $\mathbf{f}_j$. For example, suppose the beam phase (arrival time) ψ and the current I are the tagging signals, Further suppose that their mutual correlation is not zero ($\langle \psi I \rangle \neq 0$) and that they are uncorrelated with all other perturbations. In this case one can multiply the matrix $\mathbf{B}$ by the two normalized tagging vectors $(\hat{\psi}, \hat{\mathbf{I}})$, obtaining

$$\begin{pmatrix} \hat{\mathbf{I}} \\ \hat{\psi} \end{pmatrix} \mathbf{B} = \begin{pmatrix} 1 & \langle \hat{\mathbf{I}}\hat{\psi} \rangle & \dots \\ \langle \hat{\mathbf{I}}\hat{\psi} \rangle & 1 & \dots \end{pmatrix} \begin{pmatrix} \mathbf{f}_I \\ \mathbf{f}_\psi \end{pmatrix} + \mathcal{O}\left(\frac{\sigma}{\sqrt{p}}\right)\,. \tag{2.64}$$

Equation (2.64) can be inverted, up to terms of order $\sigma/\sqrt{p}$. This inversion is of great interest because it provides the explicit 'space pattern' (shape) and the magnitude of the two orbit perturbations corresponding to changes in current or phase: $\mathbf{f_I}$, and $\mathbf{f}_\psi$.

For more refined studies, the SVD technique is again of great use. For example, via SVD we can decompose the matrix $\mathbf{B}$ as

$$\mathbf{B} = \mathbf{U}(\mathbf{W}\mathbf{V}^t)\,, \tag{2.65}$$

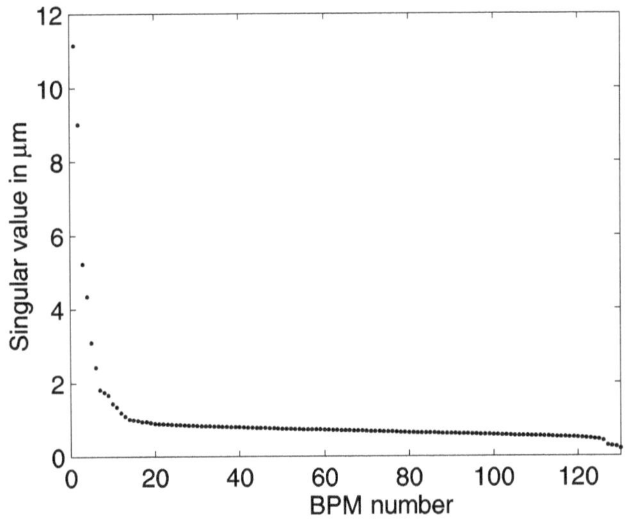

Fig. 2.20. Horizontal eigenvalues versus BPM number along the SLAC linac. In this plot 5000 pulses and 130 BPMs were analyzed

where $\mathbf{U}$ now represents the 'time patterns' (equivalent to $\mathbf{Q}$) and $(\mathbf{W}\mathbf{V}^t)$ the 'space patterns' (equivalent to $\mathbf{F}^t$) [50]. If we calculate the significant SVD eigenvalues w_j of the matrix $\mathbf{W}$ for an increasing number of BPMs (in the order of their position along the beamline), we can find locations where additional eigenvalues become large. An example is shown [52] in Fig. 2.20. The large eigenvalues indicate locations where either the trajectory measurement has become faulty or an additional orbit jitter has been introduced. The latter may be caused for by improper regulation of corrector dipole or quadrupole fields or by a structure misalignment, which translates an incoming bunch length variation into an orbit variation. Online monitoring of the number of nonzero eigenvalues may prove useful in future accelerators to, among other things, ensure functionality of the BPMs particularly those used in orbit feedback.

This method was been applied to determine the alignment of all structures along the SLAC linac [53]. Similar methods can of course be used at storage rings.

2.7 Coherent Oscillations and Nonlinear Optics

2.7.1 Beam Response to a Kick Excitation

If a beam is deflected, e.g., by a kicker, a coherent oscillation is observed on the beam-position monitors. This oscillation will ultimately decrease to zero. There are at least three effects contributing to this decrease:

- radiation damping in an electron storage ring;
- head-tail damping, if the chromaticity is positive (above transition)

- filamentation due to a spread in betatron frequencies within the bunch. If the frequency spread results from nonzero chromaticity or energy spread, the signal will recohere after one synchrotron period. Such a frequency spread may also arise from amplitude-dependent detuning, in which case there is no recoherence.

2.7.2 Coherent Damping

Beautiful data from LEP are shown in Figs. 2.21–2.24, where the beam was kicked and the resulting oscillation was fitted to a harmonic function with exponentially decreasing amplitude, $A = A_0 e^{-t/\tau}$ and amplitude-dependent frequency [54]. In LEP the decrease of centroid oscillations was dominated by the head-tail damping and by radiation damping. The damping rate could thus be written as

$$\frac{1}{\tau} = \frac{1}{\tau_{SR}} + \frac{1}{\tau_{HT}} + \dots , \tag{2.66}$$

where the head-tail damping rate $1/\tau_{HT}$ is proportional to the bunch current, the chromaticity and the real part of the impedance. The contribution from head-tail damping was examined by varying the chromaticity [54]; see Fig. 2.21. Taking data for various beam currents, and extrapolating to zero determined the radiation damping time, as illustrated in Figs. 2.21 and 2.22.

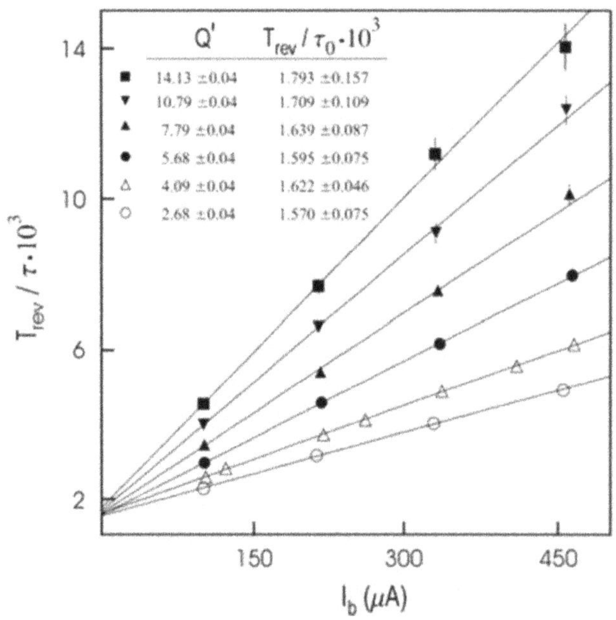

Fig. 2.21. Coherent damping rate in LEP as function of bunch current measured at a beam energy of 45.625 GeV for several chromaticities Q' [54]. The *straight lines* are fits to the individual samples. The *table* gives measured chromaticities and 'zero current' damping rates (Courtesy A.-S. Müller, 2001)

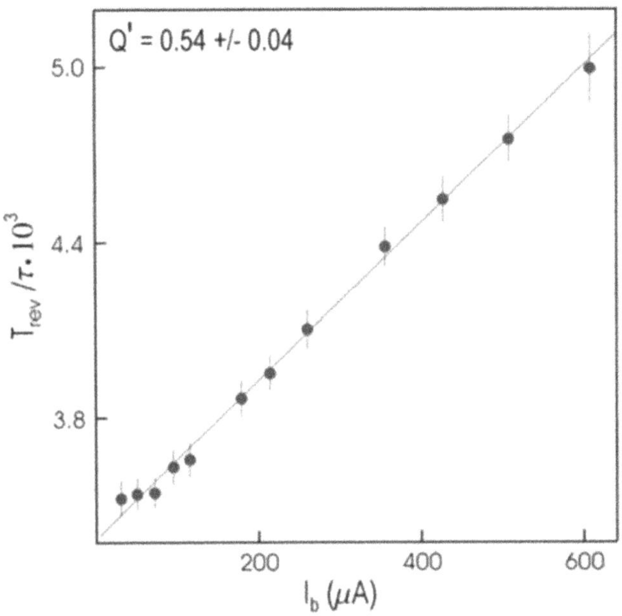

Fig. 2.22. Coherent damping rate in LEP as function of bunch current at one BPM with 60 GeV beam energy [54]. The *straight line* is a fit to the data (Courtesy A.-S. Müller, 2001)

2.7.3 Detuning with Amplitude

The tune shift with amplitude conveys information about the nonlinear fields experienced by the beam. Nonlinear fields have strong influence on beam stability and on the beam lifetime. Nonlinear fields result in an amplitude-dependent tune shift, since the average focusing experienced by a particle executing large betatron oscillations is changed.

Since, unlike filamentation, the radiation and head-tail damping change the real oscillation amplitude of individual particles, at LEP a single measurement could give the tune at various amplitudes (Fig. 2.23) and, from this, the detuning (see Fig. 2.24).

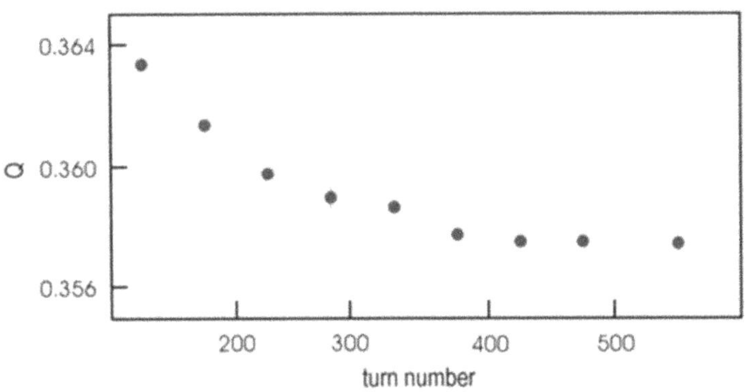

Fig. 2.23. Tune evolution inferred from a damped oscillation in LEP [54] (Courtesy A.-S. Müller, 2001)

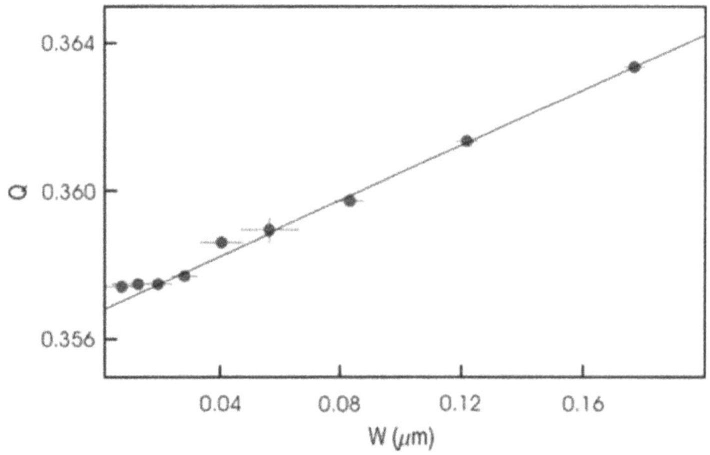

Fig. 2.24. Tune as a function of Courant-Snyder invariant with the 108/90 degree phase advance optics at an arc monitor in LEP [54]. The *straight line* is a fit to the data (Courtesy A.-S. Müller, 2001)

Earlier measurements at LEP and SPEAR also benefitted from the coherent damping, and determined the tune shift with betatron amplitude by applying an interpolated FFT with data windowing [55], as discussed in Sect. 2.1.2. In these LEP experiments, the beam was kicked in the vertical plane, and the tune was calculated over successive short time windows, of 32 turns each, as the beam oscillation damped rapidly. Figure 2.25 shows a result for LEP at 20 GeV. In Fig. 2.25, the vertical beam position is displayed as a function of time demonstrating the fast damping, over 200 turns. The shift of the horizontal and vertical betatron tune with the vertical action variable, as computed over time windows of 32 turns in length, is also shown. The vertical action was inferred from the os-

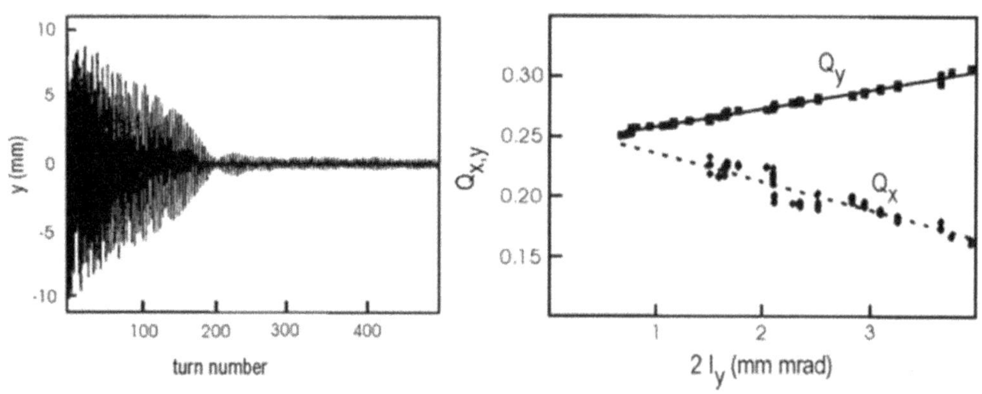

Fig. 2.25. Measurement of tune shift with amplitude in LEP at 20 GeV, using a high-precision FFT tune analysis [55]: (*left*) vertical oscillation amplitude after a kick; (*right*) horizontal and vertical betatron tunes vs. twice the vertical action variable I_y of the beam centroid motion. The observed tune shift with amplitude was consistent with the expected effect of the sextupole and octupole field components in the dipole magnets (Courtesy R. Bartolini, 1998)

cillation amplitudes in each time window. The amplitude-dependent tune shifts calculated by an off-line model and the measurement agreed to within 5%.

2.7.4 Filamentation due to Nonlinear Detuning

In other cases, notably at proton accelerators, the amplitude-dependent detuning will lead to a rapid reduction in the measured oscillation amplitude due to filamentation. In this case, to extract the amplitude-dependent detuning several measurements with different kick amplitudes may be required, and care is needed in the analysis of the signal.

An elegant mathematical formalism was developed by R.E. Meller and coworkers [56]. Considering a beam with a Gaussian transverse distribution, and further assuming that the decoherence of the centroid oscillation is due to a quadratic tune shift with amplitude a,

$$Q = Q_0 - \mu a^2 , \tag{2.67}$$

where a denotes the oscillation amplitude in units of σ and μ characterizes the strength of the nonlinear detuning, the centroid oscillation after a kick $\Delta x'$ is

$$\bar{x}(N) = \sigma_x Z A(N) \cos[2\pi Q_0 N + \Delta\bar{\phi}(N)] , \tag{2.68}$$

where N is the turn number (the kick is applied at $N = 0$), $Z = \beta \Delta x' / \sigma_x$ is the magnitude of the kick in units of σ_x, A the decoherence factor [56]

$$A(N) = \frac{1}{1+\theta^2} \exp\left[-\frac{Z^2}{2} \frac{\theta^2}{1+\theta^2} \right] \tag{2.69}$$

and $\bar{\phi}$ the centroid phase shift [56]

$$\Delta\bar{\phi}(N) = -\frac{Z^2}{2} \frac{\theta}{1+\theta^2} - 2 \arctan\theta . \tag{2.70}$$

The parameter θ contains the time dependence of the decoherence:

$$\theta = 4\pi\mu N . \tag{2.71}$$

Figure 2.26 shows the decoherence factor $A(N)$, (2.69), as a function of θ for four different normalized kick strengths. For moderate kicks (up to 1σ), we can estimate the detuning parameter μ from the number of turns, $N_{1/2}$, after which the initial signal amplitude has decreased by a factor of two, as

$$\mu \approx \frac{1}{4\pi N_{1/2}} . \tag{2.72}$$

For larger kicks, this approximation overestimates the value of μ.

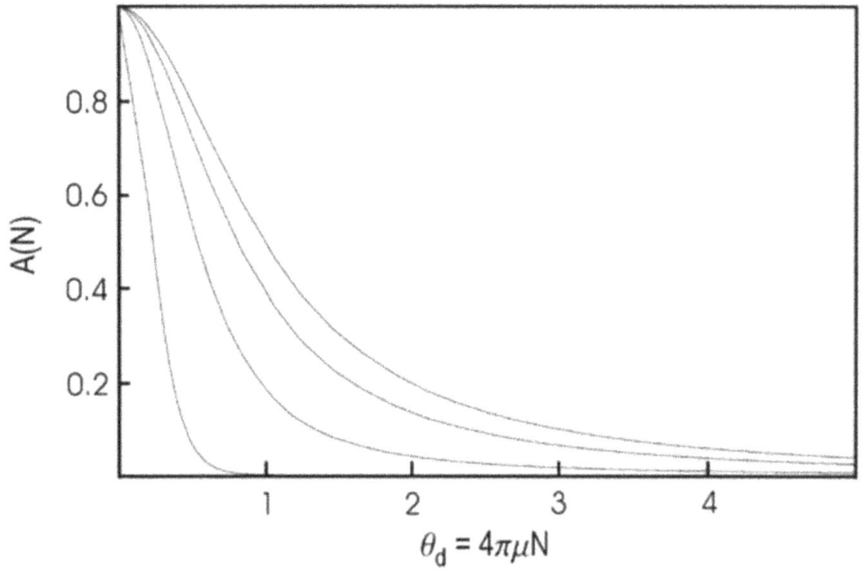

Fig. 2.26. Decoherence factor $A(N)$ as a function of $\theta = 4\pi\mu N$ for normalized kick strengths of $Z = 0.2$, 1, 2, and 5 (ordered from *top* to *bottom*)

2.7.5 Decoherence due to Chromaticity and Momentum Spread

Even without nonlinear detuning the beam signal can decohere due to the finite energy spread σ_δ and a non-vanishing chromaticity $Q' \neq 0$. In this case, the amplitude of the centroid motion after a kick, defined in (2.68), evolves as [56]

$$A(N) = \exp\left[-\frac{2\sigma_\delta^2 Q'^2}{Q_s^2}\sin^2(\pi N Q_s)\right] . \tag{2.73}$$

Note that (2.73) significantly differs from (2.69), so that the two effects can be experimentally separated. The signal described by (2.73) recoheres every full synchrotron period. From the signal modulation after a kick, one can infer the momentum spread if the chromaticity is known, and vice versa. In the case of chromaticity, the tune does not change: $\Delta\bar{\phi}(N) = 0$.

2.7.6 Resonance Driving Terms

Nonlinear magnetic fields not only induce a tune shift with amplitude, but they also excite higher-order resonances, which are visible as additional lines in the tune Fourier spectrum. A general betatron resonance is defined by the condition

$$kQ_x + lQ_y = p , \tag{2.74}$$

where k, l, and p are integers. Spectral analysis in the presence of nonlinear resonances [57] shows that in the Fourier spectrum of the horizontal coordinate $x(n)$, the above resonance gives rise to lines at the two frequencies

$(k \pm 1)Q_x + lQ_y$, and in the Fourier spectrum of the vertical signal $y(n)$ it generates lines at $kQ_x + (l \pm 1)Q_y$. Note that there is no line at $kQ_x + lQ_y$ [57], as one might have naively expected.

From amplitude, phase and frequency of the various spectral lines the dominant nonlinearities affecting the beam motion can be reconstructed [58, 59]. An example is shown in Fig. 2.27.

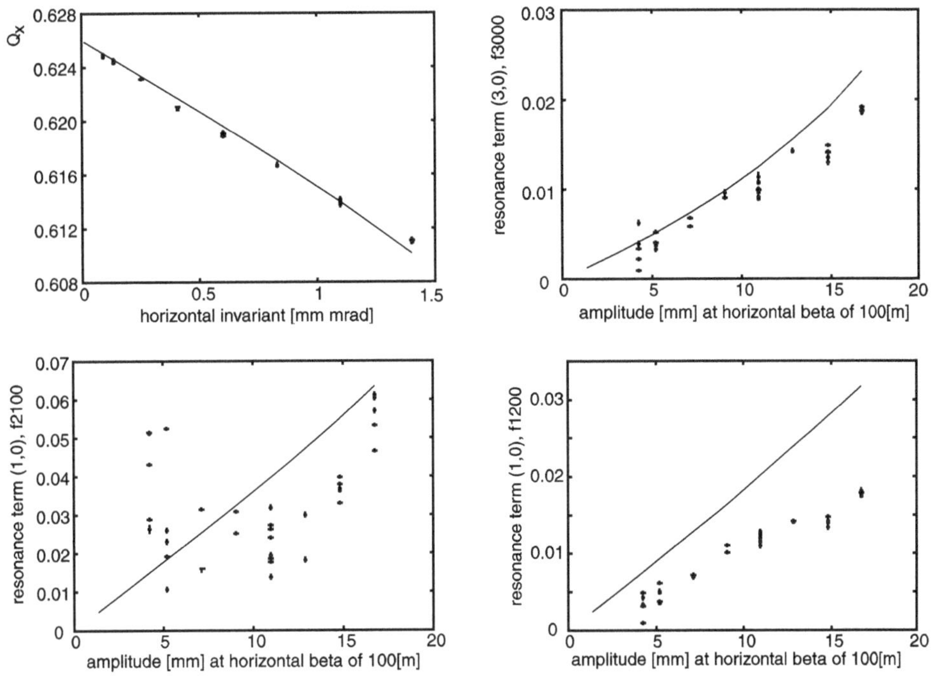

Fig. 2.27. Detuning and first-order resonance driving terms, measured at the SPS [60]. Shown are the betatron tune [*top left*] and the relative amplitude of several resonance lines with respect to the main tune line [*other*] as a function of the oscillation amplitude. Plotting symbols are experimental data; lines are from a tracking simulation. 170 turns were sampled per point (Courtesy F. Schmidt, 2000)

The nonlinearities may also be probed by adiabatically exciting large coherent betatron oscillations using an ac dipole [61]. This second method will be tested at the Relativistic Heavy Ion Collider (RHIC).

2.7.7 Tune Scans

The beam lifetime is often related to the dynamic aperture of the storage ring, where the term 'dynamic aperture' denotes the maximum stable betatron amplitude beyond which particles are lost after a certain finite number of turns. In the case of colliding beams, the lifetime is likely limited by the beam-beam interaction. Both dynamic aperture and the beam-beam interaction are sensitive to the value of the betatron tune. Measuring and plotting the beam lifetime as a function of the horizontal and vertical betatron tunes, Q_x and

Q_y, yields a tune diagram, in which higher-order resonances, given by (2.74), are evident as stripes with reduced lifetime. Figure 2.28 compares a typical beam-lifetime tune scan performed during the commissioning of the PEP-II High Energy Ring with the result of a dynamic-aperture simulation [62].

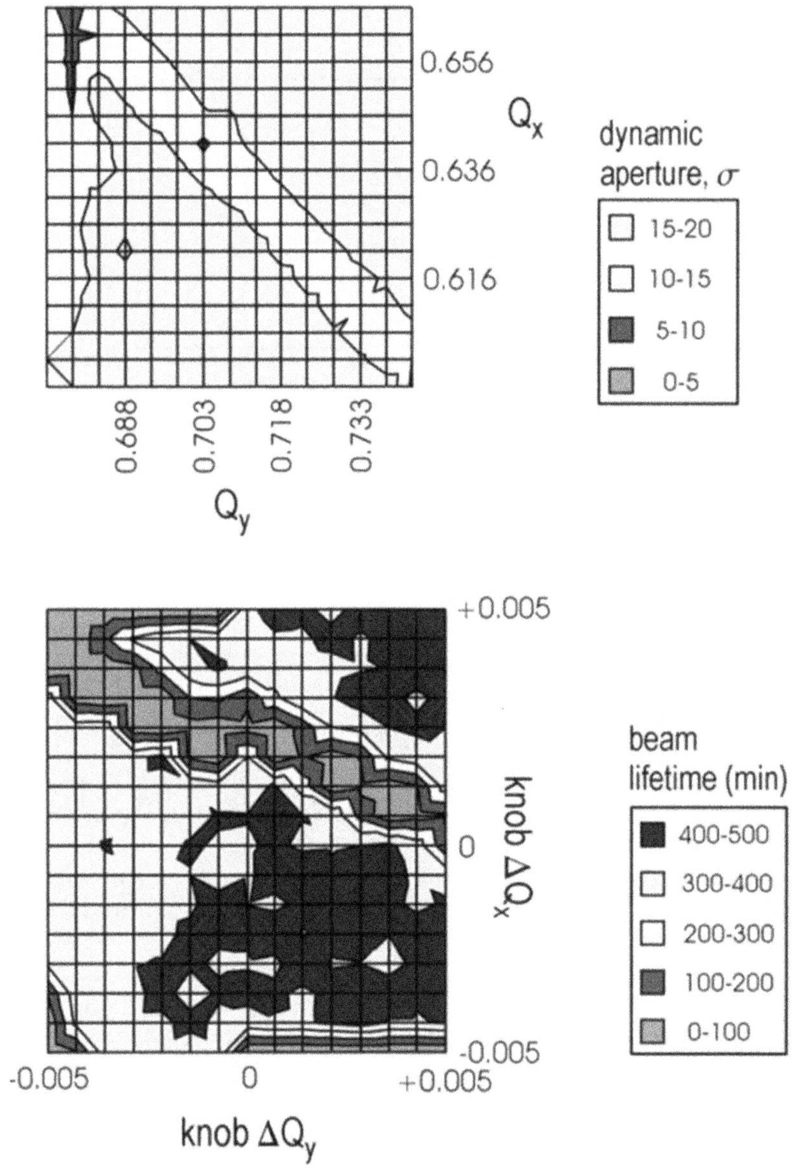

Fig. 2.28. Tune scan in PEP-II centered at $Q_x = 24.709$ and $Q_y = 23.634$ [62]: (*top*) simulated dynamic aperture (for a momentum offset $\Delta p/p = 10\ \sigma_\delta$, where δ is the rms momentum spread) as a function of the horizontal and vertical betatron tunes, Q_x and Q_y; (*bottom*) measured beam lifetime as a function of Q_x and Q_y. Total scan range is ± 0.005 on both axes. The different slope of the resonance line, as compared with the top figure, is attributed to a miscalibration of the tune knobs (Courtesy Y. Cai, 1998)

2.8 Betatron Coupling

Skew quadrupole field errors and detector solenoids generate betatron coupling between the horizontal and vertical planes of motion. Spurious betatron coupling is a concern, since it may reduce the dynamic aperture [63] and since, in electron accelerators, it contributes to the vertical equilibrium emittance. The coupling of horizontal and vertical oscillations generates two new eigenmodes of oscillation. These eigenmodes are no longer purely vertical or purely horizontal, but rather they correspond to oscillations whose reference planes are tilted and rotate with the azimuthal position s. In this case, new coupled beta functions can be defined [64, 65, 66, 67].

To illustrate the fundamental coupling phenomenon, we consider two coupled linear oscillators described by the equations

$$\frac{\mathrm{d}^2 x}{\mathrm{d}\theta^2} + Q_0^2 x = -\kappa y\,,$$

$$\frac{\mathrm{d}^2 y}{\mathrm{d}\theta^2} + Q_0^2 y = -\kappa x\,, \tag{2.75}$$

where κ is a constant describing the strength of the coupling. We can decompose the oscillation into two new normal modes. The normal-mode coordinates are

$$u = \frac{x+y}{2} \tag{2.76}$$

$$v = \frac{x-y}{2}\,. \tag{2.77}$$

Using (2.75) it is easily verified that the normal coordinates fulfill the uncoupled equations of motion

$$\frac{\mathrm{d}^2 u}{\mathrm{d}\theta^2} + (Q_0^2 + \kappa)u = 0\,, \tag{2.78}$$

$$\tag{2.79}$$

$$\frac{\mathrm{d}^2 v}{\mathrm{d}\theta^2} + (Q_0^2 - \kappa)v = 0\,. \tag{2.80}$$

The squared eigenmode frequencies are $Q_u^2 = (Q_0^2 + \kappa)$ and $Q_v^2 = (Q_0^2 - \kappa)$. Although the original oscillation frequencies were equal to Q_0 in both planes, x and y, the coupling introduces a frequency split between the two eigenmode frequencies $(Q_u^2 - Qv^2) = 2\kappa$, which is proportional to the coupling parameter κ.

2.8.1 Driving Terms

In a storage ring, the coupling between the planes is not constant around the ring, and as a consequence there are two important coupling parameters. These are the so-called two driving terms for the sum and difference resonances, which are given by [63, 67, 68]:

$$|\kappa_{\pm}| = \left| \frac{1}{2\pi} \oint ds\ k_s(s) \sqrt{\beta_x(s)\beta_y(s)}\ e^{\mathrm{i}(\phi_x \pm \phi_y - (Q_x \pm Q_y - q_{\pm})2\pi s/C)} \right| , \qquad (2.81)$$

where k_s is the normalized gradient of the skew quadrupole (in units of m^{-2}), $\phi_{x,y}$ the horizontal and vertical betatron phase, C is the circumference, $\beta_{x,y}$ are the uncoupled beta functions, and, in (2.81), we assume that the betatron tunes are near the sum or difference resonance

$$Q_x \pm Q_y + q_{\pm} = 0 , \qquad (2.82)$$

where $q_{\pm}$ is an integer. The dynamic aperture or the beam lifetime of colliding beams can be increased by measuring and minimizing the two driving terms $|\kappa_{\pm}|$.

In an electron storage ring, the vertical emittance contribution due to weak betatron coupling is [69]

$$\gamma\epsilon_y = \frac{C_q\gamma^3}{16 \oint G^2 ds} \oint \mathcal{H}_x |G^3| \left[\sum_{\pm} \frac{|W_{\pm}(s)|^2}{\sin^2 \pi(\Delta Q_{\pm})} + \frac{2\mathrm{Re}\left\{W_{+}^{*}(s)W_{-}(s)\right\}}{\sin \pi(\Delta Q_{+})\ \sin \pi(\Delta Q_{-})} \right] \mathrm{d}s , \qquad (2.83)$$

where $C_q = 3.84 \times 10^{-13}$ m, $\mathcal{H}_x$ is the horizontal dispersion invariant ($\mathcal{H}_x = (D_x^2 + (\alpha_x D_x + \beta_x D_x')^2)/\beta_x$), $G = 1/\rho$ the inverse bending radius, the star $*$ denotes the complex conjugate, Re gives the real portion of its argument, $\Delta Q_{\pm} = Q_x \pm Q_y + q_{\pm}$ characterizes the distance to the resonance (which should not be too small for the perturbation theory to be valid), and

$$W_{\pm}(s) = \int_s^{s+C} \mathrm{d}z\ k_s(s') \sqrt{\beta_x(s')\beta_y(s')} e^{\mathrm{i}[(\phi_x(s) \pm \phi_y(s)) - (\phi_x(s') \pm \phi_y(s')) + \pi(Q_x \pm Q_y)]} \qquad (2.84)$$

are the s-dependent driving terms, including all Fourier components. Note that $|W_{\pm}(0)| = |\kappa_{\pm}| 2\pi$, if there is a single skew quadrupole in the ring.

Equation (2.83) shows that, in order to minimize the vertical emittance, the driving terms $W_{\pm}(s)$, for the two nearest sum and difference resonances, should be corrected.

2.8.2 First Turn Analysis

Large coupling sources can be identified as locations where a horizontal orbit change generates a vertical kick and vice versa. In order to find such locations, the orbit is changed in one plane, by exciting steering correctors or by changing injection conditions (for example, the kicker amplitude), and the effect on the orbit in the perpendicular plane is measured. The same type of analysis can be applied to a transport line.

Large numbers of orbits and BPM data for excitations of different correctors can be fitted to determine the skew quadrupole component of each magnet in the beam line.

2.8.3 Beam Response after Kick

The driving term $|\kappa_-|$ may be measured by first kicking the beam, and then observing its response in the plane of the kick over *many turns*.

In the vicinity of the difference resonance, the envelopes of the oscillations in the horizontal and vertical plane exhibit a beating (energy exchange between the two planes) with a characteristic total modulation amplitude of [67, 70]

$$S = \frac{\hat{x}^2_{\mathrm{min}}}{\hat{x}^2_{\mathrm{max}}} \,. \tag{2.85}$$

Here $\hat{x}$ denotes the envelope of the betatron oscillation in the plane in which the kick was applied; $\hat{x}_{\mathrm{min}}$ is its minimum value, and $\hat{x}_{\mathrm{max}}$ its maximum value; these two extreme values are assumed alternately, with a modulation (or beating) period T. The driving term for the difference resonance, $|\kappa_-|$ of (2.81), is given by [70]

$$|\kappa_-| = \frac{\sqrt{1-S}}{f_{\mathrm{rev}} T} \,. \tag{2.86}$$

Thus measuring the modulation period T and the squared envelope ratio S after a kick is sufficient to infer $|\kappa_-|$.

An example from the ATF Damping Ring is shown in Fig. 2.29. The frequency spectrum from a horizontal BPM signal is viewed over a wide frequency range on a spectrum analyzer (top figure), and the frequency of the betatron signal is identified as the peak of the spectrum. The span of the spectrum analyzer is then set to zero, and its center set to the betatron frequency. This produces a signal proportional to the square of the betatron-oscillation amplitude. The output signal of the spectrum analyzer can be viewed on an oscilloscope, with results as displayed in Fig. 2.29 (bottom). The slow oscillation in this picture corresponds to synchrotron motion (the BPM is at a location with nonzero mismatched dispersion), while the fast beating reflects the transverse coupling. The picture was taken for a tune separation of $|Q_x - Q_y + q_-| \approx 0.02$. If the two tunes are separated further, the modulation period increases and the modulation amplitude decreases. Using (2.86) with $T \approx 17.6$ µs and $S \approx 0.3$–0.7, we infer a coupling term of $|\kappa_-| \approx 0.02$, consistent with other measurements [71].

It is of course possible to perform a much more detailed analysis of multi-turn BPM data. For example, one can determine the evolution of the coupled optical functions (e.g., the tilt angle of the two transverse eigenplanes) around the ring. An example may be found in [72].

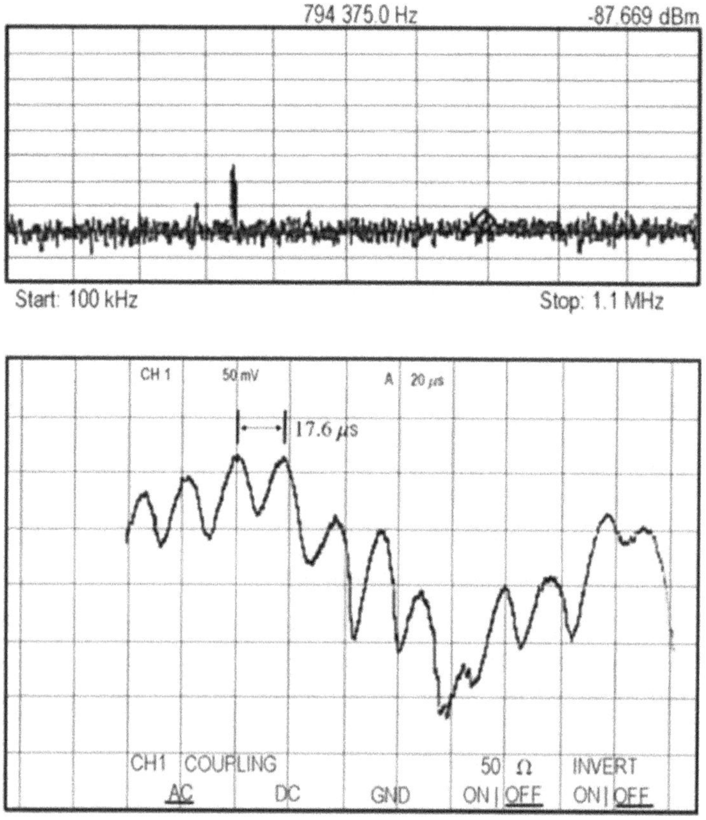

Fig. 2.29. Monitoring betatron coupling at the ATF Damping Ring [71]. (*top*) Frequency spectrum of a horizontal pick up on a spectrum analyzer; (*bottom*) evolution of the peak signal in the frequency spectrum as a function of time, as viewed on an oscilloscope; the slow variation reflects synchrotron motion, the fast beating with a period of about 17.6 μs is due to the transverse coupling; the amplitude and period of the modulation can be used to determine the driving term $|\kappa_-|$

2.8.4 Closest Tune Approach

Near the difference resonance, the tunes of the two coupled eigenmodes in the vertical plane are [67, 70]

$$Q_{I,II} = \frac{1}{2}\left(Q_x + Q_y + q \pm \sqrt{(Q_x - Q_y + q)^2 + |\kappa_-|^2}\right), \qquad (2.87)$$

where Q_x and Q_y are the tunes which one would expect without coupling. A similar formula, with the same fractional values of $Q_{I,II}$, describes the coupled tunes in the horizontal plane. Equation (2.87) shows that the measured fractional parts of Q_I and Q_{II} are never exactly equal, but can only approach each other up to a distance $|\kappa_-|$. Figure 2.30 illustrates this with an example from the PEP-II High Energy Ring (HER), which is consistent with our simple example in (2.79) and (2.80). A common technique for correcting the betatron coupling in a storage ring is to minimize the distance of closest approach using at least two skew quadrupole magnets. It is often the only correction necessary, especially if the tunes are close to the difference resonance $Q_x - Q_y + q = 0$.

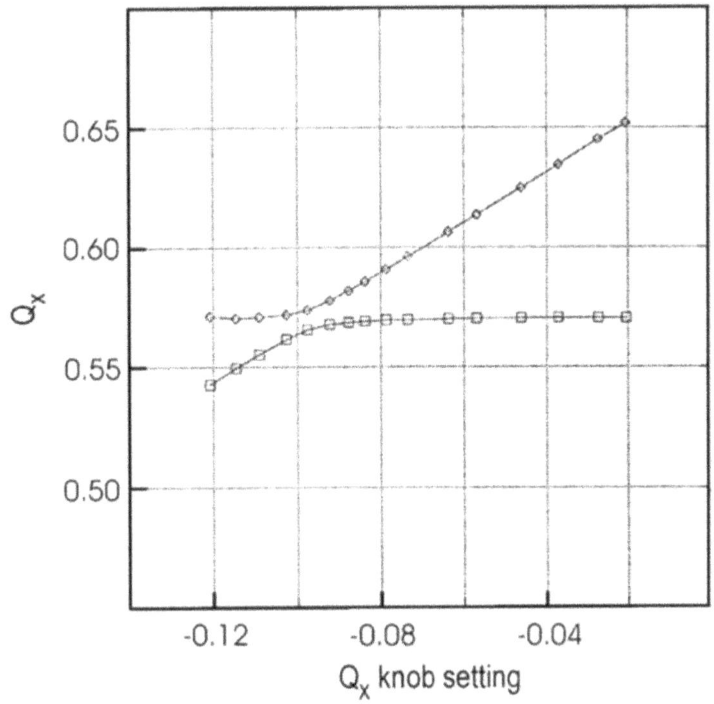

Fig. 2.30. Closest tune approach in the PEP-II HER before final correction [73]. Shown are the measured fractional tunes as a function of the horizontal tune 'knob' (which would only change Q_x if the accelerator were uncoupled), in dimensionless units. The minimum tune distance is equal to the driving term $|\kappa_-|$ of the difference resonance (Courtesy Y. Cai, 1998)

2.8.5 Compensating the Sum Resonance

In the vicinity of the difference resonance, while there is a continuous energy exchange between the two transverse planes, the beam or particle motion remains bounded. In contrast, close to the sum resonance for which $|Q_x + Q_y + q| < |\kappa_+|$, the motion is unstable. The total width of the stop band around the sum resonance is equal to twice the driving term $|\kappa_+|$ of (2.81) [67]. In practice, while perhaps more difficult, than determining the driving term $|\kappa_-|$ for the difference resonance using the closest-tune approach, the driving term for the sum resonance $|\kappa_+|$ can be compensated by adjusting two skew quadrupoles so as to minimize the stop band width.

In order to optimize the vertical emittance in an electron damping ring, in general it is necessary to correct both driving terms, $|\kappa_-|$ and $|\kappa_+|$. A minimum of six skew quadrupole correctors are needed to correct both coupling driving terms and the the vertical dispersion function since one sine-like and one cosine-like corrector are required for correction type of each aberration.

Even after perfect global coupling compensation, there can be a residual contribution of local coupling to the vertical emittance. Equation (2.83) suggests that local correction might be necessary. It is indeed employed at several light sources, the KEK ATF, and the B-factories.

2.8.6 Emittance near Difference Resonance

A third approach to determine the coupling parameter $|\kappa_-|$ is to measure the horizontal (or vertical) emittance as a function of the distance from the difference resonance [71]. Equation (2.83) does not apply close to the resonance. Instead, near the difference resonance, the horizontal emittance is described by [74]

$$\epsilon_x = \epsilon_{x0} \frac{(\Delta Q_-)^2 + |\kappa_-|^2/2}{(\Delta Q_-)^2 + |\kappa_-|^2}, \tag{2.88}$$

where ϵ_{x0} is the unperturbed horizontal emittance without coupling or far from the resonance. The tune difference $\Delta Q_- = (Q_x - Q_y - n)$ function is related to the measured tune difference $\Delta Q_{I,II} = |Q_{II} - Q_I|$ via

$$\Delta Q_- = \sqrt{\Delta Q_{I,II}^2 - |\kappa_-|^2}\,. \tag{2.89}$$

After inserting this into (2.88), the latter becomes

$$\epsilon_x = \epsilon_{x0} \left(1 - \frac{|\kappa_-|^2}{2\Delta Q_{I,II}^2}\right). \tag{2.90}$$

A more recent and different derivation [75] describes the horizontal and vertical emittances near the difference resonance in the following modified form:

$$\epsilon_x = \epsilon_{x0} - (\epsilon_{x0} - \epsilon_{y0}) \frac{|\kappa_-|^2/2}{(\Delta Q_-)^2 + |\kappa_-|^2 + \Delta Q_- \sqrt{(\Delta Q_-)^2 + |\kappa_-|^2}}, \tag{2.91}$$

$$\epsilon_y = \epsilon_{y0} + (\epsilon_{x0} - \epsilon_{y0}) \frac{|\kappa_-|^2/2}{(\Delta Q_-)^2 + |\kappa_-|^2 + \Delta Q_- \sqrt{(\Delta Q_-)^2 + |\kappa_-|^2}}. \tag{2.92}$$

In an experiment at the KEK ATF, the horizontal emittance ϵ_x was inferred from the spot size σ_x measured with the ATF interferometric monitor [76], using the formula $\epsilon_x = (\sigma_x^2 - (D_x\delta)^2)/\beta_x$, where the design beta function ($\beta_x = 0.33$ m) and dispersion ($D_x = 4$ cm) at the point of light emission, and the calculated energy spread of $\delta = 7.14 \times 10^{-4}$ were assumed.

Performing a nonlinear fit to (2.90) of the emittance ϵ_x obtained for various tune differences $\Delta Q_{I,II}$, the driving term $|\kappa_-|$ can be extracted [71]. Figure 2.31 shows the measured horizontal emittance as a function of tune separation $\Delta Q_{I,II}$. The result of the nonlinear fit is also depicted. The coupling strength of $|\kappa_-| \approx 0.037$ inferred from the fit [71] agrees well with the coupling strength $|\kappa_-| \approx 0.042$ obtained from a measurement of the closest tune approach under the same conditions [77].

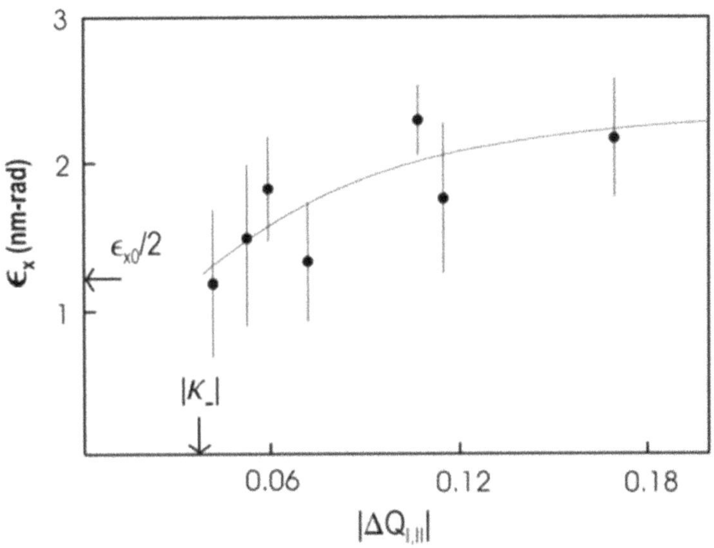

Fig. 2.31. Horizontal emittance as a function of the tune separation $\Delta Q_{I,II}$ at the ATF Damping Ring; the measured data [77] as well as the result of a nonlinear fit to (2.90) are shown; the fitted coupling strength of $|\kappa_-| \approx 0.037$ is consistent with the value $|\kappa_-| \approx 0.042$ inferred from a simultaneous measurement of the closest tune approach [77, 71]

2.8.7 Emittance near Sum Resonance

The driving term of the sum resonance κ_+ may also be inferred by measuring the emittance in the vicinity of this resonance [78]. However, we caution the reader that experimental experience with this scheme is scarce, and that there appear to be some uncertainties in the predictions by different theories and computer simulations.

Using the formulae in [74] and [79], the vertical emittance should depend on the measured distance to the sum resonance, $\Delta Q_{I,II,+} \equiv |Q_I + Q_{II} - p|$ (where p is the integer which minimizes the expression) as [78]

$$\epsilon_y \approx \epsilon_{x0} \frac{2|\kappa_+|^2}{5|\kappa_+|^2 + \Delta Q_{I,II,+}^2}, \tag{2.93}$$

while the horizontal emittance should vary as

$$\epsilon_x \approx \epsilon_{x0} \frac{3|\kappa_+|^2 + \Delta Q_{I,II,+}^2}{5|\kappa_+|^2 + \Delta Q_{I,II,+}^2}. \tag{2.94}$$

In deriving (2.93) and (2.94), we have used that the actual tune distance to the sum resonance, $\Delta Q_+ = (Q_x + Q_y - n)$, is related to the measured distance $\Delta Q_{I,II,+} = |Q_{II} + Q_I - p|$ via

$$\Delta Q_+ = \sqrt{\Delta Q_{I,II,+}^2 + |\kappa_+|^2}. \tag{2.95}$$

2.8.8 Coupling Transfer Function

A different method of measuring the coupling is through the 'coupling transfer function' [80]. Here, the beam is excited horizontally, while detecting the resulting vertical coherent motion. Such a technique was used to continually monitor and correct the coupling strength during collisions in the CERN ISR [80].

2.8.9 Excursion: Flat Versus Round Beams

Frequently the question is asked why flat beams are used in electron-positron colliders. One answer is that the beams naturally become flat due to synchrotron radiation, as will be discussed in a later chapter. However, as we have seen, one could make these beams round, by operating close to the linear difference resonance $Q_x = (Q_y + k)$, where k is an integer.

In general, whether round or flat beam collisions would give higher luminosity is not easy to answer. The luminosity is defined as the reaction rate divided by the cross section. Aside from the beam energy, it is the most important performance parameter of a high-energy collider. For a ring collider, assuming Gaussian beam distributions, the luminosity, typically measured in units of $\mathrm{cm}^{-2}\,\mathrm{s}^{-1}$, can be expressed as (compare Chap. 1)

$$L = \frac{N_b^2 n_b f_{\mathrm{rev}}}{4\pi\sigma_x\sigma_y}\,, \tag{2.96}$$

where N_b is the number of particles per bunch, n_b the number of bunches per beam, and f_{rev} the revolution frequency. The beam sizes at the interaction point (IP), $\sigma_{x,y}$, can be expressed in terms of the IP beta function and the transverse emittances as

$$\sigma_x = \sqrt{\beta_x\epsilon_x}\,, \tag{2.97}$$

$$\sigma_y = \sqrt{\beta_y\epsilon_y}\,. \tag{2.98}$$

The sum of the emittances may be held constant, equal to ϵ_{x0}, while the ratio of the horizontal and vertical emittance can be varied via the betatron coupling according to

$$\epsilon_x = \frac{\epsilon_{x0}}{1+\kappa}\,, \tag{2.99}$$

$$\epsilon_y = \kappa\frac{\epsilon_{x0}}{1+\kappa}\,. \tag{2.100}$$

An important constraint of a collider is the beam-beam tune shift, i.e., the tune shift and tune spread due to the additional focusing at the collision point by the field of the opposing beam. The beam-beam tune shift is characterized by the parameters $\xi_{x,y}$,

$$\xi_{x,y} = \frac{N_b r_e}{2\pi\gamma} \frac{\beta_x}{\sigma_{x,y}(\sigma_x + \sigma_y)} , \tag{2.101}$$

where r_e is the classical electron radius, and γ the relativistic factor. Best performance is achieved if the vertical and horizontal are both at the maximum possible value. If we assume that the maximum tune shift parameter is approximately equal for the two planes, $\xi_x = \xi_y$, it follows immediately that

$$\frac{\epsilon_x}{\epsilon_y} = \frac{\beta_x}{\beta_y} = \frac{1}{\kappa} . \tag{2.102}$$

Inserting this into (2.96), one can rewrite the luminosity as

$$L = \frac{1+\kappa}{2} \gamma\xi_{x,y} \frac{1}{r_e} \frac{N_b f_{\text{rev}} n_b}{\beta_y} . \tag{2.103}$$

This expression shows that round beams may give two times higher luminosity, thanks to the factor $(1+\kappa)$, but only if the horizontal beta function can be made as small as the vertical, $\beta_x = \beta_y/\kappa = \beta_y$.

Recently a transformation was proposed [81] which can convert a round beam created in a solenoid field into a flat beam outside of the solenoid or vice versa. As discussed in Chap. 4, the transformation which achieves the conversion can be realized by a combination of skew and regular quadrupoles. As a practical example, using this scheme the round beam from an rf photoinjector can be transformed into a beam with a flat emittance ratio $\epsilon_y/\epsilon_x \ll 1$, which is more suitable for a linear collider [82]. A first experiment at the Fermilab A0 line has demonstrated the viability of this scheme [83] (see Chap. 5). Another application is the matching of a flat electron beam to a round proton beam, e.g., in an electron cooling device.

Exercises

2.1 Schottky Signals

The longitudinal spectrum of a single particle, denoted by k, of an unbunched beam circulating in a storage ring, considering positive frequencies only, is given [84] by

$$i_k(t) = e f_{\text{rev}} + 2e f_{\text{rev}} \sum_{n=1}^{\infty} \cos(n\omega_{\text{rev}} t + \phi) , \tag{2.104}$$

where n represents the turn number and $\omega_{\text{rev}} = 2\pi f_{\text{rev}}$ is the angular revolution frequency of this particle.

a) Sketch the current spectrum for this particle in both the time and frequency domains.

b) Show that summing (2.104) over N particles and averaging in time gives the dc beam current $I_{\mathrm{dc}} = \langle \sum_{k=1}^{N} i_k \rangle_t = N e f_{\mathrm{rev}}$, assuming that different particles (index k) have slightly different angular revolution frequencies $\omega_{\mathrm{rev},k}$ and different initial phases ϕ_k, and that f_{rev} denotes the average revolution frequency.

c) The power spectrum one would measure with a spectrum analyzer is proportional to the rms current $I_{\mathrm{rms}} = \langle (\sum_k i_k)^2 \rangle_t^{\frac{1}{2}}$. Show that

$$I_{\mathrm{rms}} = 2 e f_{\mathrm{rev}} \sqrt{\frac{N}{2}} \,. \tag{2.105}$$

2.2 Betatron Tunes

a) From Fig. 2.5, what was the synchrotron tune in LEP?

b) Suppose that when the quadrupole family for the horizontal tune is increased in strength, the horizontal tune line is observed to move to the right in similar measurements. What is the fractional horizontal betatron tune?

c) After how many turns would a particle with the tunes found in a) and b) return to exactly the same place in the longitudinal or horizontal phase space?

2.3 Application of Multipole Field Expansion

A useful formula for fields produced by a magnet of order n ($n = 0$ represents a dipole) is[2]

$$B_y + \mathrm{i} B_x = B_0 \sum_{n=0} (b_n + \mathrm{i} a_n)(x + \mathrm{i} y)^n \,, \tag{2.106}$$

where for upright, or 'normal' magnets $b_n \neq 0$ and $a_n = 0$ while for magnets rotated by 45°, so-called 'skew' magnets, $b_n = 0$ and $a_n = 0$.

a) For the case of a upright, or 'normal', sextupole ($b_2 \neq 0$, $a_2 = 0$), show that the Lorentz force seen by the particle is

$$F_x \propto (x^2 - y^2) \quad \text{and} \quad F_y \propto xy \,. \tag{2.107}$$

b) For a nondispersive orbit ($x = x_{\mathrm{co}} + x_\beta$ and $y = y_{\mathrm{co}} + y_\beta$, where the subscripts co and β refer to the closed orbit and to the betatron orbit), show that a beam off-axis in x experiences a normal quadrupole field while a beam off-axis in y experiences a skew quadrupole field.

c) What transverse-to-longitudinal coupling terms arise if the dispersive contribution to the orbit ($x_\delta = D_x \delta$ and $y_\delta = D_y \delta$) are also included?

[2] We note that (2.106) refers to the American convention, and that in Europe the exponent of $(x + \mathrm{i}y)$ is chosen as $(n - 1)$, so that a sextupole field would be given by b_3 and not by b_2.

2.4 Beta-Beat

Consider a quadrupole error. We split the quadrupole into two halves each of which is parametrized by the usual matrix for a quadrupole

$$R_q = \begin{pmatrix} 1 & 0 \\ -1/(2f) & 1 \end{pmatrix} . \tag{2.108}$$

where $1/f \equiv \Delta K$ in a thin-lens approximation, ΔK being the (integrated) quadrupole strength error. Let the one turn map (OTM) with the field error be given by $R = R_q R_0 R_q$ (where for algebraic convenience we have taken as a reference point the center of the quad), where R_0 is the OTM for the ideal ring:

$$M = \begin{pmatrix} \cos\phi_0 & \beta_0 \sin\phi_0 \\ -(1/\beta_0)\sin\phi_0 & \cos\phi_0 \end{pmatrix} . \tag{2.109}$$

a) Show that the perturbed phase advance ϕ is given in terms of the unperturbed phase advance ϕ_0 by

$$\cos\phi = \cos\phi_0 - \frac{\beta_0}{2f}\sin\phi_0 . \tag{2.110}$$

b) From the measurement of phase advance errors in Fig. 2.9, estimate the amplitude of the modulation on the beam size.

2.5 Quadrupole with a Shorted Coil

So-called difference orbit measurements are taken by saving a BPM reference orbit under nominal conditions, perturbing the beam (for example, by powering a single corrector) and forming the difference orbit consisting of measurements of the perturbed orbit minus that of the nominal orbit. This difference orbit may be easily compared with model expectation. Suppose such lattice diagnostic measurements are made using a negatively charged beam and reveal that the beam is errantly deflected up and outboard (away from ring center) at a focussing quadrupole.

a) Based on these observations, which of the quadrupoles' coils is suspect of a turn-to-turn short?

b) With a position resolution of 10 μm at a BPM located $l = 1$ m downstream of the quadrupole, how well can the relative change in the current through the offending coil be determined assuming $K = Bl_q/((B\rho)a) = 1.0\,\mathrm{m}^{-1}$ with a pole tip radius of $a = 1$ cm?

2.6 Quadrupole Gradient Errors

a) Design a closed π-bump spanning a focussing quadrupole of strength K assuming perfect calibration of the corrector dipoles.

b) Using this bump, suppose that the orbit is observed not to close with a residual amplitude A measured using a BPM. Find an expression relating the measured leakage A to the gradient error ΔK.

2.7 Multiknobs

A storage ring employs two families of quadrupoles for varying the tranverse focussing. As is common, each quadrupole family has influence on both the horizontal and vertical betatron tunes, Q_x and Q_y respectively. Design a multiknob that allows independent control of each of the betatron tunes; that is, find strength coefficients for each of the quadrupole families for the cases (i) $\Delta Q_x \neq 0$, $\Delta Q_y = 0$ and (ii) $\Delta Q_x = 0$, $\Delta Q_y \neq 0$.

3 Orbit Measurement and Correction

In practice, there are many uncertainties whose presence must be appreciated when correcting the beam orbit in both linear and circular accelerators. Such uncertainties include the variations in the electronic and/or mechanical centers of the beam position monitors (BPMs), in the magnetic center of the quadrupoles (inside which the position monitors are often mounted), or in the electromagnetic center of accelerating structures. Consider the case illustrated in Fig. 3.1. In this case, the absolute beam position, with respect to a reference axis, is given by

$$x = x_d + x_b + x_m \,, \tag{3.1}$$

where x_d represents the quadrupole offset from the reference axis, x_b gives the offset of the electronic center of the BPM relative to the quadrupoles, and x_m denotes the measurement from the BPM. The reference axis may be chosen to minimize emittance dilutions in a global sense. In the ideal case for which the reference axis is straight and in the center of all quadrupoles, a typical most simple steering algorithm aimed towards zeroing the BPM readings would

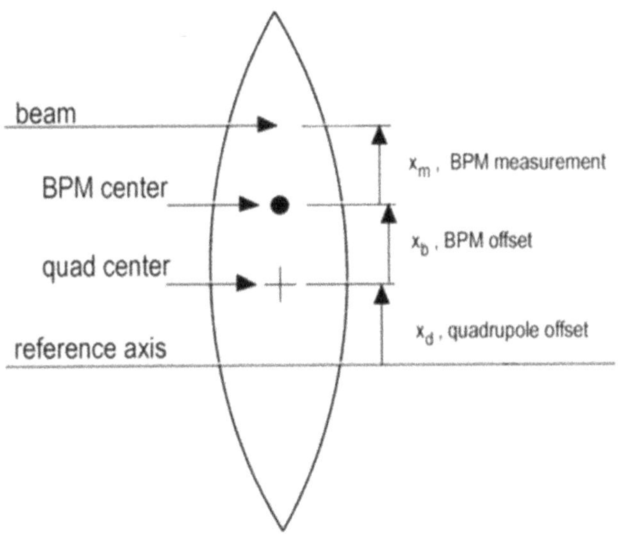

Fig. 3.1. Sketch showing relative positions of the BPM, the quadrupole, and the beam position measurement from an ideal reference axis (Courtesy C. Adolphsen, 1999)

M. G. Minty et al., *Measurement and Control of Charged Particle Beams*,
https://doi.org/10.1007/978-3-662-08581-3_3

still place the beam off-axis in the quadrupole if $x_b \neq 0$. For accelerators with tight emittance budgets and correspondingly tight alignment tolerances, clearly more sophisticated steering algorithms are needed.

If the orbit is off-center in a quadrupole magnet, dispersion is generated, and, in a ring, also the beam energy may be changed or the polarization may decrease. An orbit that is off-center in a sextupole induces skew coupling and/or beta beating. Thus it is very important to center the orbit in these magnets. The standard tools for correcting the orbit are corrector dipoles. Of course, such an orbit correction will never be perfect. Figure 3.2 shows a typical absolute orbit reading from the PEP-II HER, after moderate orbit correction during commissioning.

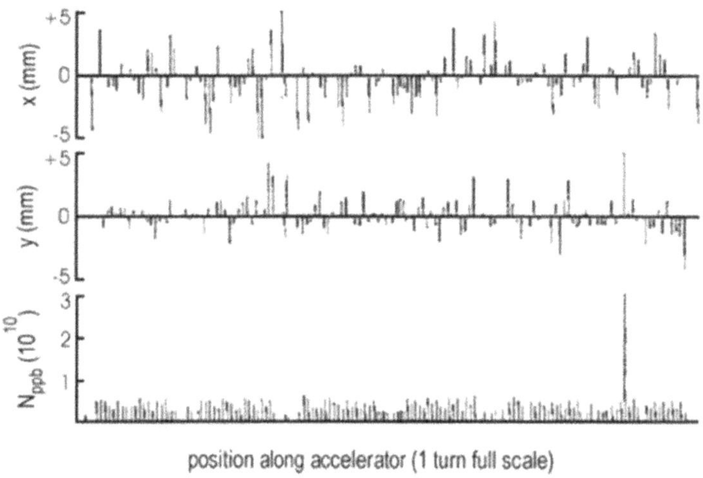

Fig. 3.2. Typical commissioning orbit in the PEP-II HER: (*top*) horizontal orbit in mm; (*center*) vertical orbit in mm; (*bottom*) intensity in 10^{10} (Courtesy U. Wienands, J. Seeman *et al*, 1998)

We will see that if the BPM offsets are not known, and possibly larger than the alignment errors, a better strategy for optimal emittance preservation is to reduce the rms strength of the steering correctors, and to pay less attention to the absolute orbit reading. In several cases, at the SLC and at the ATF, this second approach significantly reduced the magnitude of the residual vertical dispersion [1]. Sometimes other constraints are imposed on the orbit. For example, a certain orbit amplitude or a certain angle may be desired near the injection or extraction points, or near a synchrotron light beamline. In such cases, a constant orbit must be maintained at the adjacent BPMs.

Increased beam emittances may arise from beam-to-magnet or beam-to-structure position deviations as shown conceptually in Fig. 3.3. The beam passing through a single misplaced quadrupole experiences the next lower-order field namely a dipole field. The misalignment therefore generates a betatron oscillation *and* dispersion as higher energy particles are less deflected by the dipole field. Also in the case of a displaced structure a betatron os-

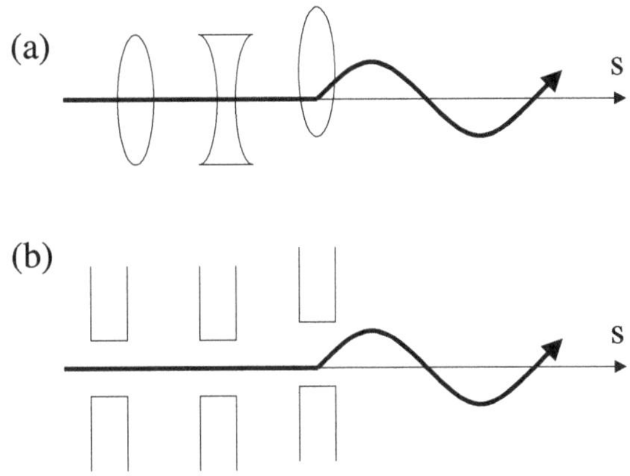

Fig. 3.3. Conceptual drawing illustrating orbit perturbations due to misaligned quadrupoles (**a**) or structures (**b**)

cillation is induced, due to the transverse wake field. The deflection caused by the wake field changes linearly with the beam displacement and with the bunch charge. In either case, the ensuing orbit is such that further emittance dilutions may result downstream of the perturbation due to the initial errors.

In this chapter we begin by reviewing beam-based alignment of single components. Such methods are particularly useful for those components with tight alignment tolerances, for example, near interaction regions. The techniques presented may be applied to the alignment of any accelerator component in general. For large accelerators, for which component-by-component alignment is too time-consuming or for which the beam trajectories must be routinely monitored, global orbit measurement and correction techniques are required. Many such algorithms will be presented in order of increasing complexity. So-called 'one-to-one' steering and R-matrix reconstruction are typically used not only to correct the orbit, but also to identify optical and/or instrumental errors. While such simple checks are absolutely mandatory, to achieve design beam parameters, more advanced tuning algorithms may be required. Common algorithms will be described including global beam-based alignment, singular value decomposition (SVD), application of wake field bumps, and dispersion-free steering.

3.1 Beam-Based Alignment

In many modern accelerators, the alignment tolerances on quadrupole and sextupole magnets are so tight that they cannot be achieved by state-of-the art surveying and installation methods. Typical residual alignment errors are in the range of 100–200 μm while the alignment tolerances may be even below 10 μm. The standard approach to meet and maintain tight tolerances is beam-based alignment.

Beam-based alignment determines the relative offset between magnet centers and nearby BPMs. If these offsets are sufficiently stable, a simple orbit correction (steering) can maintain a well-centered orbit until the optical alignment measurement is repeated at a later time (e.g., after several months).

3.1.1 Quadrupole Excitation

If the beam is not centered in a quadrupole magnet, and the strength of this quadrupole is varied, the beam receives a kick. This causes a change in the beam trajectory for single-turn measurements or a change in the closed orbit for measurements on a stored beam.

For a single-pass measurement, the dipole kick θ can easily be inferred by fitting the difference trajectory to a betatron oscillation including one additional kick at the location of the quadrupole. The dipole kick θ obtained from the fit is proportional to the quadrupole misalignment x_q and the change in the integrated quadrupole strength ΔK:

$$\theta = \Delta K \; x_q \,. \tag{3.2}$$

Note that the beam is offset from the center of the quadrupole by $-x_q$, so that there is no minus sign in (3.2) assuming $K > 0$ refers to a horizontally focusing quadrupole.

If beam-based alignment is performed using a stored beam, the additional kick of the closed orbit induced by the change in quadrupole strength is given by the sum of two components: the change in field strength and the change in the closed-orbit offset at the quadrupole. To lowest order [2],

$$\theta \approx \Delta K \; x_q - K \; \Delta x \,, \tag{3.3}$$

where x_q is the original quadrupole offset, Δx the change in closed-orbit position, K the integrated quadrupole gradient in units of m^{-1}, and a second-order term ($\Delta K \; \Delta x$) has been neglected. Applying the formula for the closed orbit distortion at the location of the dipole kick, (2.34), with $s = s_0$. Neglecting here the possible small contribution from the dispersion D_x, then

$$\Delta x = (\Delta K \; x_q - K \; \Delta x) \left(\frac{\beta}{2 \tan \pi Q} \right) , \tag{3.4}$$

which may be solved for Δx:

$$\Delta x = \Delta K \; x_q \left(\frac{\beta/(2 \tan \pi Q)}{1 + K\beta/(2 \tan \pi Q)} \right) . \tag{3.5}$$

Inserting this back into (3.3) gives the closed-orbit kick induced by a gradient change ΔK:

$$\theta = \Delta K \; x_q \left(\frac{1}{1 + K\beta/(2 \tan \pi Q)} \right) . \tag{3.6}$$

This is the stored-beam equivalent of (3.2).

The precision of this method is much improved by taking difference orbits for several quadrupole-to-beam off-sets, Δx_q, varied with a local bump [3]. One can also define a merit function

$$f(\Delta x_q) = \frac{1}{N_{\mathrm{BPM}}} \sum_{i=1}^{N_{\mathrm{BPM}}} \left[x_i(\Delta K) - x_i(-\Delta K) \right]^2 , \tag{3.7}$$

where $N_{\rm BPM}$ is the total number of BPMs in the ring and $x_i(\Delta K)$ are the BPM readings for a quadrupole strength change of ΔK. The quadrupole offset may then determined by minimizing $f(\Delta x_q)$ as a function of the bump amplitude Δx_q, using a least-squares parabolic fit. At the Advanced Light Source (ALS), this procedure allows determination of the center of the quadrupoles to within ±5 μm [2] (in case of the ALS, the orbit at the quadrupole is varied with a single corrector and not by a closed bump).

This type of measurement does not require an independent power supply for each quadrupole to be aligned, but, for several magnets in series, a simple switchable shunt resistor across each magnet will suffice. Simultaneously, such shunt resistors allow a measurement of the local beta function, according to (2.28) or (2.29) of the previous chapter.

Figure 3.4 illustrates the application of this technique at the storage ring SPEAR. The left figure shows the circuit diagram for a magnet with shunt resistor, and the right figure presents a typical alignment measurement for a SPEAR quadrupole. Plotted in the right figure is the orbit shift induced by

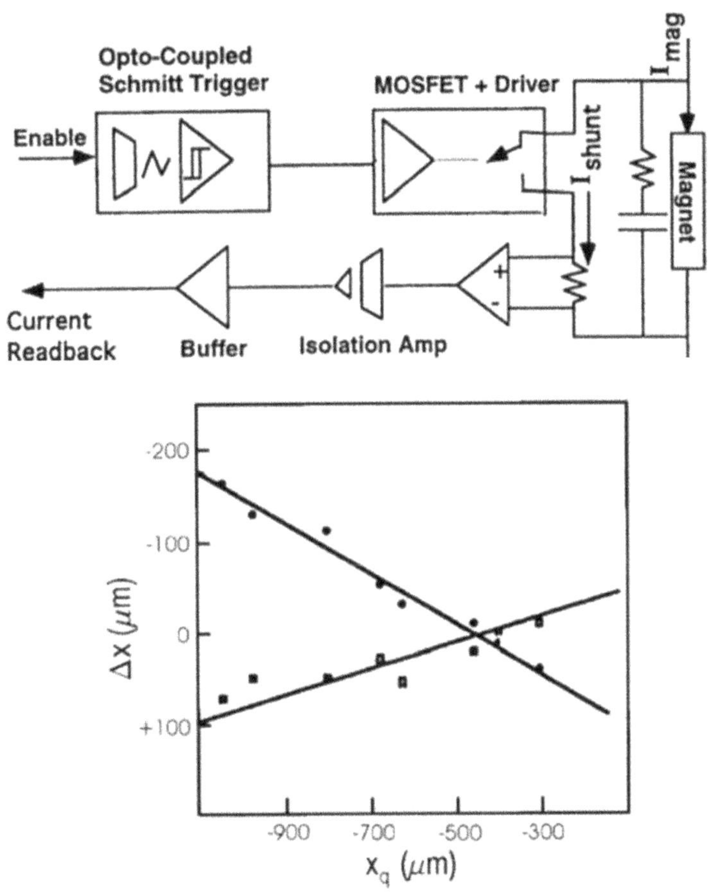

Fig. 3.4. Beam-based alignment with quadrupole shunts at SPEAR [4]: (*top*) electric circuit with shunt resistor; (*bottom*) shunt-induced orbit shift at two downstream BPMs as a function of the beam-position read back at the BPM nearest to the quadrupole being varied (Courtesy J. Corbett, 1998)

the shunt at two downstream BPMs as a function of the orbit at the shunted quadrupole, which is varied by a local bump. The orbit is centered in the quadrupole when no orbit shift is induced by the shunt (the intersection of the two lines).

If the number of BPMs is small and only groups of quadrupoles can be changed simultaneously, it is still possible to determine the quadrupole misalignments, by applying a statistical fit to a sufficiently large number of trajectories taken for different quadrupole-group excitations, different incoming conditions and different corrector settings. An interesting example of such an analysis can be found in [5].

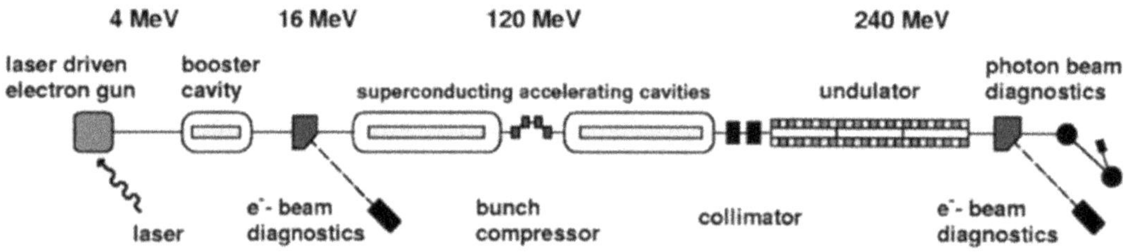

Fig. 3.5. Overview of the TTF linear accelerator including 2 accelerating modules (ACC1 and ACC2) and a bunch compressor chicane (BC2)

Another application of this method is illustrated for the Tesla Test Facility (TTF). An overview of part of the linear accelerator is shown in Fig. 3.5. In this measurement a horizontal corrector located just upstream of the bunch compressor chicane (BC2) was varied in steps for different settings of quadrupole settings. In the data shown in Fig. 3.6 the quadrupole of interest was the most upstream quadrupole in the triplet following the chicane. The next downstream beam position monitor was used for measurement of

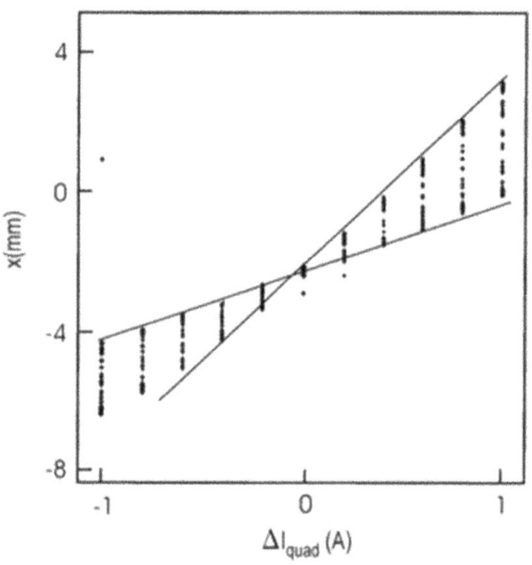

Fig. 3.6. Beam-based alignment of a single quadrupole at the TTF (Courtesy P. Castro, 2000)

the horizontal displacements. The orbit was centered when variation of the quadrupole field produced no change in the measured beam position (at the intersection of the lines).

3.1.2 Quadrupole Gradient Modulation

A scheme which allows continuous monitoring of quadrupole alignment and BPM offsets was implemented at LEP; see for example [6]. Here the strength of several quadrupoles was modulated at different frequencies in the range 0.8–15.6 Hz, and the induced oscillation amplitude, of the order of 1 μm was detected. Figure 3.7 shows the FFT over 4096 data points of this detector signal, at a time when four quadrupoles were modulated. Clearly visible are 4 peaks in the frequency sprectrum, corresponding to the four different modulation frequencies. The amplitude of the peak is proportional to the beam offset in that quadrupole.

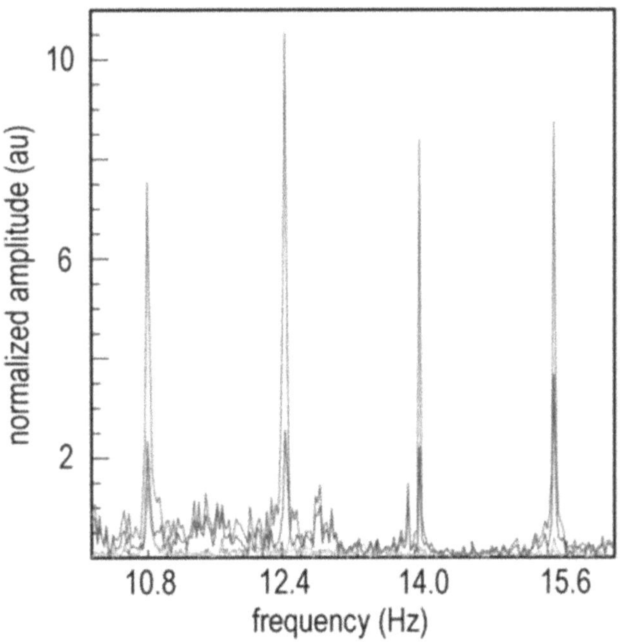

Fig. 3.7. FFT spectra with 4 modulated quadrupoles in LEP [6]. The amplitude of the peaks is proportional to the beam displacement in the 4 quadrupoles (Courtesy I. Reichel, 1998)

Using this 'K-modulation' technique, one could infer the BPM offsets from the naturally occurring beam-orbit jitter and orbit variation. This is illustrated in Fig. 3.8. The left figure shows a BPM orbit reading in LEP during several hours of a luminosity run. The reasons for the slow changes are not fully understood; the fast steps reflect corrections of the closed orbit. Making use of this natural orbit variation, one can plot the amplitude of the beam response to the quadrupole modulation as a function of the BPM

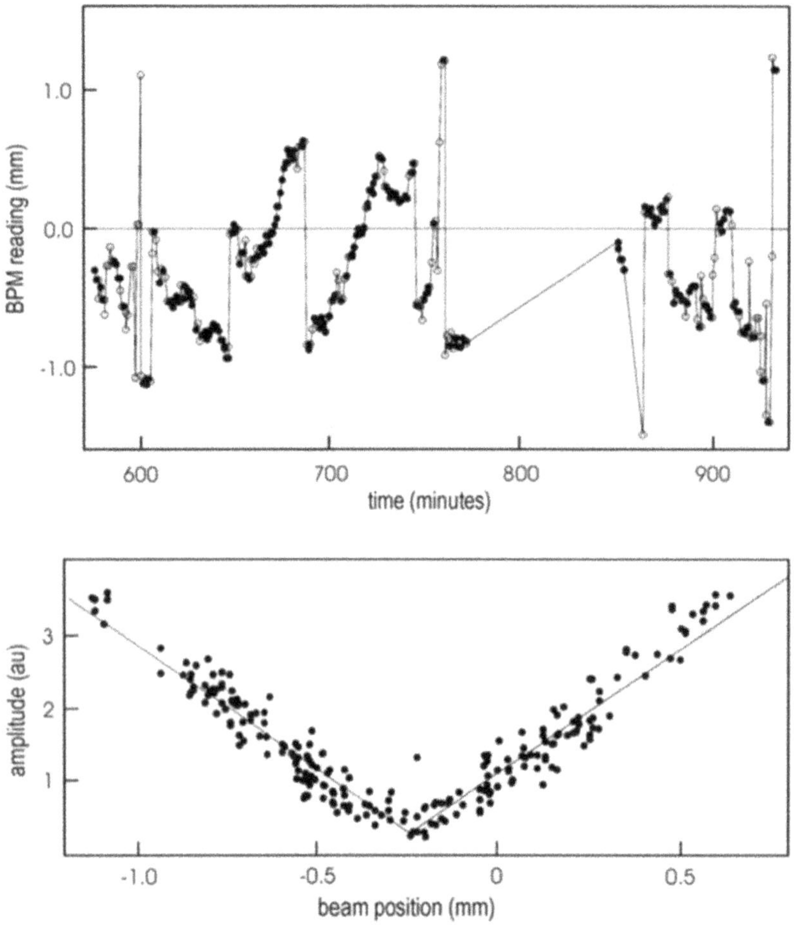

Fig. 3.8. Determination of BPM offsets using k modulation and natural orbit variation in LEP [6]: (*top*) natural orbit drifts and corrections during a LEP luminosity run at one quadrupole; (*bottom*) amplitude of beam response to k modulation vs. BPM orbit reading for the modulated quadrupole. The minimum of this plot gives the BPM offset. These data were taken continuously during 5 hours of luminosity run (Courtesy I. Reichel, 1998)

reading for the corresponding quadrupole. The result is a 'V plot', as shown in the right figure. The minimum in this plot determines the BPM reading at which the beam is centered in the quadrupole.

3.1.3 Sextupole Excitation

In present-day storage rings, it is often assumed that the sextupoles are well enough aligned with respect to the quadrupoles that only the quadrupole alignment has to be verified. An orbit off center in a sextupole will result in vertical dispersion, betatron coupling, or beta beating. Although, in principle, also the sextupoles in a storage ring can be aligned by changing their strength and measuring the induced orbit shift (which is a quadratic function of the excitation) there is little experience with such a scheme. To reach the same sensitivity as for the equivalent quadrupole alignment, the change in the sextupole gradient ΔK_s would have to be equal to

$$\Delta K_s = \frac{\Delta K_q}{2x_s}\,, \tag{3.8}$$

where x_s is the horizontal orbit offset at the sextupole, and ΔK_q the corresponding change in quadrupole gradient. A different approach, which was tested at KEK [7], is to equip the sextupole magnets with additional quadrupole trim windings for beam-based alignment. This is based on the assumption that the magnetic centers of quadrupole trim coil and sextupole will coincide. Sextupole alignment with a precision better than 50 μm was demonstrated [7].

Local orbit bumps across single sextupoles have been used for the purpose of sextupole alignment at KEK [8, 9] and DESY [10]. The strengths of all sextupoles are changed together by ΔK_s and the induced orbit change is measured. Then the measurement is repeated for a different bump amplitude. The horizontal deflection depends quadratically on the horizontal bump amplitude, while the vertical deflection is a linear function:

$$\Delta\theta_x = -0.5\,\Delta K_s(x_{\mathrm{bump}} - x_s)^2\,, \tag{3.9}$$

$$\Delta\theta_y = \Delta K_s(x_{\mathrm{bump}} - x_s)y_s\,, \tag{3.10}$$

where x_{bump} is the amplitude of the bump, and x_s, y_s are the sextupole misalignments. The advantage of this method is that it does not require individual power supplies for the sextupoles.

The vertical sextupole alignment can be determined in a similar way, but in this case a (preferably large) horizontal orbit offset x_0 at the sextupole must first be introduced, to enhance the sensitivity. The vertical deflection angle then varies linearly with the amplitude of a vertical orbit bump at the sextupole as follows:

$$\Delta\theta_y = \Delta K_s(x_0 - x_s)(y_{\mathrm{bump}} - y_s)\,, \tag{3.11}$$

where we have also included a residual horizontal misalignment x_s. After measuring the vertical misalignment y_s, the original horizontal orbit can be restored.

Alternative approaches are conceivable: one could vary multiple sextupoles at once, and fit for multiple kicks. Also, one could vary the sextupole strength and measure the induced tune variation or the tune separation near the difference resonance [11].

In the final-focus systems of linear colliders, sextupole alignment is essential. At the SLC final focus, the orbit in the sextupoles was frequently measured and adjusted to maintain high luminosity. The SLC sextupole alignment procedure was based on varying the sextupole strength and detecting the induced optics (not orbit) change [12]. If the orbit is off center, the first order effect of the sextupole excitation is a waist shift (change in the beta function), skew coupling, or dispersion at the interaction point. These optics changes can be quantified easily by reoptimizing the spot-size at the collision point, after a change in the sextupole strength. This reoptimization is done by scanning a group of quadrupole and skew quadrupole magnets excited together so that they only affect one optical parameter, e.g., either dispersion, skew coupling or waist shift. Such combinations of changes in magnet settings are called multiknobs; see the discussion in Chap. 2. For each setting of the multiknob, the IP spot size is remeasured with beam-beam deflection scans, and the magnets are finally set to the value for which the beam size is minimum. The change in the optimum settings of the multiknobs controlling waist, dispersion, etc., as a function of the sextupole excitation is proportional to the orbit offset at the sextupole. The measured offsets are corrected by means of closed bumps.

An interesting feature of the SLC final focus is that it comprises 2 pairs of interleaved sextupoles. The sextupoles in each pair, connected to the same power supply, are separated by an optical $-I$ transform. Thus, the alignment procedure actually consists in generating symmetric or antisymmetric orbit bumps for each sextupole pair, in response to the amount of waist motion or dispersion etc., induced by a change in the sextupole-pair strength [13].

3.1.4 Sextupole Movement

It is also possible to align the sextupole magnets by detecting the second-order effect of the sextupole excitation: the induced orbit kick. This method works well when the sextupoles are installed on precision movers, which can be used

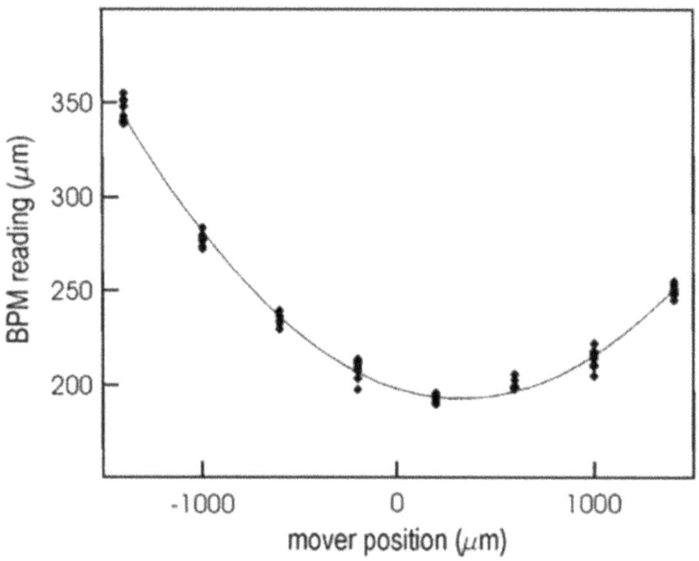

Fig. 3.9. Sextupole alignment in the Final Focus Test Beam (FFTB) [14]. The downstream orbit variation is measured as a function of sextupole mover position; the sextupole is aligned at the minimum of the parabola (Courtesy P. Tenenbaum, 1998)

for both the measurement and the alignment. The measurement principle is straightforward. Measuring the orbit change downstream as a function of horizontal or vertical sextupole-mover position results in a parabolic curve. The sextupole is aligned when the mover position is set to the minimum of this curve. A sample measurement from the FFTB [14] is displayed in Fig. 3.9.

3.1.5 Structure Alignment Using Beam-Induced Signals

For future high-gradient linear accelerators operating at even higher accelerating frequencies it becomes even more essential to center the beam orbit in the accelerating structures, thus minimizing the effect of transverse wake fields on the beam. Alignment techniques have been studied on test structures for the Next Linear Collider, which were installed in the SLAC linac as part of the ASSET experiment [15]. These studies demonstrated that the beam-induced dipole-mode signals can be used to center the beam to the level of 40 μm [16]. The result in Fig. 3.10 shows the amplitude and phase

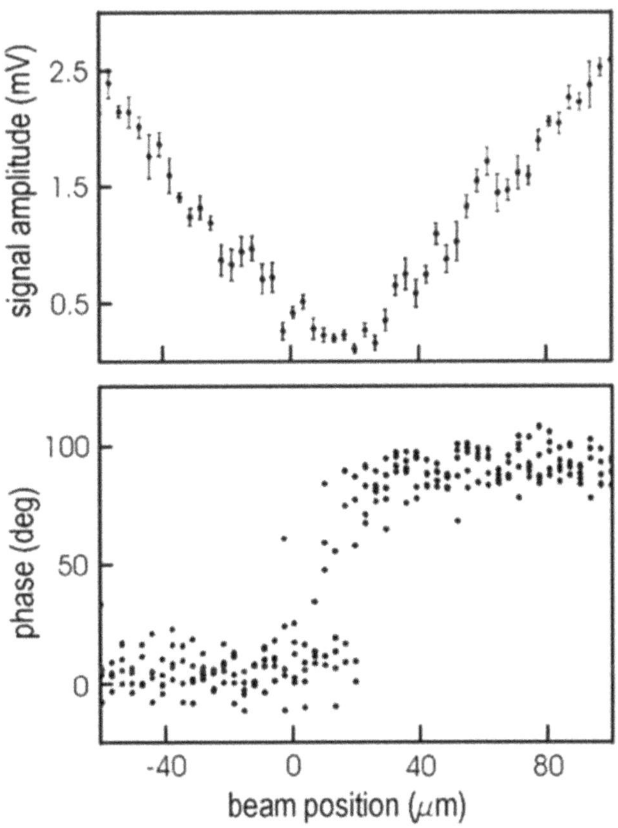

Fig. 3.10. Amplitude (*top*) and phase (*bottom*) of the beam-induced dipole mode signal in an X-band accelerating structure versus the nominal beam position (arbitrary zero), which was varied by steering correctors [16] (Courtesy M. Seidel, 1998)

(with respect to a reference phase derived from a BPM signal) of a 15-MHz wide slice of the beam-induced dipole mode signal, centered near 15 GHz, as a function of the nominal beam position. The beam position was varied with dipole steering magnets. Clearly visible is a minimum in the amplitude along with a 180 degree phase jump. Steering the beam to the position with minimum signal successfully centered the orbit in the structure as was verified by detecting the deflection experienced by a subsequent witness bunch.

We now turn our attention to global steering algorithms designed to efficiently align the beam relative to offset magnets and accelerating structures relying heavily on computational techniques. The algorithms are presented in order of increased complexity. Not by chance, this order coincides to a large degree with the order of implementation as an accelerator is being commissioned and the beam properties are refined.

3.2 One-to-One Steering

This algorithm aims to steer the beam so that the transverse displacements measured by beam position monitors (BPMs) are minimized. The BPMs are typically mounted near the center of quadrupoles since their sensitivity is highest at large β-function. A conceptual orbit steered one-to-one is shown in Fig. 3.11. The beam is successfully deflected to pass through the magnet center and, assuming that the BPM is not offset with respect to the quadrupole, the BPM would show zero displacement. Notice however that one-to-one steering generates dispersion will contribute to emittance dilutions.

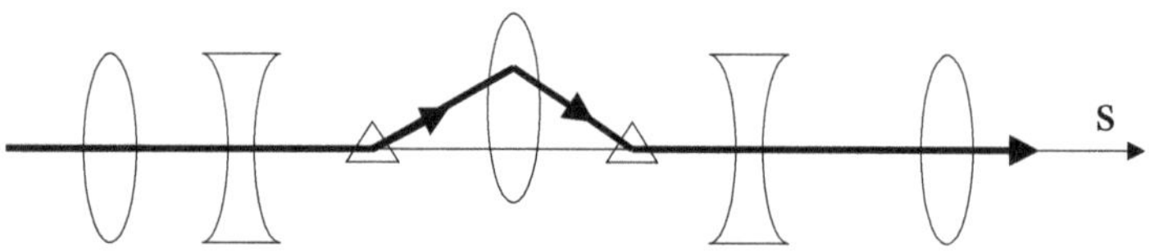

Fig. 3.11. Conceptual illustration of a closed bump that would minimize the BPM reading after one-to-one steering

In a transport line the beam centroid position measured downstream at location $s = j$ obeys

$$x_j = \sum_{i=0}^{j} \sqrt{\beta_i \beta_j} \theta_i \sin(\theta_j - \theta_i) , \tag{3.12}$$

which has contributions from each dipole kick θ_i and depends on the β-functions at the location of the initial disturbance (i) and at the observation point (j). The corrector magnet fields to be applied to minimize the BPM readings will be solved for assuming linear transport; that is, assuming that there are no nonlinear magnetic fields and that the measurements are made at low bunch current so that nonlinear wake field effects may be ignored.

In matrix form

$$\boldsymbol{x} = M\boldsymbol{\theta}\,, \tag{3.13}$$

where $\boldsymbol{x}$ is the set of measurements from m BPMs, $\boldsymbol{\theta}$ is the set of deflection angles to be applied by n correctors, and M contains the transfer matrix elements between the correctors and the BPMs:

$$x^t = (x_0, x_1, ..., x_m)\,, \tag{3.14}$$

$$\theta^t = (\theta_0, \theta_1, ..., \theta_n)\,, \tag{3.15}$$

$$M_{ij} = \sqrt{\beta_i\beta_j}\sin(\phi_j - \phi_i)\,. \tag{3.16}$$

Solving (3.13), the kick angles to be applied for minimizing the BPM readings are obtained:

$$M^t\boldsymbol{x} = M^t M\boldsymbol{\theta} \quad \text{or} \quad \boldsymbol{\theta} = (M^t M)^{-1}M^t\boldsymbol{x}\,. \tag{3.17}$$

If the number of correctors equals the number of BPMs, then M is a square matrix. In this case, (3.17) reduces to simply $\boldsymbol{\theta} = M^{-1}\boldsymbol{x}$. Otherwise the general form is taken. If $m > n$ the matrix is overdetermined. For $m < n$ the number of unknown corrector strengths exceeds the number of measurements, so that an independent measurement should be made after changing some parameter, for example, the beam energy. In a linear accelerator, (3.12) must be modified [17] to include the energy scaling factor $\sqrt{E_i/E_j}$, which reflects the transverse damping due to the longitudinal acceleration. This introduces n additional unknown variables, so that additional measurements are required to constrain the solution.

As motivation for the algorithms to be used below in the discussion of beam-based alignment and dispersion-free steering, the solution (3.17) can be equivalently formulated in terms of a minimization procedure, which is well adapted to computational evaluation. The function to be minimized, given by (3.13), is

$$\sum_j \Big[x_j - \sum_i M_{ji}\theta_i\Big]^2\,, \tag{3.18}$$

where x_j again represents the BPM measurements and the fitting function $\sum_i M_{ji}\theta_i$ contains the unknowns θ_i. The minimization procedure demands

$$\begin{aligned} 0 &= \frac{\partial}{\partial\theta_k}\Big[\sum_j (x_j - \sum_i M_{ji}\theta_i)^2\Big] \\ &= 2\sum_j \Big[x_j - \sum_i M_{ji}\theta_i\Big] M_{jk}\,, \end{aligned} \tag{3.19}$$

or

$$\sum_j M_{jk}x_j = \sum_j \sum_i M_{ji}M_{jk}\theta_i\,, \tag{3.20}$$

which is identical to (3.17).

One-to-one steering is used routinely during initial commissioning of an accelerator as it is by far the simplest of all steering algorithms. The procedure relies on having reasonably well functioning beam position monitors and may diverge in the presence of either large monitor errors or, possibly, with multiple smaller errors. A complementary algorithm, presented next, is useful to help identify BPM errors (polarity, gain, offset, etc.) and to help localize optical errors as well.

3.3 Lattice Diagnostics and R Matrix Reconstruction

Consider a beam line, without any elements coupling horizontal and vertical motion, as shown in Fig. 3.12 consisting of dipole and quadrupole magnets, BPMs and corrector magnet dipoles. The point-to-point transfer map between any two points (1) and (2) is given by

$$\begin{pmatrix} x \\ x' \end{pmatrix}_2 = \begin{pmatrix} R_{11} & R_{12} \\ R_{21} & R_{22} \end{pmatrix} \begin{pmatrix} x \\ x' \end{pmatrix}_1 . \tag{3.21}$$

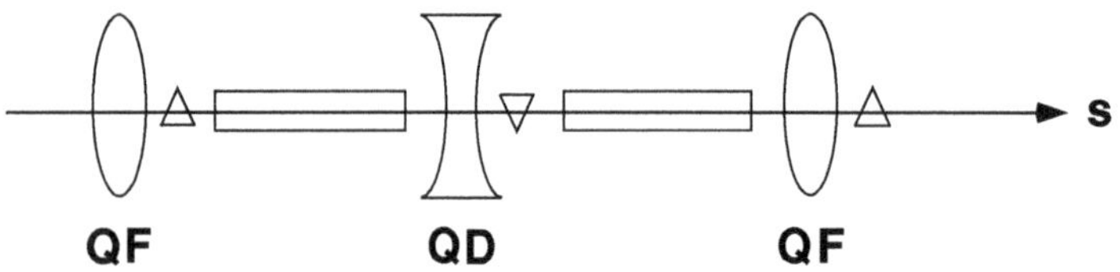

Fig. 3.12. Simple FODO lattice. Shown are the focussing quadrupoles (QF), the defocussing quadrupoles (QD), and dipoles (rectangles). The BPMs are usually mounted within the quadrupoles. Corrector magnets are denoted by triangles

Let the initial point (1) be at a corrector and the final point (2) be at a BPM. Two measurements are required to determine R_{12}: x_{BPM} with the nominal beam trajectory and the nominal initial values $(x, x')_{\mathrm{COR}}$, and with x_{BPM} for the initial position and slope $(x, x' + \theta)_{\mathrm{COR}}$, i.e., after the beam is kicked by an angle θ. The difference in x_{BPM} between these two measurements is $R_{12} = \Delta x/\theta$. In practice to decrease sensitivity to measurement error, one introduces a series of large betatron oscillations by varying the corrector in steps. The response of the BPM readings to these perturbations [18] is then measured. The resulting dependence of position on the kick angle θ is fitted with a straight line as shown in Fig. 3.13.

Shown in Fig. 3.14 is an example of a beam trajectory after excitation by a corrector magnet and an amplitude fit using many downstream BPMs. For good viewing conditions the measured trajectory is plotted for the case of a maximum kick angle $\theta = \theta_{max}$. The solid line connects the measurements from each BPM. The dashed line represents the fitted positions $x_i = R_{12}^{\theta \to i}\theta + x_{o,i}$ evaluated at θ_{max} where $x_{o,i}$ is an offset at the ith BPM.

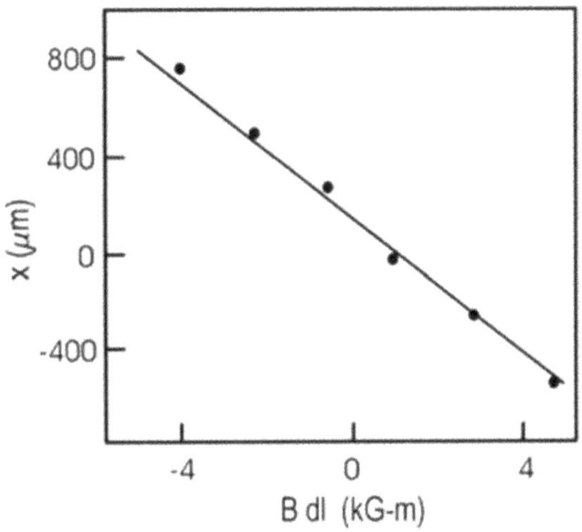

Fig. 3.13. Example of an R_{12} measurement: transverse position x versus $\int B\ \mathrm{d}l$, where the kick angle, in terms of the magnetic rigidity $B\rho$, is $\theta = \int B\ \mathrm{d}l/(B\rho)$

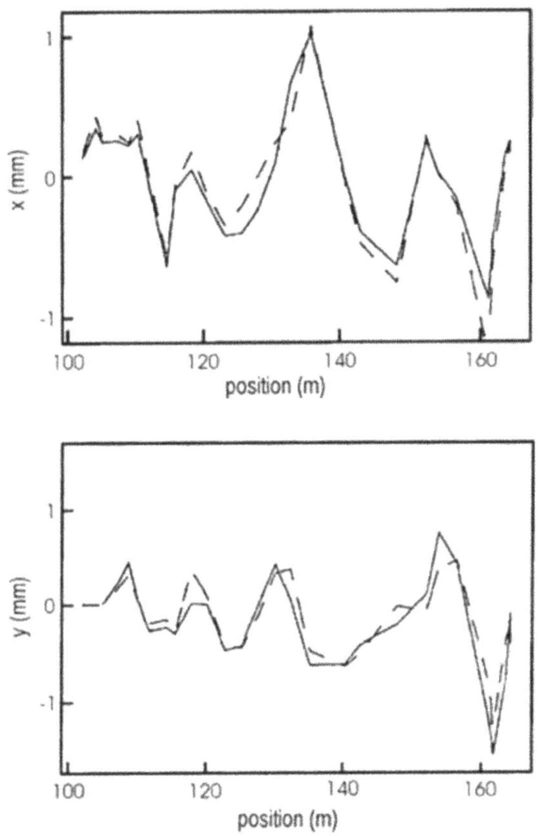

Fig. 3.14. Comparison of measured data (*solid curve*) and amplitude fit (*dashed curve*). Plotted are the horizontal and vertical beam positions as measured as a function of distance along the transport line. See also [19]

in the linear fit. A single value of θ and the values of m BPM offsets $x_{o,i}$ are fitted, so as to minimize the difference between the measured BPM readings and the model prediction, i.e., $\sum_i (x_{\mathrm{BPM},i} - R_{12}^{\theta\to i}\theta - x_{o,i})^2$. Here $R_{12}^{\theta\to i}$ is the (1,2) transport-matrix element between the corrector and the ith BPM computed by the optics model. To probe all magnetic elements in the beamline, a second measurement is required using a second corrector dipole separated by about 90° in betatron phase advance. Discrepancies between the measurement, $x_{\mathrm{BPM},i}$ and the fit, x_i, are used to reveal phase errors, which could result from a shorted quadrupole magnet, and/or "bad" BPMs for example. Identification and exclusion of bad BPMs is required for good convergence of steering algorithms.

Assuming the linear transport matrices between the different correctors and the transport matrices between adjacent BPMs are known with sufficient accuracy, the R matrix between the correctors and the BPMs can be determined by a simple least squares fit [18]. Including the additional constraint that the R matrix has to be symplectic eliminates several degrees of freedom, but then the problem must be solved by non-linear regression [18]. Reference [18] describes how a rigorous error analysis allows an estimate of the unknown systematic errors.

Suppose the agreement between the measured data and fit is unsatisfactory. In the absence of hardware errors, this may result from systematic measurement errors or from an incomplete model. Accuracy of the model is vital for basic optics checkout and requires, for example, accurate representations of magnetic field strengths.

In a linear accelerator, it may be necessary to take into account the energy dependence of the point-to-point transfer matrix elements. The change in the betatron phase $\Delta\psi$ relative to the expected phase ψ is given by

$$\frac{\Delta\psi}{\psi} = \delta\xi\,, \tag{3.22}$$

where δ is the relative energy deviation and ξ is the chromaticity (in a storage ring ξ is related to Q' via $\xi \equiv Q'/Q$, where Q denotes the betatron tune). With $\xi = -1/\pi$ for a 90° FODO cell, then a 1% energy error corresponds to a change in phase advance of $\Delta\psi = -0.003\psi$ per cell.

To take into account the energy dependence of the transport matrix, the matrix elements may be expanded in a Taylor series. Keeping the linear term only, the transport matrix from the entrance ('1') to the end of the ('2') of the linac can be written as

$$\begin{pmatrix} x \\ x' \end{pmatrix}_2 = \prod_{k=1}^{N_s} \begin{pmatrix} R_{1,1_k} + \frac{\mathrm{d}R_{1,1_k}}{\mathrm{d}E_k}\Delta E_k & R_{1,2_k} + \frac{\mathrm{d}R_{1,2_k}}{\mathrm{d}E_k}\Delta E_k \\ R_{2,1_k} + \frac{\mathrm{d}R_{2,1_k}}{\mathrm{d}E_k}\Delta E_k & R_{2,2_k} + \frac{\mathrm{d}R_{2,2_k}}{\mathrm{d}E_k}\Delta E_k \end{pmatrix} \begin{pmatrix} x_0 \\ x'_0 \end{pmatrix}_1 , \tag{3.23}$$

where N_s is the number of regions ('sectors', typically taken to be one sector per power source) into which the accelerator has been subdivided, and the total R matrix is expressed as a product of the R matrices for the different sectors. A possible procedure now consists of

1. measuring the transverse position x_m after the beam has been deflected,
2. selecting a set of E_k's,
3. comparing the measured position x_m with the expected, or calculated, position x_c by computing

$$\chi^2 = \sum (x_m - x_c)^2 , \tag{3.24}$$

where the sum is over all the BPMs used in the measurement, and
4. iterating steps 2 and 3 to minimize the χ^2.

An example of this approach is shown in Fig. 3.15 for the SLAC linear accelerator [20]. In the top plot an amplitude fit was used without taking into account any energy deviations (as in Fig. 3.14). In the bottom plot, also the values of E_k were fitted. The minimum χ^2 corresponded to an energy error of about 30%, which far exceeded the estimated uncertainty in the energy. Possible reasons for the discrepancy might include calibration errors of the quadrupole strengths, random errors in the BPM gains, or wake fields, which all are not included in the model. Nonetheless, the dynamics were made more predictable by incorporating the fitted energy errors into the accelerator model, despite the fact that the actual source of the error (perhaps wake fields) was different from the assumed one (energy errors).

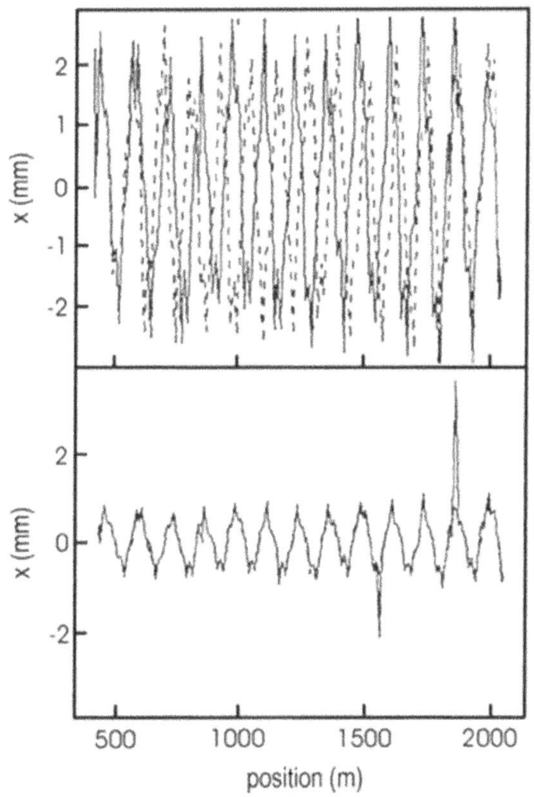

Fig. 3.15. Comparison of measured data and fit to betatron amplitude only (*top*) or fit to both betatron amplitude and energy in the different linac sections (*bottom*). Plotted is the horizontal beam position as a function of distance along the linear accelerator (Courtesy T. Himel, 1999)

3.4 Global Beam-Based Steering

We next present a more global approach to beam-based alignment than that discussed previously. We consider beam-based steering algorithms as ones which provide information on magnet, BPM, or structure misalignments using measurements with the beam. With this definition, one-to-one steering may also be considered a beam-based alignment algorithm since the applied kicks θ are related to the quadrupole (or BPM) displacements Δx by $\theta = K\Delta x$, where K is the integrated quadrupole focussing field.

More generally, we take in this example into account that the electrical zero of the BPMs may not be coincident with the magnetic center of the quadrupoles and that the quadrupoles themselves may be displaced with respect to the reference axis. The coordinate system used is sketched in Fig. 3.16. Here the beam position x measured with respect to some reference axis, which is common to all magnets, is given as a sum of the quadrupole displacement x_q, the difference in location of the electrical center of the BPM and the magnetic center of the quadrupole x_{bpm}, and the measured BPM value x_m.

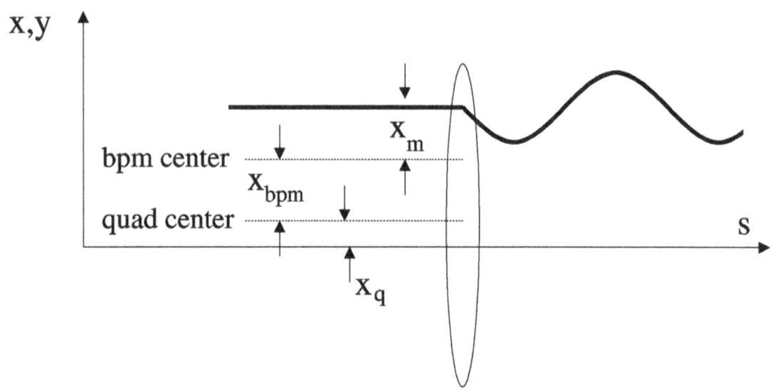

Fig. 3.16. Coordinate system used in the example of beam-based alignment (Courtesy C. Adolphsen, 2000)

The beam position x_k and angle $x_k{}'$ at quadrupole k, defined with respect to the reference axis, are given by [21]

$$\begin{pmatrix} x_k \\ x_k{}' \\ 1 \end{pmatrix} = \sum_{j=1}^{k-1} R_{j+1,k} \left\{ R_{j,j+1} \left[\begin{pmatrix} x \\ x' \\ 1 \end{pmatrix}_j + \begin{pmatrix} -x_{q,j} \\ 0 \\ 0 \end{pmatrix} \right] + \begin{pmatrix} x_{q,j} \\ 0 \\ 0 \end{pmatrix} \right\} , \tag{3.25}$$

where $()_j$ gives the beam position and angle with respect to the quad center, the term in $[]$ is the beam position with respect to the reference axis, $R_{j,j+1}[]$ is the beam position with respect to the reference axis transported between quad j and quad $j+1$, and the term in $\{\}$ is the beam position and angle with respect to the quad center transported between quads j and $j+1$. Rearranging terms gives

$$\begin{pmatrix} x_k \\ x_k{}' \\ 1 \end{pmatrix} = R_{0,k} \begin{pmatrix} x \\ x' \\ 1 \end{pmatrix}_0 + \sum_{j=1}^{k-1} (R_{j+1,k} - R_{j,k}) \begin{pmatrix} x_{q,j} \\ 0 \\ 0 \end{pmatrix} , \tag{3.26}$$

where the sum is taken over upstream quadrupoles. The function to be minimized is then

$$\sum_k \Big[x_m - (x_k - x_q - x_{\rm bpm})\Big]^2 , \tag{3.27}$$

where x_m are the measurements and $(x_k - x_q - x_{\rm bpm})$ is the fitting function with unknowns $x_q, x_{\rm bpm}$ and the initial position and angle x_0 and $x_0{}'$.

The number of measurements is about half the number of unknowns. So the system is underconstrained. To constrain the solution, two independent measurements are required. An independent set of data may be obtained by scaling all the quadrupoles and correctors by a common factor and repeating the measurements. Multiple such scalings may be used to overdetermine the system which reduces the sensitivity of the solution to statistical errors.

3.5 Singular Value Decomposition

A common situation is that the BPM offsets are known fairly well and the orbit already fulfills a number of constraints, but many of the corrector magnets are strongly excited with some of them 'fighting' (compensating) each other. Fortunately, there exists a very powerful technique to reduce the rms strength of the orbit correctors, while maintaining a set of constraints. This technique is sometimes called 'corrector ironing' [22] and it is based on the 'singular value decomposition' (SVD) [23].

The linear equation to be solved is

$$\Delta\mathbf{x} = \mathbf{A} \cdot \theta , \tag{3.28}$$

where the vector $\Delta\mathbf{x} = (\Delta x_1, \ldots \Delta x_m)$ may describe the desired correction (or constraint) at m BPMs, and $\theta = (\theta_1, \ldots, \theta_n)$ are the excitation strengths of n correctors, that we want to determine. If $m \geq n$, we can decompose the matrix $\mathbf{A}$ as

$$\mathbf{A} = \mathbf{U} \cdot \begin{pmatrix} w_1 & 0 & \ldots & 0 \\ 0 & w_2 & \ldots & 0 \\ & \ldots\ldots & & \\ 0 & 0 & \ldots & w_n \end{pmatrix} \cdot \mathbf{V}^t . \tag{3.29}$$

The column vectors of the $m \times n$ matrix $\mathbf{U}$ and the $n \times n$ matrix $\mathbf{V}$ are orthonormal,

$$\mathbf{U}^t \cdot \mathbf{U} = \mathbf{I}_n , \tag{3.30}$$

$$\mathbf{V}^t \cdot \mathbf{V} = \mathbf{I}_n , \tag{3.31}$$

where $\mathbf{I}_n$ denotes the $n \times n$ unity matrix. The decomposition of (3.29) can be performed, for example, using the FORTRAN subroutine described in [23]. An SVD decomposition is also provided in a convenient form by many mathematical analysis packages, such as MATLAB [24].

We now consider three different cases: First, we suppose the number of correctors is equal to the number of BPMs. In this case the matrix $\mathbf{A}$ is square. The formal solution is given by

$$\theta = \mathbf{A}^{-1} \cdot \Delta\mathbf{x} = \mathbf{V} \cdot \begin{pmatrix} 1/w_1 & 0 & \dots & 0 \\ 0 & 1/w_2 & \dots & 0 \\ & \dots & \dots & \\ 0 & 0 & \dots & 1/w_n \end{pmatrix} \cdot \mathbf{U}^t \cdot \Delta\mathbf{x}\,. \tag{3.32}$$

If none of the w_i is zero, this is the unique solution to the problem. If one or more of the w_i are zero, the equation may not have an exact solution, but for these w_i one can simply replace $1/w_i$ by 0, and with this replacement (3.32) still gives the solution in a least squares sense. This means it minimizes the distance $r = |\mathbf{A}\cdot\theta - \Delta\mathbf{x}|$. Furthermore, the solution vector θ so obtained is the (either least-squares or exact) solution with the smallest possible length $|\theta|^2$. In other words, the solution derived from the SVD decomposition also minimizes the rms strength of the correctors.

In addition, it is worthwhile to note that the columns of $\mathbf{U}$ whose same numbered w_i are nonzero form an orthonormal set of basis vectors that span the range of the matrix $\mathbf{A}$ while the columns of $\mathbf{V}$ whose same-numbered elements w_j are zero form an orthonormal set for the nullspace of $\mathbf{A}$.

Next, we consider the case that there are fewer equations than correctors. In this case, we can simply add rows with zeroes to the vectors and matrices of (3.28) until the matrix is square, and then apply the SVD formalism, as described above. In this case, there is (at least) one zero eigenvalue w_j for every row of zeroes added.

Finally, in the case of more BPM constraints than unknown correctors ($m > n$), SVD works just as well. In general the w_j will not be zero, and the SVD solution will agree with the result of a least-square fit. If there are still some small values w_j, these indicate a degeneracy in $\mathbf{A}$ and the corresponding $1/w_j$ should be set to zero as before. The corresponding column in $\mathbf{V}$ deserves attention, since it describes a linear combination of corrector excitations, which does not affect the constraints.

The SVD steering algorithm has been used successfully at many accelerators, for example, at the synchrotron light source SPEAR [25], throughout the SLC, and most recently at LEP. Applications include not only the minimization of corrector strengths and the identification of strongly excited correctors 'fighting' each other, but also orbit feedback, dispersion-free steering, and the computation of optical tuning knobs, e.g., for the beta function, dispersion, or coupling. Multiknobs were discussed in Sect. 2.5.

An Example of SVD. As a more concrete illustration, let us consider two steering correctors, separated by a betatron phase advance of π, which are followed by two BPMs at locations 1 and 2, which are also a phase of π apart from each other.

After applying a convenient normalization, the matrix equation relating the corrector strengths θ_1 and θ_2 with the BPM readings x_1 and x_2 is

$$\begin{pmatrix} x_1 \\ x_2 \end{pmatrix} = \begin{pmatrix} 1 & -1 \\ -1 & 1 \end{pmatrix} \begin{pmatrix} \theta_1 \\ \theta_2 \end{pmatrix} . \tag{3.33}$$

The 2×2 matrix $\mathbf{A}$ can be decomposed as follows:

$$\mathbf{A} \equiv \begin{pmatrix} 1 & -1 \\ -1 & 1 \end{pmatrix} \tag{3.34}$$

$$= \begin{pmatrix} \frac{1}{\sqrt{2}} & \frac{1}{\sqrt{2}} \\ -\frac{1}{\sqrt{2}} & \frac{1}{\sqrt{2}} \end{pmatrix} \begin{pmatrix} 2 & 0 \\ 0 & 0 \end{pmatrix} \begin{pmatrix} \frac{1}{\sqrt{2}} & -\frac{1}{\sqrt{2}} \\ \frac{1}{\sqrt{2}} & \frac{1}{\sqrt{2}} \end{pmatrix} \tag{3.35}$$

$$\equiv \mathbf{U}\mathbf{W}\mathbf{V}^t , \tag{3.36}$$

where W denotes the diagonal matrix. Suppose now we want to steer to $x_{t,1} = 1$ and $x_{t,2} = 0$. An exact solution does not exist, since there is no corrector between the two BPMs and the latter are a phase advance of π apart. However, there is an SVD solution:

$$\begin{pmatrix} \theta_1 \\ \theta_2 \end{pmatrix} = \mathbf{V}\mathbf{W}^{-1}\mathbf{U}^t \begin{pmatrix} 1 \\ 0 \end{pmatrix} \tag{3.37}$$

$$= \begin{pmatrix} \frac{1}{\sqrt{2}} & \frac{1}{\sqrt{2}} \\ -\frac{1}{\sqrt{2}} & \frac{1}{\sqrt{2}} \end{pmatrix} \begin{pmatrix} \frac{1}{2} & 0 \\ 0 & 0 \end{pmatrix} \begin{pmatrix} \frac{1}{\sqrt{2}} & -\frac{1}{\sqrt{2}} \\ \frac{1}{\sqrt{2}} & \frac{1}{\sqrt{2}} \end{pmatrix} = \begin{pmatrix} \frac{1}{4} \\ -\frac{1}{4} \end{pmatrix} . \tag{3.38}$$

Hence, the SVD solution equals $\theta_1 = 1/4$ and $\theta_2 = -1/4$. For these corrector settings one obtains the BPM readings $x_1 = 1/2$ and $x_2 = -1/2$, where indeed the quadratic distance to the target values, $\sum_i (x_{t,i} - x_i)^2$, assumes a minimum. Not only does SVD determine a pair of corrector settings for which the BPM readings are closest to the solution looked for, but in addition, among all possible such solutions (adding any constant to the above values of θ_1 and θ_2 would give the same minimum value of $\sum_i (x_{t,i} - x_i)^2$), it finds that one which also minimizes the corrector strengths, namely the sum of the squares $\sum \theta_i^2$.

3.6 'Wake Field Bumps'

Through the early 1990's emittance dilutions in the SLC linac were controlled by imposing tight tolerances on injection errors as a precursor to BNS damping [26, 27], steering using both one-to-one correction and localized beam-based alignment [21, 28], and by invoking BNS damping. The name BNS damping refers to V. Balakin, A. Novokhatsky, and V. Smirnov, who in the early 1980s proposed this type of scheme, in which the tail of the bunch is focussed more weakly than the bunch head in order to compensate for the defocusing effect of the single-bunch wake field [29]. At the SLC the variation of focusing across the bunch was achieved by introducing an energy difference. More details will be discussed in Chap. 4.

As the beam currents were increased, a more localized emittance preservation technique was developed in which empirically determined trajectory oscillations ('bumps') were used to cancel emittance dilutions from transverse wake fields and dispersive errors. While the origins of the disturbances could not be easily localized longitudinally along the linac, the accumulated effects could be cancelled using such bumps and the emittance dilution could be reduced by a factor of almost ten [30]. The effect on the beam was determined by emittance measurements near the end of the linac.

Two trajectories in both x and y are shown [30] in Fig. 3.17. Both trajectories produced about the same small emittance measured near the end of the linac. Notice the vertical scale which shows excursions of nearly 750 μm peak-to-peak. While wake field bumps were used for many years, it became clear as the currents were increased that this technique was inherently unstable; small (e.g., thermal) changes in the reference line phase, for example, changed the phase advance over the bump range so that even this more localized correction scheme was not sufficiently local to be stable against realistic variations in the accelerator. Both trajectories in Fig. 3.17 resulted in about the same

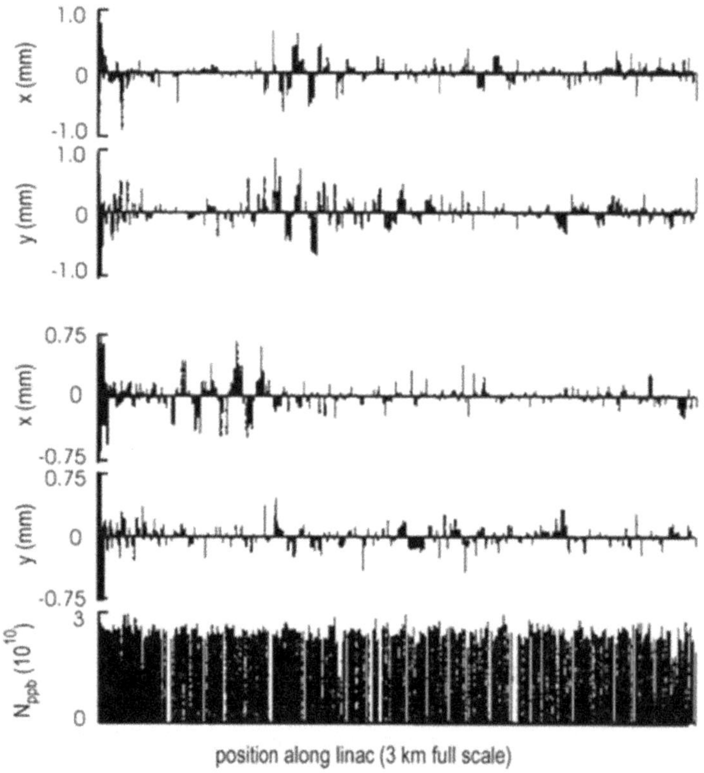

Fig. 3.17. Two measured orbits with empirically determined coherent betatron oscillations used to cancel accumulated wake field and dispersion errors (Courtesy J. Seeman, 2000)

final beam emittances indicating that the procedure did not converge to a unique solution.

Physical insights were gained by simulations carried out using the program LIAR [31]. The model included representative amplitudes of the wake field bumps which minimized the relative emittance growth at the locations of the measurements. The simulated relative growth in normalized emittance [32, 33] is shown in Fig. 3.18 as a function of position along the linac. Several important conclusions were drawn from this and other simulation results:

- At the first emittance measurement, the optimization had been made in a location where the energy spread of the beam was large; that is, a compromise was made using wake field bumps between correction of dispersive and wake field-induced errors.
- Between the emittance measurement stations, there was uncontrolled emittance growth.
- Between the final emittance measurement station and extraction of the beam from the linac, there was significant emittance growth.
- Most importantly, being nonlocal in nature, small changes in the phase advance could destroy this delicate cancellation. In practice this caused significant time-dependent variations in the measured emittances [34, 35, 36, 37].

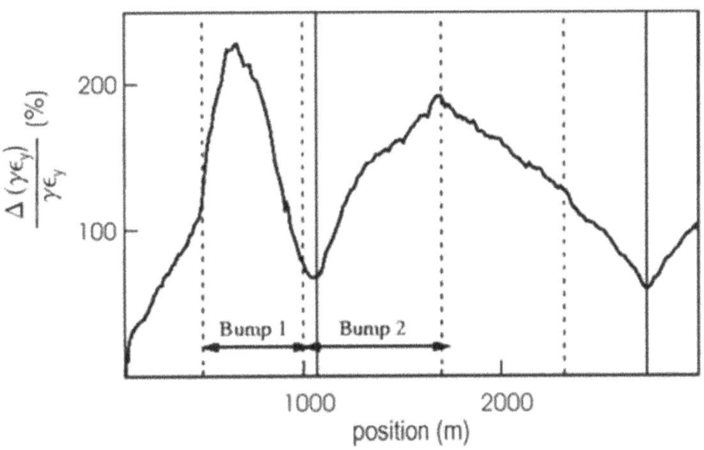

Fig. 3.18. Simulated emittance growth as a function of position along the linac. The locations of the feedback loops, which controlled the amplitude of the wake field bumps, are shown along with the locations of the emittance measurements (Courtesy R. Assmann, 2000)

3.7 Dispersion-Free Steering

So far we have described *one-to-one steering* which is a first step in orbit optimization but is imperfect as minimization of the BPM reading in a displaced quadrupole generates dispersive errors, *beam-based alignment* of quadrupole displacements which works beautifully at low beam currents where there is no wake field-generated dispersion, and wake field bumps which while more local than BNS damping is highly sensitive to small perturbations in the electromagnetic optics. With perfect implementation of either procedure, dispersive emittance dilutions may still result. As an example, consider a closed trajectory bump of the kind illustrated in Fig. 3.3. It has been shown [38] using LIAR and realistic optical parameters of the SLC linac that a closed 100 µm π-bump at a quadrupole located early in the linac generates nearly 0.5 mm dispersion at the end of the linac. Naively, about 6 such bumps acting independently would produce a dispersive emittance contribution equal to the final emittances typically achieved at the SLC.

Dispersion-free steering [35, 36, 38] is an algorithm which corrects even more locally the dispersive errors from misaligned quadrupoles *and* the dispersive errors arising from transverse wake fields. For mostly technical reasons (e.g., data acquisition and processing time) implementation was unduly delayed at the SLC. Dispersion-free steering (and rf phase stability [39, 40, 41]) proved crucial for maintaining stable linac emittances at the SLC.

The centroid trajectory[1] is given by

$$x_j = \sum_{i=1}^{j-1} \sqrt{\frac{E_i}{E_j}} \sqrt{\beta_i \beta_j} \theta_i \sin(\phi_j - \phi_i) \tag{3.39}$$

$$= \sum_{i=1}^{j-1} R_{12}^{ij} \theta_i \,, \tag{3.40}$$

where the damping factor $\sqrt{E_i/E_j}$ has been included. To constrain the system, one can equivalently change the beam energy (which may be difficult in practice) or scale the strength of the lattice, i.e., the strengths of the quadrupoles. Then the orbit difference is measured for a deflection applied in the original lattice and in the scaled lattice. This orbit difference amounts to

$$\begin{aligned} \Delta x_j &\approx \sum_{i=1}^{j-1} \left[R_{12}^{ij} - \sqrt{\frac{E_i}{E_j}} \sqrt{\beta_{i,\kappa} \beta_{j,\kappa}} \sin(\phi_{j,\kappa} - \phi_{i,\kappa}) \right] \theta_i \\ &= \sum_{i=1}^{j-1} R_{12,\kappa}^{ij} \theta_i \,, \end{aligned} \tag{3.41}$$

where $\beta_{i,\kappa}$ and $\phi_{i,\kappa}$ denote the beta function and betatron phase in the scaled lattice, and the change in lattice focussing is expressed by the parameter

$$\kappa = \frac{\Delta K}{K} + 1 \,, \tag{3.42}$$

where K is the quadrupole strength.

[1] a warning: intrabunch position-energy correlations, when projected, may result in measured centroid displacements which underestimate the contributions from off-axis bunch tails

The function to be minimized is

$$\sum_j \Big[x_j - \sum_i M_{ji}\theta_i \Big]^2 , \tag{3.43}$$

where x_j is an $m \times 1$ vector containing the difference measurements and the fitting function is given by $\sum_i M_{ji}\theta_i$ where θ_i is an $n \times 1$ vector of unknowns. In the simplest case, the coefficients M_{ji} represent an $m \times n$ matrix containing the transfer matrix elements $M_{ji} = R_{12}^{ij}$.

In practice it is not difficult to minimize not only the absolute orbit, but simultaneously several (k) difference orbits Δx. In this case x_j is a $(k+1) \times 1$ vector containing the difference measurements and the absolute orbit, M_{ji} is a $(k+1) \times n$ matrix and θ remains an $n \times 1$ matrix. This approach was used at the SLC where in addition, to overconstrain the solution and minimize systematic errors arising from magnet hysteresis, the measurements were performed for $k = 4$ or 5 values of κ corresponding to energy variations of +5% to −30%. In later years, the problem of magnet hysteresis was eliminated and the application became noninvasive as two independent measurements could be obtained *without* changing the lattice by measuring independently the orbits of the electrons and positrons which passed through the same lattice. The opposite charge of electrons and positrons was equivalent to a lattice scaling by −200%.

Shown in Fig. 3.19 are absolute ($\kappa = 1.0$) and difference trajectories ($\kappa = 0.9$, 0.8, and 0.7) measured after trajectory steering of the SLC linac [38] using 20-pulse BPM averaging. With an equivalent energy change of 30% ($\kappa = 0.7$), a difference trajectory of up to 1.5 mm was observed. Similar measurements made after iteration of dispersion-free steering [38] are given in Fig. 3.20. Iteration proved useful to reduce sensitivity to errors in the

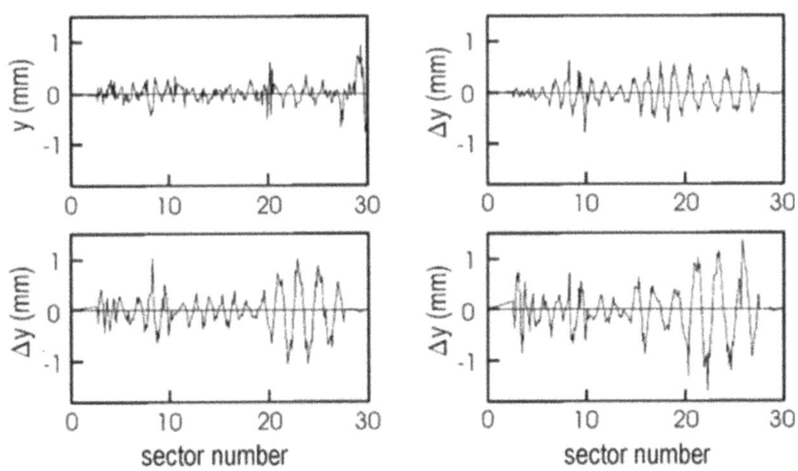

Fig. 3.19. Absolute and difference vertical trajectories measured after trajectory steering before dispersion-free steering of the SLC linac (Courtesy R. Assmann, 2000)

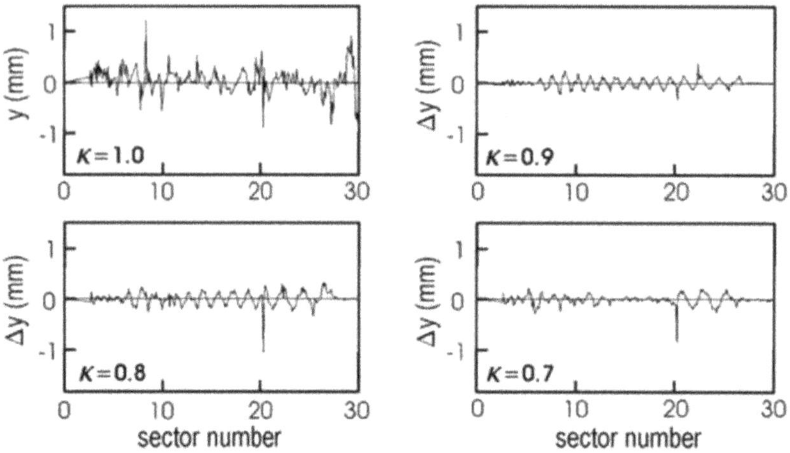

Fig. 3.20. Absolute and difference vertical trajectories measured after dispersion-free steering in the SLC linac (Courtesy R. Assmann, 2000)

assumed optics even though experience showed that the first iteration yielded the largest improvements. With $\kappa = 0.3$, neglecting the errant point due possibly to a bad BPM near sector 20, the maximum orbit difference after dispersion-free steering was reduced from 1.5 mm to less than 200 μm. Notice that the rms of the measurements of the absolute orbit are actually larger following dispersion-free steering. This suggests that significant BPM and/or quadrupole misalignment errors were still present.

3.8 Errors

For simplicity of expression, measurement errors have been neglected up to now. At the SLC, error sources and their typical rms contributions included BPM resolution errors $\sigma(x_j) < 10$ μm, BPM misalignments $\sigma_{\mathrm{bpm}} \sim 100$ μm, and systematic errors arising from beam jitter and/or slow drifts $\sigma_{\mathrm{sys}} \sim$ 20 μm. To propagate the measurement errors used in the minimization procedures, a weighting function may be defined as

$$w_j = \frac{1}{\sum_j \sigma_{m,j}{}^2}, \tag{3.44}$$

where the subscripts m give the different error sources and j is a sum over the BPM measurements. The functions to be minimized then are (c.f. (3.18), (3.27), and (3.43))

$$\sum_j \left[\frac{x_j - \sum_i M_{ij}\theta_i}{\sum_m \sigma_{m,j}{}^2}\right]^2 \quad \text{(one}-\text{to}-\text{one steering)},$$

$$\sum_k \left[\frac{x_m - (x_k - x_q - x_{\mathrm{bpm}})}{\sum_m \sigma_{m,j}{}^2}\right]^2 \quad \text{(beam}-\text{based alignment)},$$

$$\sum_j \left[\frac{x_j - \sum_i M_{ij}\theta_i}{\sum_m \sigma_{m,j}^2} \right]^2 \qquad \text{(dispersion}-\text{free steering)} . \tag{3.45}$$

A goodness of fit parameter, or χ−squared may be correspondingly constructed. In the dispersion-free steering example given above for which both the trajectory and the trajectory differences were to be simultaneously minimized,

$$\chi^2 = \sum_j \left[\frac{x_j^2}{\sigma_{\text{bpm}}^2} + \sum_\kappa \frac{\Delta x_{j,\kappa}^2}{\sigma_{\text{sys}}^2} \right] , \tag{3.46}$$

where the second summation over κ corresponds to the difference orbits for the different energy scalings under which the measurements were made. The errors from BPM resolution were assumed to be negligible and the summation over errors has been simplified to reflect the dominating errors; that is, the systematic errors contribute less than the alignment errors in the measurements of the absolute trajectories, while in the measurements of difference trajectories the BPM misalignments cancel and the associated errors are therefore set to zero.

3.9 Orbit Feedback

Automated feedback systems that continually stabilize the beam orbit are becoming more common in accelerators. A comprehensive overview of orbit feedback system design can be found in [42]. A simple orbit feedback maintains a constant orbit by adjusting the strength of 2 or 4 steering correctors based on BPM readings. Many orbit feedback systems employ an SVD algorithm which flattens the orbit while at the same time minimizing the strength of the correctors.

Slightly more complicated feedback loops are designed so that they maintain both the beam orbit and the beam energy. Orbit and energy can be separated using BPMs at dispersive locations. The orbit is corrected via steering correctors; the beam energy by adjustments to some upstream rf phase.

The effectiveness of a feedback can be tested by measuring its response to a step change. An example in Fig. 3.21 shows the response of an SLC feedback loop to a sudden step change in energy. The picture illustrates the improvement achieved by increasing the number of feedback BPMs to better constrain the fit. Specifically, 2 BPMs separated by $-I$ were included so that the sum of the (difference) signals from the two BPMs gave the dispersive contribution to the particle orbit independent of the betatron component. In addition to having a well understood accelerator model, for accelerators with high repetition frequency (with intra-pulse spacing comparable to the the response time of the correction magnets) it was found necessary to carefully match the response times of correction magnets [43].

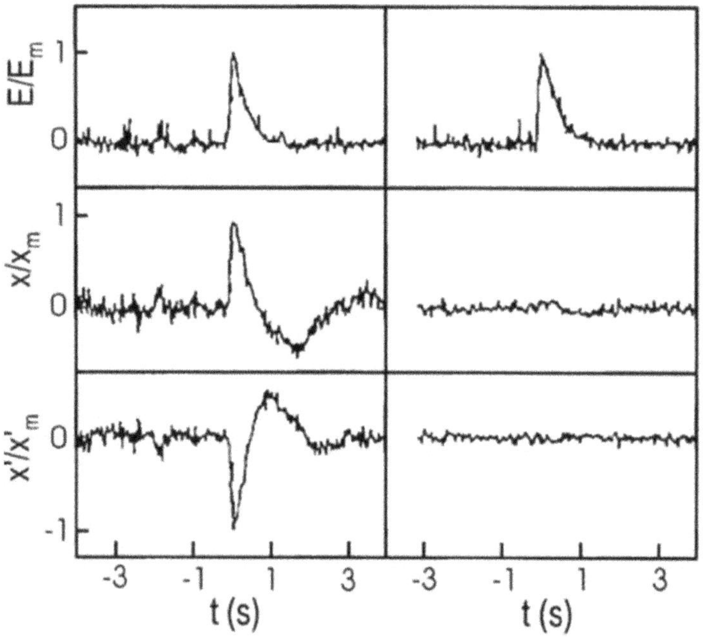

Fig. 3.21. Response of the orbit and energy feedback in the ring-to-linac transfer line of the SLC to a fast step change in energy [43]: (*left*) before and (*right*) after additional BPMs were included in the feedback loop

There are different techniques to calibrate the local transport matrices between correctors and BPMs within each feedback loop, which are used to continually compute the excitation of the feedback steering correctors. For example, the induced change in orbit position and angle can be measured as a function of the individual feedback corrector strengths.

If there are successive feedback loops on a beam line, these loops could interfere with each other, and cause unwanted orbit oscillations due to "double-compensation". This interference can be avoided by one of 4 different approaches [42]: (1) orthogonality of correction, (2) different feedback response time, (3) inter-loop communication (feedback "cascades") and (4) integration of loops into one global loop. The orbit feedbacks in the SLAC linac are connected by a so-called adaptive cascade, where each feedback passes its information to the loop downstream, in order to avoid multiple corrections of the same perturbation. The linear transport matrix between successive loops is monitored and updated continuously using the naturally occurring beam-orbit variation.

3.10 Excursion – AC Dipole

Not always is there a well defined closed orbit in a storage ring which is stable from turn to turn. As an example of a case without regular closed orbit, we consider a dipole field that is modulated at a frequency near the fractional part of the betatron tune. We assume that the field changes on successive turns, but has a sinusoidal variation, so that it repeats after a larger number of turns. In this case there is a new dynamic 'closed orbit' which returns to its starting point after a full modulation period. The amplitude of this periodic orbit is [44]

$$\hat{x} = \frac{\beta_x (B_{\mathrm{mod}} l_{\mathrm{dip}})}{4\pi (B\rho) \Delta Q}, \tag{3.47}$$

where B_{mod} is the amplitude of the dipole-field modulation, l_{dip} the length of the dipole, $(B\rho)$ the magnetic rigidity, $\Delta Q = |Q_x - Q_m - k|$, Q_x the betatron tune, Q_m the modulation tune, and k an integer which minimizes ΔQ. The amplitude of the dynamic orbit increases inversely with the frequency difference of modulation and betatron oscillations. Such a dynamic closed orbit may be used to increase the strength of spin resonances in order to preserve the polarization during acceleration [45] (see also the discussion of polarization in Chap. 9) or it can aid in measuring lattice nonlinearities [44] without diluting the transverse emittance. First tests were successfully performed at the BNL AGS [44, 45] and the CERN SPS [46].

Exercises

3.1 Design of an Orbit Feedback Loop

Write an algorithm for orbit correction in one plane assuming uncoupled, linear transport. Let the beam position and angle be detected using two BPMs and use two fast corrector dipoles for implementing the desired deflections. Introduce (assumed known) relative phase advances and beta functions as needed to take into account phase differences between the correctors, between the BPMs, and between the correctors and BPMs. Comment on the optimum phase advances between the correctors and BPMs and comment on the optimum beta functions at the locations of the measurements and corrections. Consider

a) a transport line,

b) a circular accelerator.

3.2 Linac Dispersion and Orbit Correction

a) Dispersion for a free betatron oscillation. Consider a beam deflected by an angle θ at $s = 0$. In a smooth approximation, and ignoring wake field effects, the betatron motion of a single particle with relative momentum deviation δ_1 is

$$x_1''(s) + \frac{k_\beta^2}{1+\delta_1} x_1(s) = \frac{\theta}{1+\delta_1} \delta(s), \tag{3.48}$$

where the last $\delta(s)$ is the Dirac delta function, indicating a single deflection of strength θ at location $s = 0$, and $k_\beta = \sqrt{K}$ is the wave number of the betatron oscillation. Equation (3.48) describes the motion of an individual off-momentum particle in the bunch, as well as of the bunch centroid if the latter experiences a momentum offset δ_1. The equality of the particle and centroid motion (for small bunch charges) is an advantage, since the single-particle dispersion can be measured by observing the response of the centroid motion to an energy error. Solve (3.48), linearize the solution in δ_1 and determine the dispersion for $k_\beta s \delta_1 \ll 1$.

b) Dispersion behind a π bump. A one-to-one orbit correction can be thought of as a superposition of π bumps. Calculate the dispersion generated by a single π bump represented by two kicks θ, at $s_1 = 0$ and $s_2 = \pi/k_\beta$.

4 Transverse Beam Emittance Measurement and Control

The beam emittance ϵ_{xyz} represents the volume of the beam occupied in the six dimensional phase space $(x, x', y, y', \phi, \delta)$, where x and y are the transverse positions, x' and y' are the transverse angles, ϕ is the time-like variable representing the relative phase of the beam, and δ is the relative beam momentum error. Using the notation of the beam matrix Σ_{beam} introduced in Chap. 1, the 6-dimensional emittance is

$$\epsilon_{xyz} = \det\, \Sigma_{\text{beam}}^{xyz} \,. \tag{4.1}$$

Considering now only the horizontal plane, the corresponding 2-dimensional horizontal emittance is obtained from

$$\epsilon_x = \sqrt{\langle x^2\rangle\langle x'^2\rangle - \langle xx'\rangle^2}\,, \tag{4.2}$$

where the first moments have been subtracted, and the average $(\langle\ldots\rangle)$ is taken over the distribution function of the beam; recall also (1.27–1.29). An analoguous expression holds for the vertical plane. For a coupled system, the general form of (4.1) must be taken.

Control of the beam emittance and prevention of emittance dilutions are mandatory for achieving high brilliance in light sources and high luminosity in colliding beam accelerators. Some sources of emittance dilution, e.g., beam-gas scattering, are inevitable and can be reduced only via hardware improvements. In the same vane, processes involving space-charge dilutions, which constitute a predominant limitation for low-energy ion or proton beams, have after-the-observation been treated by fundamental changes in the acceleration optics, as illustrated, e.g., by the FNAL linac upgrade, and a new booster ring at BNL. Another class of dilutions involving man-made sources, e.g., component vibration, ground motion, power supply regulation, etc., may be curable using sophisticated measurement devices and feedback or feedforward schemes. In this case, the crucial ingredient leading to improved performance is the detection of the offending presence, e.g., using model-independent analysis, as discussed in Chap. 2, or more commonly analyses in the frequency domain.

An interesting example of the latter case is shown in Fig. 4.1. Plotted is the Fourier transform of pulse-by-pulse BPM measurements from the SLC linac. The data evidence a strong component at about 1 Hz which seemed

M. G. Minty et al., *Measurement and Control of Charged Particle Beams*,
https://doi.org/10.1007/978-3-662-08581-3_4

to partly explain the 1 Hz 'wave' in the backgrounds recorded by the SLD detector. However, at this time, the data acquisition was limited to 30 Hz (with a beam repetition frequency of 120 Hz). As a result, the sampled data were aliased; that is, the Fourier line appearing at 1 Hz corresponded actually to 59 Hz as verified later with a faster (120 Hz) data acquisition speed. The source of these excitations was eventually traced back to asynchronous (to the beam) operation of the cooling water pumps for the linac quadrupoles. Once the disturbance was identified[1], corresponding measures were implemented – in this case, damping of the vibrations resulting from the turbulent water flow and modifications to the pump impellers – and the impact on the beam was correspondingly reduced. Identification of the error sources evidenced by the remaining peaks in Fig. 4.1 was also partially successful [1].

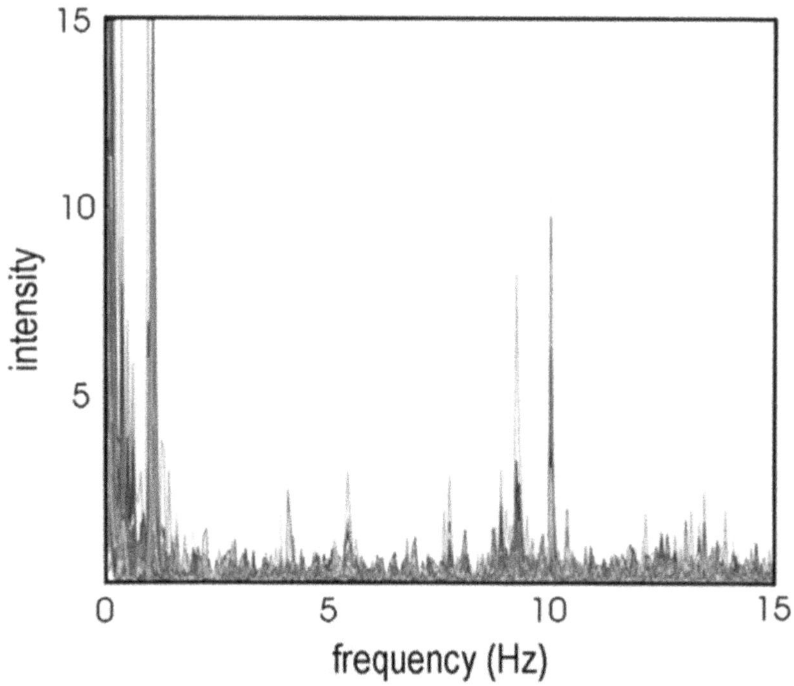

Fig. 4.1. Frequency spectrum of beam motion in the SLC linac, revealing the contribution of water pumps operating at 59 Hz [1]

Other common situations which lead to emittance dilutions arise from poor setting of the accelerator including poor "matching" between accelerator subsystems, residual betatron coupling (which may also be caused by unknown field errors), or spurious dispersion which arises either directly from magnet misalignments or from imperfect attempts to correct for alignment errors by one-to-one steering. The latter may arise from insufficient knowledge of the accelerator optics (as limited by the BPM resolution, for example) or

[1] which proved difficult since in fact there actually was at that time a 1 Hz component on the beam introduced incorrectly by the orbit feedback loops as these also reacted on the aliased frequency

may arise intrinsically from the alignment procedure itself (as is the case, e.g., of using the one-to-one steering algorithm).

Finally, there are processes which are in practice hard to predict that also lead to emittance dilutions. Such processes, predominant especially with high beam intensities, include, at high single-bunch charge both transverse and longitudinal wakefields effects leading to beam instabilities in circular accelerators and to beam centroid excursions in linear accelerators, and at high single-bunch charge and high total current also beam instabilities arising from ions or electron clouds.

In all the above cases, the diagnosis is precedented by an accurate measurement of the beam emittance. In this chapter, standard emittance measurement techniques will be outlined in a step-by-step manner. Then some methods used to minimize the beam emittance by changing the underlying accelerator optics will be reviewed.

4.1 Beam Emittance Measurements

In this section we describe not single particle transport, but transport of the beam as a whole. The beam quality will be characterized by the beam emittance which is often measured with reference to a particular plane of interest; i.e., the horizontal, vertical, or longitudinal emittance. Methods for measuring the beam emittance and for parametrizing the degree of mismatch will be outlined.

We note that many novel measurement techniques have recently been developed for measuring very small beam sizes including interferometric methods, applied in the beamline at the SLAC final focus test facility [2] or in the ATF damping ring [3], or tomographic phase space reconstruction applied at the TTF linac [4]. A recent review of similar topics is given in [5]. In this section we restrict the discussion to emittance measurement techniques, which are based on commonly used and relatively simple hardware. For illustration we describe measurements and present experimental data obtained using wire scanners although the measurement principles have been similarly applied with the use of fluorescent screens.

4.1.1 Single Wire Measurement

The beam emittance can be measured by varying the field strength of a quadrupole located upstream of a single wire or screen. In general this will lead to trajectory and beam-size changes downstream, and, hence, is referred to as an "invasive" measurement. The measurement could be made less invasive by simultaneously adjusting another quadrupole downstream of the wire scanner, so as to compensate for the change in the beta function induced by the first quadrupole. This has been rarely done in the past, but is a proposed scheme for future linear colliders.

The total transfer matrix of interest here is $R = SQ$, where S denotes the known transfer matrix between the quadrupole and the wire, and Q is the transfer matrix of the quadrupole:

$$Q = \begin{pmatrix} 1 & 0 \\ K & 1 \end{pmatrix}, \tag{4.3}$$

where we have invoked the thin-lens approximation, which is valid if the length of the quadrupole is short compared to its focal length $f = 1/K$. After matrix multiplication, one obtains

$$R = \begin{pmatrix} S_{11} + KS_{12} & S_{12} \\ S_{21} + KS_{22} & S_{22} \end{pmatrix}, \tag{4.4}$$

where the coefficients S_{ij} are the components of the matrix S. Expanding the matrix product for the 2×2 beam matrix $\Sigma^x_{\text{beam}} = (SQ)\Sigma^x_{\text{beam},0}(SQ)^T$ and equating the (1,1) elements on both sides, the square of the horizontal beam size follows as

$$\begin{aligned} \Sigma_{11}(= \langle x^2 \rangle) = {} & ({S_{11}}^2 \Sigma_{11_0} + 2S_{11}S_{12}\Sigma_{12_0} + {S_{12}}^2 \Sigma_{22_0}) \\ & + (2S_{11}S_{12}\Sigma_{11_0} + 2{S_{12}}^2\Sigma_{12_0})K + {S_{12}}^2 \Sigma_{11} K^2 , \end{aligned} \tag{4.5}$$

which is quadratic in the field parameter K.

To make use of these results in an emittance measurement, the following procedure is often employed:
1. For each value of quadrupole field strength K, the wire is scanned and the amplitude of the response measured by a detector is obtained as a function of wire position.
2. For each wire scan at fixed K, the distribution is fitted with a Gaussian of the form

$$f(x) = f_0 + f_{\max} e^{-\frac{(x - \langle x \rangle)^2}{2\langle x^2 \rangle}} , \tag{4.6}$$

where f_0 is the baseline level offset and $f_{\max}$ is the peak value of the Gaussian distribution.
3. The fitted beam size $\langle x^2 \rangle$ is plotted as a function of K.
4. The result is fitted with a parabola. One parametrization for the fit [6] is

$$\begin{aligned} \Sigma_{11} &= A(K - B)^2 + C \\ &= AK^2 - 2ABK + (C + AB^2) . \end{aligned} \tag{4.7}$$

5. The Σ matrix is reconstructed by equating coefficients of (4.5) and (4.7):

$$A = S_{12}^2 \Sigma_{11} , \tag{4.8}$$

$$-2AB = 2S_{11}S_{12}\Sigma_{11} + 2S_{12}^2\Sigma_{12} , \tag{4.9}$$

$$C + AB^2 = {S_{11}}^2 \Sigma_{11} + 2S_{11}S_{12}\Sigma_{12} + {S_{12}}^2 \Sigma_{22} , \tag{4.10}$$

and solving for Σ_{11}, Σ_{12} $(= \Sigma_{21})$, and Σ_{22}. The results are

$$\Sigma_{11} = A/{S_{12}}^2 \,, \tag{4.11}$$

$$\Sigma_{12} = -\frac{A}{{S_{12}}^2}\left(B + \frac{S_{11}}{S_{12}}\right) \,, \tag{4.12}$$

$$\Sigma_{22} = \frac{1}{{S_{12}}^2}\left[(AB^2 + C) + 2AB\left(\frac{S_{11}}{S_{12}}\right) + A\left(\frac{S_{11}}{S_{12}}\right)^2\right] \,. \tag{4.13}$$

6. The beam emittance is then calculated from the determinant of the beam matrix $\epsilon_x = \sqrt{\det\, \Sigma^x_{\text{beam}}}$ and the errors are propagated:

$$\det\, \Sigma^x_{\text{beam}} = \Sigma_{11}\Sigma_{22} - {\Sigma_{12}}^2 \tag{4.14}$$

$$= AC/S^4_{12} \,, \tag{4.15}$$

so that

$$\epsilon_x = \sqrt{AC}/S^2_{12} \,. \tag{4.16}$$

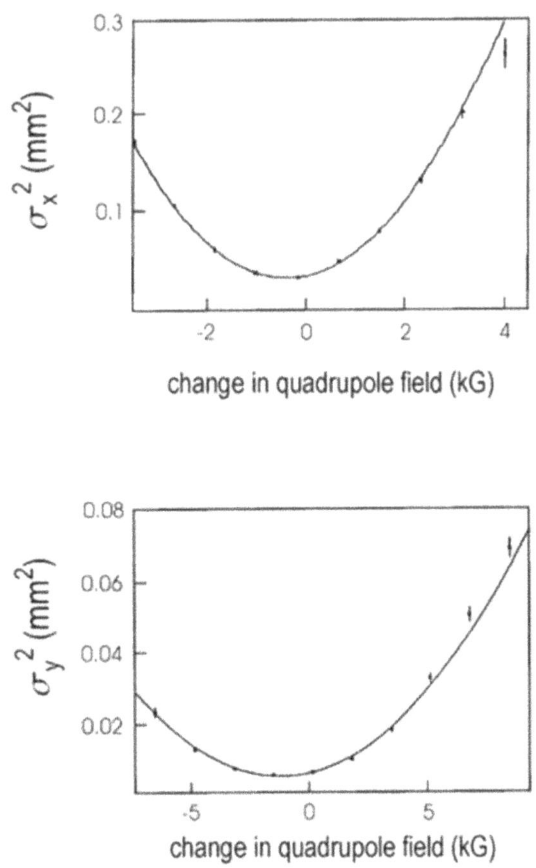

Fig. 4.2. Example transverse beam emittance measurements from the SLC prior to injection into the main linac using a single wire. The fit parameters for the horizontal emittance (*top*) were $A = 3494 \pm 52$, $B = -118.8 \pm 0.03$, $C = 3.2 \times 10^4 \pm 297$ with $\chi^2/\text{dof} = 1.5$ giving $\epsilon_x = 12.9 \pm 0.2$ nm-rad or $\gamma\epsilon_x = 30.1 \pm 0.4$ µm-rad. In the vertical plane (*bottom*), $A = 158.5 \pm 3.5$, $B = -129.3 \pm 0.08$, and $C = 5.1 \times 10^3 \pm 103$ with $\chi^2/\text{dof} = 4.6$ giving $\epsilon_y = 1.71 \pm 0.02$ nm-rad or $\gamma\epsilon_x = 3.98 \pm 0.04$ µm-rad

The above results also give the beam-ellipse parameters α_x, β_x, and γ_x:

$$\beta_x = \frac{\Sigma_{11}}{\epsilon} = \sqrt{\frac{A}{C}}, \tag{4.17}$$

$$\alpha_x = -\frac{\Sigma_{12}}{\epsilon} = \sqrt{\frac{A}{C}} \left(B + \frac{S_{11}}{S_{12}} \right), \tag{4.18}$$

$$\gamma_x = \frac{{S_{12}}^2}{\sqrt{AC}} \left[(AB^2 + C) + 2AB \left(\frac{S_{11}}{S_{12}} \right) + A \left(\frac{S_{11}}{S_{12}} \right)^2 \right]. \tag{4.19}$$

As a useful check, the beam-ellipse parameters should satisfy $(\beta_x \gamma_x - 1) = \alpha_x^2$.

An example of emittance measurements in the two transverse planes x and y is shown in Fig. 4.2. Notice that the optics at the wire scanners has been optimally chosen to allow simultaneous measurement of the beam waists in both transverse planes.

4.1.2 Multiple Wire Measurement

The beam emittance in a transport line or linac may be measured (in many applications noninvasively) using multiple wire scanners. Here, the quadrupole gradients are fixed, but the R matrices between s_0 and the different wire scanners (or other beam-size monitors) are different. If there are no coupling elements, three measurements using three wire scanners are required. With coupling, four wire scanners are needed (in this case each wire scanner should be equipped with several wires oriented at different angles in the transverse plane, e.g., a horizontal, a vertical and a wire oriented at 45° in the case of 'round beams' with equal horizontal and vertical beam size). The optimum wire locations for maximum sensitivity (without coupling) are such that the separation between wires corresponds to a difference in betatron phase advance $\Delta\mu$ of $90°/N_w$, where N_w is the number of wires used in the measurement.

The matrix equation to be solved is

$$\begin{pmatrix} ({\sigma_x}^{(1)})^2 \\ ({\sigma_x}^{(2)})^2 \\ ({\sigma_x}^{(3)})^2 \\ \cdots \\ ({\sigma_x}^{n})^2 \end{pmatrix} = \begin{pmatrix} ({R_{11}}^{(1)})^2 & 2R_{11}^{(1)} R_{12}^{(1)} & ({R_{12}}^{(1)})^2 \\ ({R_{11}}^{(2)})^2 & 2R_{11}^{(2)} R_{12}^{(2)} & ({R_{12}}^{(2)})^2 \\ ({R_{11}}^{(3)})^2 & 2R_{11}^{(3)} R_{12}^{(3)} & ({R_{12}}^{(3)})^2 \\ \cdots & & \\ ({R_{11}}^{(n)})^2 & R_{11}^{(n)} R_{12}^{(n)} & ({R_{12}}^{(n)})^2 \end{pmatrix} \begin{pmatrix} \beta(s_0)\epsilon \\ -\alpha(s_0)\epsilon \\ \gamma(s_0)\epsilon \end{pmatrix}. \tag{4.20}$$

This equation is applicable for both a multiple wire measurement or for a quadrupole scan. The superindex within parenthesis refers to the different different measurements; i.e., it either corresponds to the setting of some quadrupole magnet, in the case of a quadrupole scan, or to a different wire scanner or monitor, in the case of a multi-wire emittance measurement. The superindex '2' is meant to denote the square of the quantity. At least 3 measurements are required in order to solve for the three independent parameters ϵ, $\beta(s_0)$ and $\alpha(s_0)$.

To simplify the notation, let us denote the $n \times 3$ matrix on the right-hand side of (4.20) as $\mathbf{B}$, the n-component vector on the left side by $\Sigma_x = (\sigma_x^{(1)2}, ..., \sigma_x^{(n)2})$, and the 3-component vector on the far right by

$$\mathbf{o} = (\beta(s_0)\epsilon, -\alpha(s_0)\epsilon, \gamma(s_0)\epsilon) . \tag{4.21}$$

The equation is then

$$\Sigma_x = \mathbf{B} \cdot \mathbf{o} . \tag{4.22}$$

The problem of determining the elements of the vector $\mathbf{o}$ can be solved by a simple least-squares fit. We have to minimize the sum

$$\chi^2 = \sum_{l=1}^{n} \frac{1}{\sigma^2_{\Sigma_x^{(l)}}} \left(\Sigma_x^{(l)} - \sum_{i=1}^{3} B_{li} o_i \right)^2 , \tag{4.23}$$

where $\sigma_{\Sigma_x^{(l)}}$ denotes the rms error of $\Sigma_x^{(l)} = \sigma_x^{(l)2}$. This error is obtained from the fit to the lth wire scan which determines the rms beam size $\sigma_x^{(l)}$.

We find it convenient to normalize the coordinates $\Sigma^{(l)}$ so that the rms error is 1, introducing

$$\hat{\Sigma}_x^{(l)} = \frac{\Sigma_x^{(l)}}{\sigma_{\Sigma_x^{(l)}}} , \tag{4.24}$$

and

$$\hat{B}_{li} = \frac{B_{li}}{\sigma_{\Sigma_x^{(l)}}} . \tag{4.25}$$

Forming a symmetric $n \times n$ covariance matrix

$$\mathbf{T} = (\hat{\mathbf{B}}^t \cdot \hat{\mathbf{B}})^{-1} , \tag{4.26}$$

the least-squares solution to (4.22) is

$$\mathbf{o} = \mathbf{T} \cdot \hat{\mathbf{B}}^t \cdot \hat{\Sigma}_x , \tag{4.27}$$

and the error of any scalar function $f(\mathbf{o})$ is given by

$$\sigma(f)^2 = (\nabla_{\mathbf{o}} f)^t \cdot \mathbf{T} \cdot (\nabla_{\mathbf{o}} f) . \tag{4.28}$$

In particular, the errors of the parameters $\mathbf{o}$ themselves are

$$\sigma_{o_i} = \sqrt{T_{ii}} . \tag{4.29}$$

Once the components of $\mathbf{o}$ are known, we still need to perform a simple nonlinear transformation to infer ϵ, β, and α:

$$\epsilon = \sqrt{o_1 o_3 - o_2^2}\,, \tag{4.30}$$

$$\beta = o_1/\epsilon\,, \text{ and} \tag{4.31}$$

$$\alpha = -o_2/\epsilon\,. \tag{4.32}$$

The error propagation is straightforward, using (4.28).

A possible procedure for the multiple wire emittance measurement is as follows:

1. Each wire is scanned to obtain detector counts versus wire position x.
2. For each wire scan, the distribution is fit with a Gaussian using (4.6).
3. The Σ matrix is reconstructed using (4.20), the transfer matrix elements R_i from the model, and the σ_i from the measurements.
4. The emittance is calculated $\epsilon = \sqrt{\det \Sigma_{\text{beam}}}$.
5. The ellipse parameters $\alpha = -\Sigma_{12}/\epsilon$, $\beta = \Sigma_{11}/\epsilon$, and $\gamma = \Sigma_{22}/\epsilon$ are calculated, if desired.

4.1.3 Graphics

Increased operational efficiency may be obtained from a meaningful graphical representation of the experimental data. In the multiple wire emittance measurement it is useful to project the measurements to a single point along the accelerator and to plot the normalized phase space. The emittance ϵ, multiplied by π, corresponds to the area of the ellipse parametrized by

$$\epsilon = \gamma_x x^2 + 2\alpha_x x x' + \beta_x x'^2\,. \tag{4.33}$$

Since $\beta\gamma = 1 + \alpha^2$,

$$\epsilon = \frac{1}{\beta_x}[x^2 + (\alpha_x x + \beta_x x')^2] = \frac{1}{\beta_x}(x^2 + {p_x}^2)\,, \tag{4.34}$$

where $p_x = \alpha_x x + \beta_x x'$ is the coordinate orthogonal to x.

A succinct representation of the measured beam emittances is obtained by the following procedure (applied at the SLC), which displays the data in the normalized phase space, so that deviations from the design values are immediately obvious. The wire orientations are also plotted to indicate the phase space coverage provided by the wires:

1. Plot the design rms ellipse in the coordinates

$$\left(\frac{x}{\sqrt{\beta_x}}, \frac{\alpha_x x + \beta_x x'}{\sqrt{\beta_x}}\right) \tag{4.35}$$

at some reference point s along the trajectory. This results in a circle. Normalize the design ellipse to unit radius.

2. Plot also the ellipse obtained from the measurements of the ellipse parameters transposed to the reference point. Apply the same normalization as in step 1.

3. Using the lattice model, project each beam-size measurement back to the reference point and add the result as a line to the figure; that is, for each point $(\sigma_{x,w}, x'_w)$, where $\sigma_{x,w}$ is the beam size measured at the wire, apply an inverse mapping to the reference point:

$$\begin{pmatrix} x \\ x' \end{pmatrix}_{\text{ref point}} = R^{-1} \begin{pmatrix} \sigma_{x,w} \\ x'_w \end{pmatrix} , \tag{4.36}$$

where R is the 2×2 transport matrix from the reference point to the location of the wire. Here x'_w represents the undetermined angle variable (divergence) at the wire which parametrizes the location along the straight line in the phase space defined by (4.35). The slope of this line is related to the phase advance between the reference point and the wire.

An accompanying display of numbers should summarize the measurements which might include the measured and expected beam widths at each of the wires, the measured and design beam emittances, as well as the beam intensity. In addition, a measure of the degree of "mismatch" is useful. This will be further discussed in the next section.

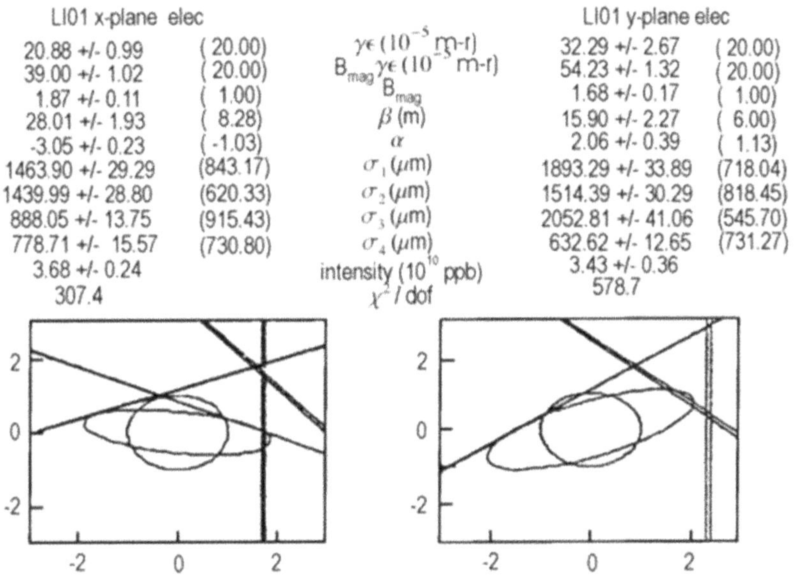

Fig. 4.3. Graphics output of multiple-wire transverse beam emittance measurement in the injector linac at the SLC

An example of this graphics is shown in Fig. 4.3. The corresponding raw data are given in Fig. 4.4. From Fig. 4.3 it is immediately obvious that while the measured ellipse has roughly the same emittance as the design circle (the horizontal emittance is 208.8 ± 9.9 [mm-mrad] compared to the design of 200 [mm-mrad], the vertical emittance is 323 ± 26.7 [mm-mrad] compared to the design of 200 [mm-mrad]), the ellipse orientation is incorrect. As will be shown in the next section, if this beam were allowed to propagate uncorrected,

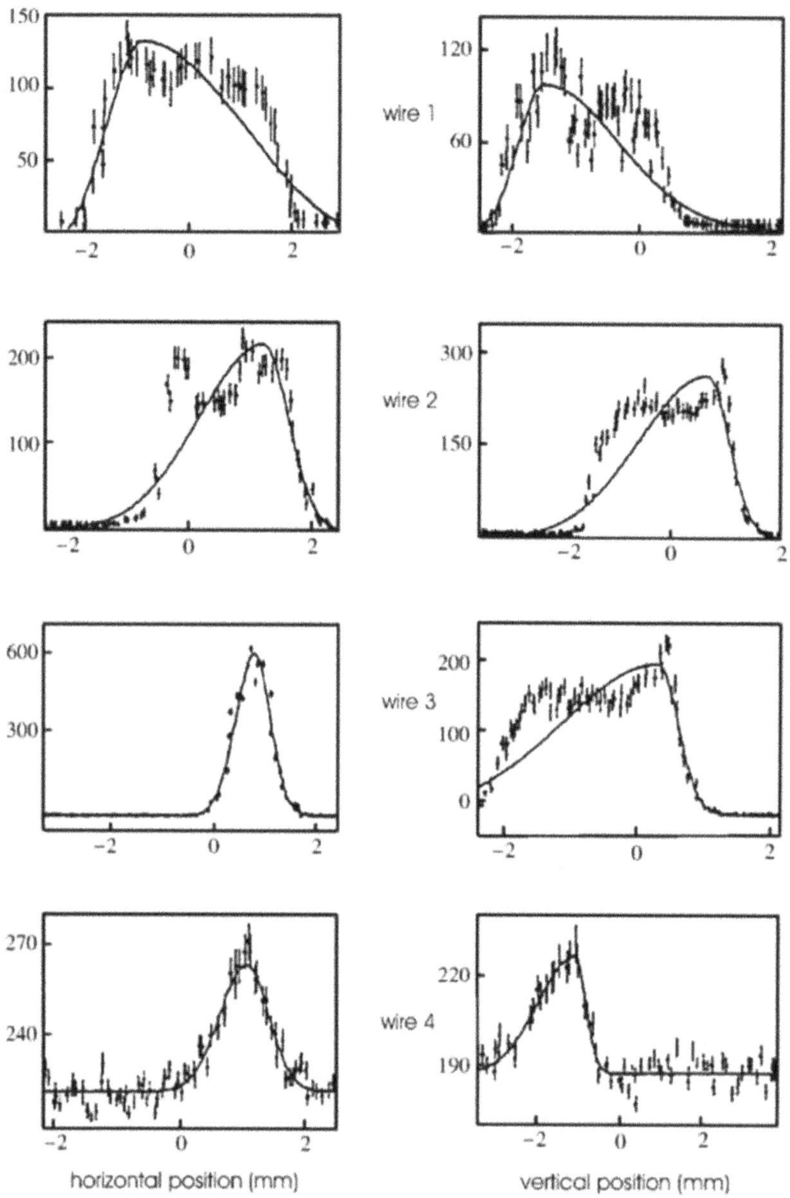

Fig. 4.4. Raw data showing individual wire scans used in the summary display of Fig. 4.3 and the "asymmetric Gaussian" fits of (4.37)

the final emittance titled $\mathrm{B_{mag}}\epsilon$ would be 390.0 ± 10.2 [mm-mrad] in x and 543.3 ± 13.2 [mm-mrad] in y. The emittance dilution factor $\mathrm{B_{mag}}$ represents the degree of the mismatch. From Fig. 4.3 can be deduced immediately the degree of phase space coverage spanned by the wires. In the horizontal plane, for example, the wire orientations are about $0°$, $-45°$, $-22.5°$, and $-67.5°$, which is ideal for a 4-wire emittance measurement.

The "measured ellipse", that is the ellipse that was reconstructed from the beam widths obtained by Gaussian fits to the individual wire scans, does not always represent the true rms emittance of a Gaussian beam distribution.

This can be seen by inspecting the raw data shown in Fig. 4.4. For more complex beam distributions, a somewhat better characterization is achieved by applying an "asymmetric Gaussian" fit to the wire-scan measurement, in which the left and right hand sides of the measured beam profile are approximated independently by two separate Gaussians. For example, the fitting function used at the SLC [7] was

$$f(x) = f_0 + f_{\max} \exp\left[-\frac{(x - \langle x \rangle)^2}{2\langle x^2 \rangle (1 + \alpha[\mathrm{sign}(x - \langle x \rangle)]}\right], \tag{4.37}$$

where α represents an asymmetry factor and is zero for a perfect Gaussian[2]. The σ for the left and right hand sides of the fitted distribution are $\sigma = \langle x^2 \rangle (1 \pm \alpha)$. For the ellipse reconstruction the average σ was used. When large tails are present in the raw data this more accurately represents the beam distribution. Based on the raw data it is clear, however, that even with the more sophisticated fitting algorithm, the fit only marginally represents the actual distributions.

For reasonably well "matched" beams, the graphical summary display is most useful for quick evaluation of the beam. In this example the deviations between the design and measured ellipse in the graphics suggest that a closer inspection of the raw data may be warranted. The "double-humps" in the single-wire measurements are characteristic of an upstream error; a beam, if kicked transversely, will filament, i.e., it loses coherency due to the natural spread in the phase advance between particles, resulting in an increased emittance and the characteristic double humps.

If a wire is mounted at 45° with respect to x and y (a "u-plane" wire), then it is also possible to measure the coupling between the horizontal and vertical plane. The (4×4) Σ-matrix is

$$\Sigma^{xy}_{\mathrm{beam}} = \begin{pmatrix} \Sigma_{11} & \Sigma_{12} & \Sigma_{13} & \Sigma_{14} \\ \Sigma_{21} & \Sigma_{22} & \Sigma_{23} & \Sigma_{24} \\ \Sigma_{31} & \Sigma_{32} & \Sigma_{33} & \Sigma_{34} \\ \Sigma_{41} & \Sigma_{42} & \Sigma_{43} & \Sigma_{44} \end{pmatrix}, \tag{4.38}$$

where, for example, Σ_{14} represents the correlation between x and y'. Notice that $\Sigma_{14} \neq \Sigma_{23}$. Whereas for the single plane uncoupled beam matrix reconstruction a minimum of 3 measurements are required, to fully reconstruct the coupled beam matrix a total of 10 measurements is needed. This includes the 3 measurements in the x plane, 3 in the y plane, and 4 in the u plane. The raw data used in such a coupled emittance measurement is presented in Figs. 4.5–4.7. In this case the raw data are well fitted using Gaussian fits. From these fits, using the uncoupled analysis (with only the x and y wires) presented previously, $\gamma\epsilon_x = (2.20 + / - 0.01) \times 10^{-5}$ m-r and

[2] with high resolution scanners this parameter may also prove useful for characterizing 'banana' beams in future linear colliders

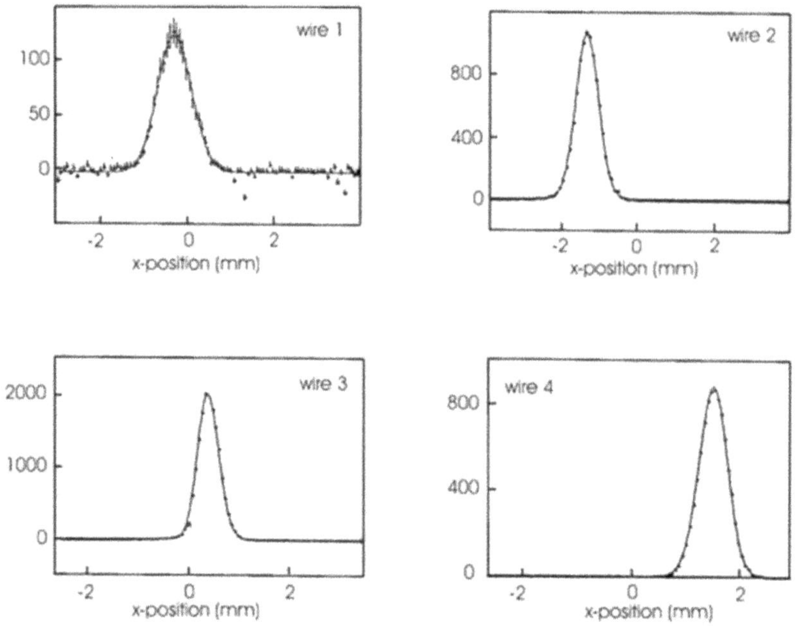

Fig. 4.5. Raw data in the x-plane for emittance measurement with $x-y$ coupling

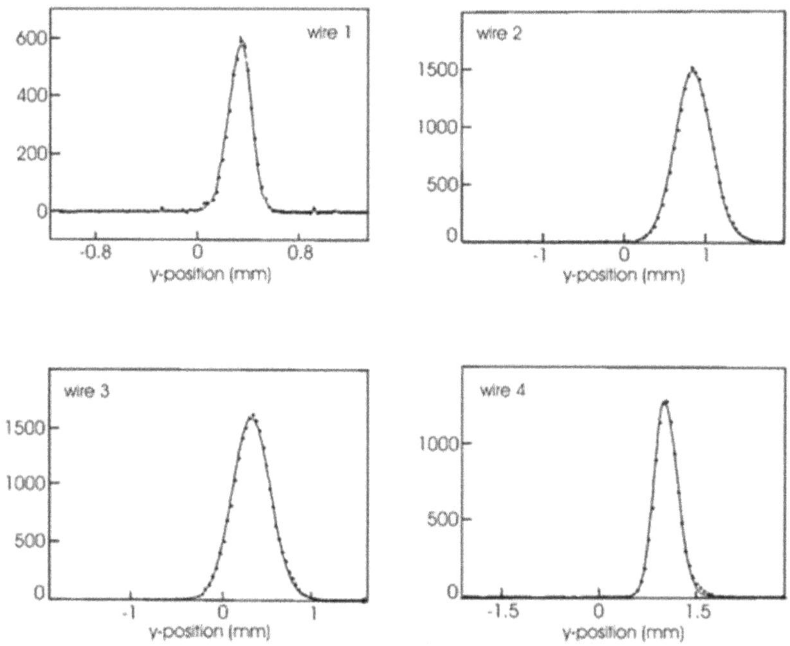

Fig. 4.6. Raw data in the y-plane for emittance measurement with $x-y$ coupling

$\gamma\epsilon_y = (1.89 +/- 0.02) \times 10^{-5}$ m-r. Using the complete data set, and projecting the distributions onto the eigenplanes of the tilted beam ellipses, the fits yielded $\gamma\epsilon_1 = 2.13 \times 10^{-5}$ m-r and $\gamma\epsilon_2 = 1.71 \times 10^{-5}$ m-r, which is in reasonable agreement with the results obtained excluding coupling effects indicating that the coupling is small. However, the error bars were not fully propagated. Recent experience at the ATF has shown that proper wire orientations are

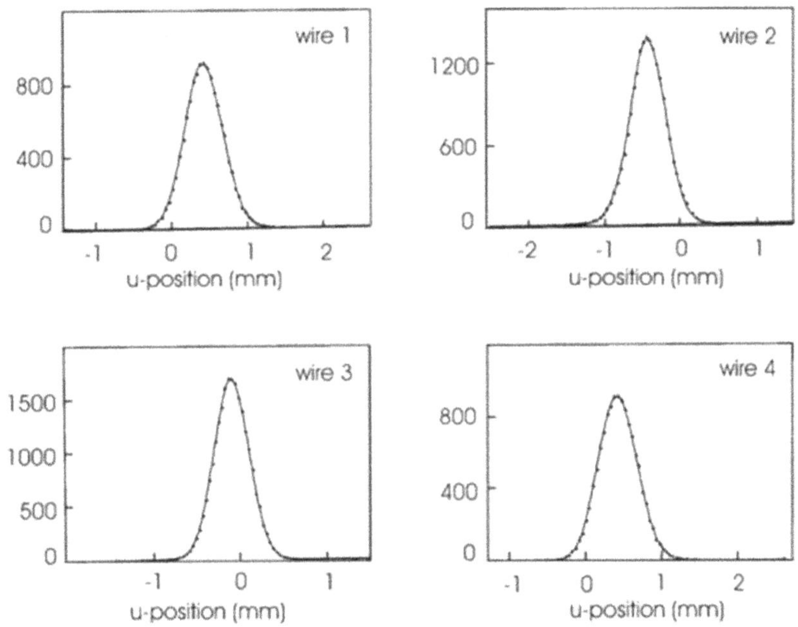

Fig. 4.7. Raw data in the u-plane for emittance measurement with $x-y$ coupling

critical for maximizing sensitivity particularly for flat beam measurements. Moreover, measurement errors can strongly influence the value of the emittance [8] interpreted from the approach given above. More robust tuning procedures for minimizing the linear coupling are given in [8].

4.1.4 Emittance Mismatch

In this section we begin by explicitly computing, in two dimensions, the transverse position and angle using the general form of the beam transfer matrix for a periodic lattice. This result is used to calculate the individual elements of the beam transfer matrix and to derive an expression for the mismatch parameter $\mathrm{B_{mag}}$ [9, 10].

The parameter $\mathrm{B_{mag}}$ has an important physical meaning. If a beam is injected into a ring or linac with a mismatch, the beam will filament until its distribution approaches a shape that is matched to the ring or the linac lattice. However, the filamentation causes the beam emittance to increase, such that, after complete filamentation, the emittance is given by the product of $\mathrm{B_{mag}}$ and the initial value of ϵ. The mismatch parameter is well suited for analysis in circular machines for which the periodicity is implicit. We will see that the same formalism is useful in describing emittance transport in linear accelerators and transport lines as well.

Derivation of Beam Matrix Elements. According to (1.27) and after subtracting the mean values from all coordinates, the beam matrix is

$$\Sigma^x_{\mathrm{beam}} = \begin{pmatrix} \langle x^2\rangle & \langle xx'\rangle \\ \langle xx'\rangle & \langle x'^2\rangle \end{pmatrix} . \qquad (4.39)$$

From (1.25), the point-to-point transfer matrix is

$$R_{fi} = \begin{pmatrix} \sqrt{\frac{\beta_f}{\beta_i}}(\cos\phi_{fi} + \alpha_i \sin\phi_{fi}) & \sqrt{\beta_f\beta_i}\sin\phi_{fi} \\ \frac{\alpha_i-\alpha_f}{\sqrt{\beta_f\beta_i}}\cos\phi_{fi} - \frac{1+\alpha_f\alpha_i}{\sqrt{\beta_f\beta_i}}\sin\phi_{fi} & \sqrt{\frac{\beta_i}{\beta_f}}(\cos\phi_{fi} - \alpha_f\sin\phi_{fi}) \end{pmatrix}, \tag{4.40}$$

where α_f and β_f are the ellipse parameters at a (final) observation point f downstream of a reference point, which is denoted by the subscript i. Here ϕ_{fi} is the phase advance between the reference point and the observation point and is equal to

$$\phi_{fi} = \int_{s_i}^{s_f} \frac{\mathrm{d}s}{\beta}\,. \tag{4.41}$$

For a periodic lattice for which $\alpha = \alpha_0$ and $\beta = \beta_0$, the periodic, one turn map, transfer matrix R_{otm} is given by (1.26):

$$R_{\mathrm{otm}} = \begin{pmatrix} \cos\mu + \alpha\sin\mu & \beta\sin\mu \\ -\gamma\sin\mu & \cos\mu - \alpha\sin\mu \end{pmatrix}. \tag{4.42}$$

The beam matrix elements after 1 iteration through the periodic lattice are

$$\langle x^2\rangle_n = \langle x^2\rangle_0 {R_{11}}^2 + 2\langle xx'\rangle_0 R_{11}R_{12} + \langle {x'}^2\rangle_0 {R_{12}}^2\,, \tag{4.43}$$

$$\begin{aligned}\langle xx'\rangle_n &= \langle x^2\rangle_0 R_{11}R_{21} + \langle xx'\rangle_0[R_{11}R_{22} + R_{12}R_{21}] \\ &\quad + \langle {x'}^2\rangle_0 R_{12}R_{22}\,,\end{aligned} \tag{4.44}$$

$$\langle {x'}^2\rangle_n = \langle x^2\rangle_0 {R_{21}}^2 + 2\langle xx'\rangle_0 R_{21}R_{22} + \langle {x'}^2\rangle_0 {R_{22}}^2\,. \tag{4.45}$$

After substitution of the matrix elements of (4.42) into (4.44–4.45), and using $\cos 2\psi = \cos^2\psi - \sin^2\psi$ and $\sin 2\psi = 2\sin\psi\cos\psi$,

$$\begin{aligned}\langle x^2\rangle_n &= \frac{1}{2}[\langle x^2\rangle_0 + \langle(\alpha x_0 + \beta x_0')^2\rangle] \\ &\quad - \frac{1}{2}[\langle x^2\rangle_0\,(\alpha^2-1) + 2\beta\alpha\langle xx'\rangle_0 + \beta^2\langle {x'}^2\rangle_0]\cos 2\psi \\ &\quad + [\alpha\langle x^2\rangle_0 + \beta\langle xx'\rangle_0]\sin 2\psi\,,\end{aligned} \tag{4.46}$$

$$\begin{aligned}\langle {x'}^2\rangle_n &= \frac{1}{2}[\langle {x'}^2\rangle_0 + \langle(\alpha x' + \gamma x)^2\rangle_0] \\ &\quad + \frac{1}{2}[\langle {x'}^2\rangle_0 - \langle(\alpha x' + \gamma x)^2\rangle_0]\cos 2\psi \\ &\quad - [\alpha\langle {x'}^2\rangle_0 + \gamma\langle xx'\rangle_0]\sin 2\psi\,,\end{aligned} \tag{4.47}$$

$$\begin{aligned}\langle xx'\rangle_n &= \frac{1}{2}[-\alpha\gamma\langle x^2\rangle_0 - 2\alpha\langle xx'\rangle_0 - \alpha\beta\langle {x'}^2\rangle_0] \\ &\quad + \left[\frac{\alpha\gamma}{2}\langle x^2\rangle_0 + (1+\alpha^2)\langle xx'\rangle_0 + \frac{\alpha\beta}{2}\langle {x'}^2\rangle_0\right]\cos 2\psi \\ &\quad + \frac{1}{2}[-\gamma\langle x^2\rangle_0 + \beta\langle {x'}^2\rangle_0]\sin 2\psi\,.\end{aligned} \tag{4.48}$$

Next, let

$$a = \frac{\beta}{2}[\gamma\langle x^2\rangle_0 + 2\alpha\langle xx'\rangle_0 + \beta\langle x'^2\rangle_0]\,, \tag{4.49}$$

$$b = \frac{\gamma}{\beta}a\,, \tag{4.50}$$

$$c = -\frac{\alpha}{\beta}a\,, \tag{4.51}$$

and use

$$c_1\cos 2\psi + c_2\sin 2\psi = \sqrt{{c_1}^2 + {c_2}^2}\cos(2\psi - \chi),\ \text{with}\ \chi = \tan^{-1}\left(\frac{c_2}{c_1}\right). \tag{4.52}$$

Then,

$$\langle x^2\rangle_n = a + \sqrt{a^2 - \beta^2(\langle x^2\rangle_0\langle x'^2\rangle_0 - {\langle xx'\rangle_0}^2)}\cos(2\psi + \chi_{\langle x^2\rangle_0})\,, \tag{4.53}$$

$$\langle x'^2\rangle_n = b + \sqrt{b^2 - \gamma^2(\langle x^2\rangle_0\langle x'^2\rangle_0 - {\langle xx'\rangle_0}^2)}\cos(2\psi + \chi_{\langle x'^2\rangle_0})\,, \tag{4.54}$$

and

$$\langle xx'\rangle_n = c + \sqrt{(\langle xx'\rangle_0 - c)^2 + \left(-\frac{\gamma}{2}\langle x^2\rangle_0 + \frac{\beta}{2}\langle x'^2\rangle_0\right)^2}\cos(2\psi + \chi_{\langle xx'\rangle_0})\,, \tag{4.55}$$

where the angles $\chi_{\langle x^2\rangle_0}$, $\chi_{\langle x'^2\rangle_0}$, and $\chi_{\langle xx'\rangle_0}$, follow from (4.52), (4.46), (4.47), and (4.48), respectively. Note that since $\epsilon = \sqrt{\det\sigma}$ is an invariant in the absence of filamentation,

$$\langle x^2\rangle_n\langle x'^2\rangle_n - \langle xx'\rangle_n^2 = \langle x^2\rangle_0\langle x'^2\rangle_0 - \langle xx'\rangle_0^2\,. \tag{4.56}$$

The Mismatch Parameter [9] $\mathrm{B_{mag}}$. Dividing both sides of (4.53) by $\beta\epsilon_0 = \beta\sqrt{\langle x^2\rangle_0\langle x'^2\rangle_0 - \langle xx'\rangle_0}$, we have

$$\begin{aligned}\frac{\langle x^2\rangle_n}{\beta\epsilon_0} &= \frac{a/\beta}{\sqrt{\langle x^2\rangle_0\langle x'^2\rangle_0 - {\langle xx'\rangle_0}^2}} \\ &\quad + \sqrt{\left(\frac{a/\beta}{\sqrt{\langle x^2\rangle_0\langle x'^2\rangle_0 - {\langle xx'\rangle_0}^2}}\right)^2 - 1}\ \cos(2\psi - \chi_{\langle x^2\rangle_0}) \\ &= \mathrm{B_{mag}} + \sqrt{\mathrm{B_{mag}^2} - 1}\cos(2\psi - \chi_{\langle x^2\rangle_0})\,,\end{aligned} \tag{4.57}$$

where the mismatch parameter $\mathrm{B_{mag}}$ is defined as

$$\mathrm{B_{mag}} = \frac{a/\beta}{\sqrt{\langle x^2\rangle_0\langle x'^2\rangle_0 - \langle xx'\rangle_0}}\,. \tag{4.58}$$

With the ellipse parameters α, β, and γ referring to the steady-state or equilibrium beam distribution, we have

$$\mathrm{B}_{\mathrm{mag}} = \frac{1}{2}\frac{[\gamma\langle x^2\rangle_0 + 2\alpha\langle xx'\rangle_0 + \beta\langle x'^2\rangle_0]}{\sqrt{\langle x^2\rangle_0\langle x'^2\rangle_0 - \langle xx'\rangle_0}} . \tag{4.59}$$

Thus, $\mathrm{B}_{\mathrm{mag}}$ is the ratio of the area of the decohered beam to the area of the injected beam. The factor of 2 results from the numerator representing an rms area.

Examples of Emittance Dilution due to Mismatch. Emittance dilution results if $\mathrm{B}_{\mathrm{mag}} \neq 1$ due to the difference in the transverse phase advance of the particles within the bunch. There are multiple sources of such phase advance variations. The two most commonly considered sources depend on the chromaticity or on the amplitude of the betatron oscillations. The chromaticity ξ ($\equiv Q'/Q$) $= (\Delta\psi/\psi)/\delta$ characterizes the energy dependence of the phase advance where $\Delta\psi$ is the difference in the phase advance of a particle from the mean phase advance of the bunch and δ is the relative energy deviation of that particle compared to the mean energy of the bunch. The amplitude dependence of the phase advance due to sextupole or octupolar magnetic fields is approximately described by

$$2\pi\psi = 2\pi\psi_0 - \mu a^2 , \tag{4.60}$$

where ψ_0 is the phase advance for a reference particle on the closed orbit, μ characterizes the strength of the sextupolar or octupolar fields, and a is the betatron oscillation amplitude of the particle. Less commonly considered sources for phase advance variations include wakefield focussing or space-charge defocussing for high current beams, focussing due to ions or electron clouds, and focussing due to the beam-beam tune shift in colliding beam accelerators.

a) Periodic Lattice

Let the length of the lattice period be L. Then, as shown in Fig. 4.8, for $\mathrm{B}_{\mathrm{mag}} = 1$, the beam always fills the same area in phase space after each lattice period. The rms area of the ellipse after n turns is

$$\langle x^2\rangle_n = \beta\epsilon_0 , \tag{4.61}$$

and the beam is said to be matched. Under these conditions, no emittance dilution will occur. In particular, since $\langle x^2\rangle_n$ is independent of the phase advance ψ, the phase space area is unchanged even if the phase of each of the particles in the beam advances differently.

For $\mathrm{B}_{\mathrm{mag}} > 1$, then

$$\begin{aligned}\epsilon_n &= \frac{\langle x^2\rangle_n}{\beta} \\ &= \epsilon_0\left[\mathrm{B}_{\mathrm{mag}} + \sqrt{\mathrm{B}_{\mathrm{mag}}{}^2 - 1}\cos(2\psi - \chi_{\langle x^2\rangle_0})\right] ,\end{aligned} \tag{4.62}$$

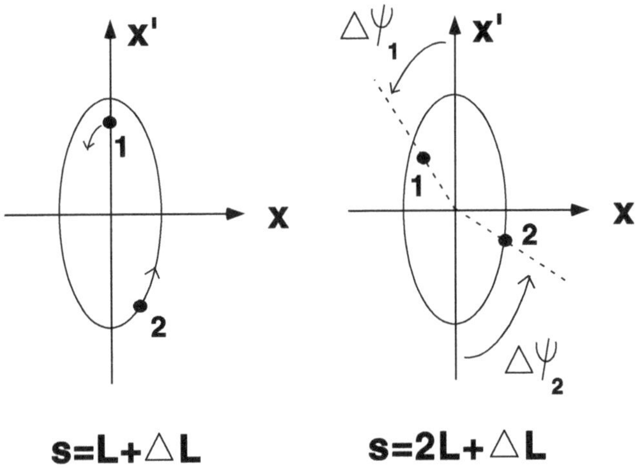

Fig. 4.8. Horizontal phase space for a matched beam in a periodic lattice. The emittance is preserved even if the phase advance is different for different particles (denoted here by the numbers 1 and 2)

as shown in Fig. 4.9. The solid, small ellipse represents the matched ellipse for which $B_{mag} = 1$. The shaded ellipse represents the (1σ) rms distribution of the mismatched beam. During the first few traversals of identical lattice segments, the phase advance variations of the different particles may not be obvious. As n approaches infinity, however, the phase advance variations lead to a smearing in the transverse phase space resulting in a larger emittance. This is represented by the area occupied by the hatched ellipse as given by (4.57).

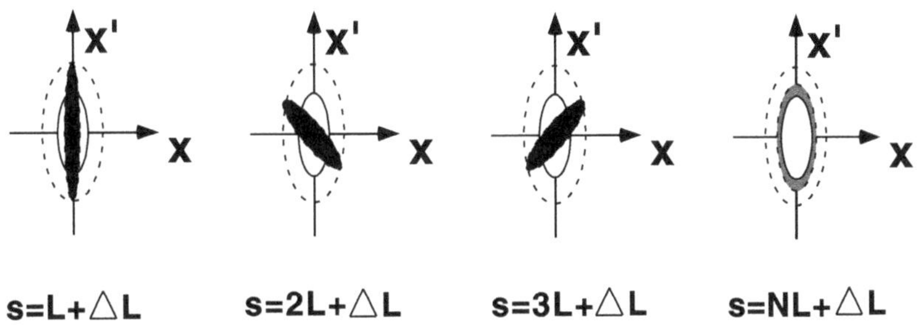

Fig. 4.9. Schematic of the horizontal phase space for a mismatched beam in a periodic linear lattice after 1, 2, 3 and $N \gg 1$ periods. The emittance is not preserved: the dilution is given by the ratio of the areas of the hatched ellipse to the design ellipse (for $N \to \infty$)

b) Circular Lattice

For a circular accelerator, the periodicity is usually taken to be not the superperiodicity of the machine (i.e., the number of identical lattice sections), but the revolution period. The index n therefore represents the turn number. The mismatch B_{mag} most often arises from improper orientation of the beam ellipse at injection. Neglecting the constant phase offset $\chi_{\langle x^2 \rangle_0}$ in (4.57), the equilibrium emittance is

$$\epsilon_n = \frac{\langle x^2 \rangle_n}{\beta} = \epsilon_0 \left[\mathrm{B_{mag}} + \sqrt{\mathrm{B_{mag}}^2 - 1} \cos(4\pi\nu) \right] , \tag{4.63}$$

where ν is the phase advance per turn. Shown in Fig. 4.10 is the evolution of the transverse phase space for $\mathrm{B_{mag}}$ along with the projections onto the horizontal axis. With a turn-by-turn beam size monitor, the mismatch can be measured directly by detecting the beam size changes at every turn. An example is given in Sect. 9.6.

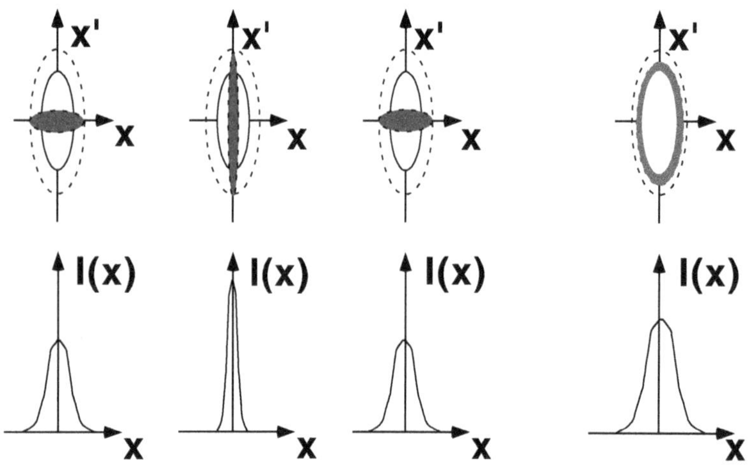

Fig. 4.10. Horizontal phase space and x-projection for a mismatched beam in a circular accelerator

4.2 Beta Matching in a Transport Line or Linac

The beam size (squared) at the location s can be expressed in terms of the α and β functions and the emittance at an upstream location s_0 as

$$\langle x^2(s) \rangle = R_{11}^2 \beta(s_0)\epsilon - 2R_{12}R_{11}\alpha(s_0)\epsilon + R_{12}^2\gamma(s_0)\epsilon \, . \tag{4.64}$$

In a quadrupole scan, the transfer matrix elements R_{11} and R_{12} are varied, by changing the strength of a quadrupole between s_0 and s. Beam-size measurements for at least 3 different quadrupole settings are required in order to solve for the three independent unknown parameters: ϵ, $\beta(s_0)$ and $\alpha(s_0)$. The fourth parameter, $\gamma(s_0)$ is not free, but determined by $\alpha(s_0)$ and $\beta(s_0)$: $\gamma = (1 + \alpha^2)/\beta$.

The deviation of the β, α, and γ from the design parameters β_D, α_D and γ_D is often characterized in terms of the '$\mathrm{B_{mag}}$' (β matching) parameter [9, 10] of (4.59), which can also be written as

$$B_{\rm mag} = \frac{1}{2}\,(\beta\gamma_D - 2\alpha\alpha_D + \gamma\beta_D)\,. \tag{4.65}$$

Once the values of β and α are known, quadrupole magnets can be adjusted so as to match the optical functions at a selected point to their design value, which is equivalent to $B_{\rm mag} = 1$.

The SLC had more than 10 multi-wire emittance measurement stations, which monitored the beam emittances in various parts of the machine in hourly intervals, and were indispensable for emittance control and tuning. As will be shown in Sect. 4.4.3, in the SLC linac transverse orbit bumps were intentionally induced as a global correction which cancelled the accumulated local effects of dispersion or wakefields. The bumps were optimized by minimizing the emittance downstream, as calculated by this measurement technique.

Example. To illustrate the beta matching method, Fig. 4.11 shows an example from the KEK/ATF beam transport line (BT), connecting the S-band linac and the ATF damping ring. The top picture shows the result of a typical quadrupole scan at the end of the BT. Plotted is the square of the vertical beam size versus the strength of an upstream quadrupole, as well as a quadratic fit to the data. The Twiss parameters deduced from such a fit can be propagated through the BT, using a model derived from the actual or the design magnet settings. The bottom picture displays the inferred beta functions compared to the design optics.

4.3 Equilibrium Emittance

We now discuss different methods for changing and controlling the equilibrium emittance in electron or positron storage rings. In these rings, synchrotron radiation gives rise to an equilibrium beam size, which is independent of the beam emittance at injection. While at high beam currents, collective effects and intrabeam scattering may be important as well, at low beam intensity the value of the equilibrium emittance is determined solely by the ring optics and the beam energy.

The discreteness and the random character of the synchrotron radiation increases the beam emittance. The expression for the transverse emittance growth in the plane u ($u = x$ or y) due to quantum excitation is [12]:

$$\frac{{\rm d}\epsilon_u}{{\rm d}t} = cC_Q E^5 \left\langle \frac{\mathcal{H}_u}{\rho^3} \right\rangle , \tag{4.66}$$

where the function $\mathcal{H}_u$, introduced by Sands [13], is

$$\mathcal{H}_u(s) = \frac{1}{\beta_u}\left\{D_u^2 + (\beta_u D_u' + \alpha_u D_u)^2\right\} , \tag{4.67}$$

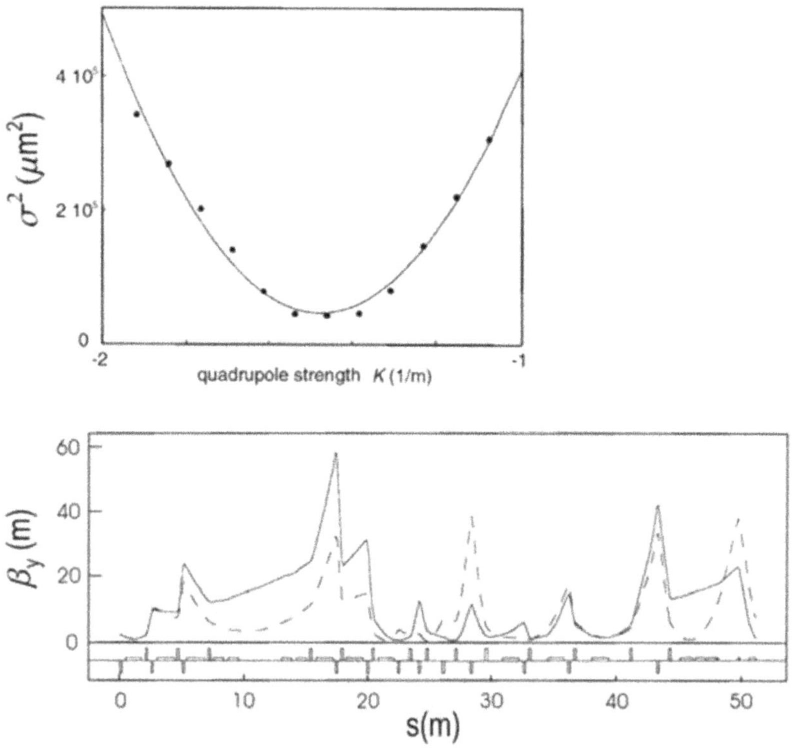

Fig. 4.11. Beta matching in the KEK/ATF beam transport line (BT) [11]: (*top*) quadrupole-scan emittance measurement; shown is the square of the vertical beam size measured using a fluorescent screen vs. the strength of an upstream quadrupole; (*bottom*) the vertical beta function obtained by propagating the measured Twiss parameters (*solid*) through the actual BT optics is compared with the beta function expected for the design optics (*dashed*)

the coefficient C_Q is

$$C_Q = \frac{55}{48\sqrt{3}} \frac{r_e \hbar c}{(m_e c^2)^6} \approx 2 \times 10^{-11} \text{ m}^2 \text{ GeV}^{-5} . \tag{4.68}$$

and ρ is the bending radius. The angular brackets denote an average over the ring.

On the other hand, the average energy loss due to the synchrotron radiation usually leads to damping in all three degrees of freedom. The emittance decrease due to radiation damping is described by

$$\frac{d\epsilon_u}{dt} = -2\epsilon_u C_d J_u E^3 \left\langle \frac{1}{\rho^2} \right\rangle , \tag{4.69}$$

where ϵ is the beam emittance,

$$C_d = \frac{c}{3} \frac{r_e}{(m_e c^2)^3} = 2.1 \times 10^3 \text{ m}^2 \text{ GeV}^{-3} \text{ s}^{-1} , \tag{4.70}$$

and J_u is the damping partition number.

The equilibrium emittance is reached when the quantum excitation and the damping are equal. It is

$$\epsilon_{u,\infty} = C_q \frac{\gamma^2}{J_u} \frac{\langle \mathcal{H}_u/\rho^3 \rangle}{\langle 1/\rho^2 \rangle} , \tag{4.71}$$

where the new constant C_q is defined by

$$C_q = \frac{55}{32\sqrt{3}} \frac{\hbar c}{m_e c^2} = 3.84 \times 10^{-13} \text{ m} . \tag{4.72}$$

Thus ϵ_u is inversely proportional to the transverse damping partition number J_u. Similarly, the longitudinal emittance is inversely proportional to J_z.

The exponential amplitude-damping time τ_u is obtained from the equation

$$\frac{1}{\tau_u} \equiv \frac{1}{2\epsilon_u} \frac{\mathrm{d}\epsilon_u}{\mathrm{d}t} . \tag{4.73}$$

Including also the longitudinal degree of freedom, the exponential damping times for all three oscillation modes can be written as [13]

$$\tau_i = \frac{2E_0}{\langle P_\gamma \rangle J_i} , \tag{4.74}$$

where E_0 is the nominal energy, and $\langle P_\gamma \rangle$ is the average rate of energy loss. The latter is given by

$$\langle P_\gamma \rangle = \frac{cC_\gamma}{2\pi} E_0^4 \left\langle \frac{1}{\rho^2} \right\rangle , \tag{4.75}$$

where yet another constant is introduced, namely

$$C_\gamma = \frac{4\pi r_e}{3(m_e c^2)^3} \approx 8.877 \times 10^{-5} \text{ m GeV}^{-3} , \tag{4.76}$$

as defined by Sands [13].

The factor J_i in (4.74) is the damping partition number for the ith degree of freedom. A general theorem by Robinson [14] states that the sum of the three partition numbers is a constant:

$$J_x + J_y + J_z = 4 . \tag{4.77}$$

A general proof of this theorem is given in Chap. 11. If the ring only contains horizontal, but no vertical bending magnets, then $J_y = 1$ and the partition numbers in the other two planes are related by a term $\mathcal{D}$ [13]

$$J_x = 1 - \mathcal{D} , \tag{4.78}$$

$$J_z = 2 + \mathcal{D} , \tag{4.79}$$

where

$$\mathcal{D} = \frac{\oint D_x/\rho \left(1/\rho^2 + 2k\right) \mathrm{d}s}{\oint 1/\rho^2 \ \mathrm{d}s} . \tag{4.80}$$

For separated function magnets $k/\rho = 0$ and the value of $\mathcal{D}$ is typically much smaller than 1.

It is often desirable to increase one of the damping rates or to vary the horizontal emittance. For example, in linear collider applications, a fast horizontal damping and a small horizontal emittance are advantageous, whereas the longitudinal emittance is of less concern. In storage-ring colliders one may instead want to increase the horizontal emittance near the beam-beam limit[3].

The damping rate and the equilibrium emittance can be changed by adjusting the value of $\mathcal{D}$ and/or by adding wiggler magnets. Depending on the application, there are various possibilities to do so. In the following, we describe the effect of a change in the ring circumference, an almost equivalent application in which the accelerating frequency is changed, and two different applications of wigglers.

4.3.1 Circumference Change

If the geometric circumference of the ring is changed by moving the magnet centers outwards by a step Δx^{mag} while holding the ring rf frequency fixed (so as to maintain synchronization with other systems) the quantity $\mathcal{D}$ changes by

$$\Delta \mathcal{D} \approx -\left(\sum_q k_q^2 D_{x,q} L_q\right) \frac{2\rho^2}{C} \ \Delta x^{\mathrm{mag}} , \tag{4.82}$$

where k_q is the non-integrated quadrupole gradient, L_q the quadrupole length, $D_{x,q}$ the dispersion function at the quadrupole, C the ring circumference, and ρ the bending radius of the dipole magnets. The minus sign arises because the orbit moves inwards with respect to the quadrupole magnets (the orbit shift is opposite to the displacement: $\Delta x = -\Delta x^{\mathrm{mag}}$). Note that the contributions from focusing and defocusing magnets add, because the effect is quadratic in $k_{1,q}$. Note that $-2 < \Delta\mathcal{D} < 1$ else beam loss may occur at

[3] Near the beam-beam limit, experimental observations [15] have shown that attempts to raise the luminosity by increasing the bunch charge leads to an increase in the beam size such that the beam-beam tune shift, from (2.101),

$$\xi_x = \xi_y = \frac{r_e}{2\pi\gamma} \frac{N_b}{\epsilon_x(1+\kappa)} \tag{4.81}$$

(with $\kappa = \epsilon_y/\epsilon_x$ the emittance ratio), remains approximately constant. However, since $L \propto {N_b}^2/\epsilon$, such an increase in bunch charge and emittance still can result in a higher luminosity the damping poles. The maximum tolerable shift Δx is determined by the available aperture, and by the beam size at injection.

In 1992, the magnet support girders of the SLC North Damping Ring were pulled outwards by about $\Delta x^{\mathrm{mag}} = 1.5$ mm, increasing the geometric ring circumference by 9 mm. As a result the measured horizontal damping time decreased [16] from 4.11 ± 0.11 ms to 3.41 ± 0.09 ms, consistent with predictions.

4.3.2 RF Frequency Change

An equivalent change in $\mathcal{D}$ can be achieved with a shift of the rf frequency by

$$\frac{\Delta f_{\mathrm{rf}}}{f_{\mathrm{rf}}} = \frac{2\pi \Delta x^{\mathrm{mag}}}{C} . \tag{4.83}$$

More accurately, the orbit shift in the quadrupole is proportional to the local dispersion function

$$\Delta x(s) = -\frac{D_x}{\alpha_x} \frac{\Delta f_{\mathrm{rf}}}{f_{\mathrm{rf}}} , \tag{4.84}$$

with α_c the momentum compaction factor. The change in the partition number is

$$\Delta \mathcal{D} \approx \frac{\oint 2k^2 D_x^2 \,\mathrm{d}s}{\oint \mathrm{d}s/\rho^2} \frac{\Delta p}{p} \equiv C_0 \frac{\Delta p}{p} . \tag{4.85}$$

However, in practice the rf frequency must be locked to the rf of the injection (or extraction) system. Therefore, at the SLC damping rings in addition to the static circumference change a dynamic rf frequency shift was implemented [17]. The dynamic rf frequency shift by up to 100 kHz started about 1.33 ms after injection, and was stopped 200 µs before extraction, in order to stabilize the injected beam and to minimize emittance and extraction jitter, respectively [18]. The total store time was 8.33 ms, equal to about 2.5 nominal damping times. For a dynamic frequency shift of 62.5 kHz the normalized emittance of the extracted beam decreased from 3.30 ± 0.07 m to 2.66 ± 0.06 m. The 20% reduction agreed with SAD calculations [17]. This example is discussed further in Sect. 8.10 after describing the influence of heavy beam loading on the rf system.

Emittance control via the accelerating rf has been used already before [19]. More recently it was applied at LEP [20] and in the HERA electron ring [21]. At HERA the associated increase in beam energy spread was compensated by a larger rf bucket height. With limited rf power, this was achieved by increasing the transverse focussing thereby reducing the dispersion and hence the momentum compaction factor.

We note an interesting side effect, remarked by Wiedemann [12]. From (4.85) the partition number changes with the particle momentum. If a particle performs synchrotron oscillations

$$\frac{\Delta p}{p} = \delta_{\max} \sin \Omega_s t\,, \tag{4.86}$$

its horizontal partition number and damping time vary with the synchrotron period:

$$\frac{1}{\tau} = \frac{1}{\tau_0}(1 - C_0 \delta_{\max} \sin \Omega_s t)\,, \tag{4.87}$$

The equation for the horizontal equilibrium emittance of such particles is then time-dependent according to [12]

$$\epsilon_x(t) = \epsilon_{x,\infty} \exp\left[\frac{2\delta_{\max} C_0}{\Omega \tau_0}(\cos \Omega t - 1)\right]. \tag{4.88}$$

The effect is largest for particles with large synchrotron oscillations.

4.3.3 Wigglers

A wiggler magnet generates additional synchrotron radiation and, thus, can enhance the radiation damping and/or change the equilibrium emittance. The damping time is modified as

$$\tau_{u,w} = \tau_{u,0} \frac{1}{1 + \langle 1/\rho^2 \rangle_w / \langle 1/\rho^2 \rangle_0}\,, \tag{4.89}$$

where $\tau_{u,x}$ is the damping time in the plane u including the effect of the wiggler, $\tau_{u,0}$ on the right is the damping time for the ring proper, while on the right-hand side of the equality, the subindex 0 indicates an average over the ring without wiggler magnets, while the subindex w indicates the contribution from the wiggler magnets. On the left, $\tau_{u,x}$ is the damping time in the plane u, including the effect of the wiggler, and $\tau_{u,0}$ on the right is the damping time for the ring proper.

Similarily, the relative emittance increase due to the presence of the wiggler is

$$\frac{\epsilon_{u,w}}{\epsilon_{u,0}} = \frac{1 + \langle \mathcal{H}_u/\rho^3 \rangle_w / \langle \mathcal{H}_u/\rho^3 \rangle_0}{1 + \langle 1/\rho^2 \rangle_w / \langle 1/\rho^2 \rangle_0}\,, \tag{4.90}$$

where the averages are given, for example, by

$$\left\langle \frac{1}{\rho^2} \right\rangle_w = \frac{1}{C} \oint \frac{1}{\rho_w^2}\, \mathrm{d}s\,, \tag{4.91}$$

with C the circumference, and ρ_w the bending radius in the wiggler.

In addition to changing the emittance, wigglers also affect the energy spread [12]:

$$\frac{\sigma_{\delta,w}^2}{\sigma_{\delta,0}^2} = \frac{1 + \langle 1/\rho^3 \rangle_w / \langle 1/\rho^3 \rangle_0}{1 + \langle 1/\rho^2 \rangle_w / \langle 1/\rho^2 \rangle_0}\,. \tag{4.92}$$

Damping Wigglers. If we place a wiggler in a region with no dispersion, $D_x = 0$, we might expect that the equilibrium emittance decreases according to (4.90) with $\langle H_w \rangle = 0$. However this is not completely correct, because the wiggler itself generates dispersion. As an example, we consider a sinusoidal wiggler with field

$$B(z) = B_w \cos k_p z \,, \tag{4.93}$$

where $k_p = 2\pi/\lambda_p$ and λ_p denotes the length of the wiggler period. The differential equation for the dispersion function reads:

$$D_u''(z) = \frac{1}{\rho_w} \cos k_p z \,, \tag{4.94}$$

which, assuming $D_u(0) = D_u'(0) = 0$, can be solved as

$$D_u(z) = \frac{1}{k_p^2 \rho_w}(1 - \cos k_p z) \,. \tag{4.95}$$

Using

$$\frac{1}{\rho} = \frac{1}{\rho_w} |\cos k_p z| \,, \tag{4.96}$$

for each half period of the wiggler we find [12]

$$\int_0^{\lambda_p/2} \frac{\mathcal{H}_u}{|\rho|^3} \, \mathrm{d}z = \frac{36}{15} \frac{1}{\beta_u} \frac{1}{k_p^5 \rho_w^5} + \frac{4}{15} \frac{\beta_u}{k_p^3 \rho_w^5} \approx \frac{4}{15} \frac{\beta_u}{k_p^3 \rho_w^5} \,, \tag{4.97}$$

where β is the beta function and we assumed that $\lambda_p \ll \beta$. Introducing the deflection angle per wiggler pole $\theta_w = 1/(\rho_w k_p)$, and the number of wiggler periods N_w, we can rewrite (4.97) as

$$\int_w \frac{\mathcal{H}_u}{\rho^3} \, \mathrm{d}z \approx N_w \frac{8}{15} \frac{\beta}{\rho_w^2} . \theta_w^3 \tag{4.98}$$

Similarily, we find

$$\int_w \frac{1}{\rho^2} \, \mathrm{d}z \approx \pi N_w \frac{\theta_w}{\rho_w} \,. \tag{4.99}$$

Finally the emittance ratio, (4.90), becomes

$$\frac{\epsilon_{u,w}}{\epsilon_{u,0}} = \frac{1 + \frac{8}{30\pi} N_w \frac{\beta_u}{\langle \mathcal{H}_u \rangle_0} \frac{\rho_0^2}{\rho_w^2} \theta_w^3}{1 + \frac{1}{2} N_w \frac{\rho_0}{\rho_w} \theta_w} \,, \tag{4.100}$$

where $\langle \mathcal{H}_u \rangle_0$ is the average value of $\mathcal{H}_u$ in the ring magnets, excluding the wiggler magnets. The latter can be re-expressed in terms of the emittance ϵ_{x0} to yield, e.g., with a vertical wiggler field, the horizontal emittance

$$\frac{\epsilon_{x,w}}{\epsilon_{x,0}} = \frac{1 + \frac{8 C_q}{30\pi J_x} N_w \frac{\beta_x}{\epsilon_{x0} \rho_w} \gamma^2 \frac{\rho_0}{\rho_w} \theta_w^3}{1 + \frac{1}{2} N_w \frac{\rho_0}{\rho_w} \theta_w} \,. \tag{4.101}$$

The emittance is reduced by the wiggler magnet if

$$\frac{8}{15\pi}\frac{C_q}{J_x}\frac{\beta_x}{\epsilon_0\rho_w}\gamma^2\theta_w^2 \leq 1\,. \tag{4.102}$$

In a wiggler-dominated ring the minimum emittance which one might hope to achieve is still limited. Ignoring the contributions from the arcs, after traversing a large number of wiggler periods (with intermittent re-acceleration) the emittance reaches an asymptotic value

$$\epsilon_{x,w} \to \frac{16}{30\pi}\frac{C_q\beta_x}{\rho_w}\gamma^2\theta_w^2\,. \tag{4.103}$$

The horizontal damping time with a wiggler can be written as

$$\tau_{x,w} = \tau_{x,0}\frac{1}{1+\frac{1}{2}N_w\frac{\rho_0}{\rho_w}\theta_w}\,. \tag{4.104}$$

In the limit of an extremely wiggler-dominated ring or a very long wiggler channel, again assuming intermittent re-acceleration, this simplifies to

$$\tau_{x,w} \approx \frac{2\rho_w^2}{C_d J_x E^3}\,. \tag{4.105}$$

Equations (4.103) and (4.105) set lower bounds on the emittance and damping times that can be attained in the damping ring of a future linear collider.

Robinson wiggler. A "Robinson wiggler" is a wiggler consisting of a series of combined function magnets, arranged such as to increase the horizontal partition number. Such a magnet was first used at the CEA to convert the synchrotron (which because $\mathcal{D} > 1$ was horizontally unstable) into a stable storage ring with $0 < \mathcal{D} < 1$ [14]. Such a wiggler will change the partition number according to [22]

$$\Delta\mathcal{D} \approx \frac{\bar{D}_x L_{\mathrm{Rob}} k}{2\pi(1+F_\omega)}\frac{\rho_0}{\rho_{\mathrm{Rob}}}\,, \tag{4.106}$$

where L_{Rob} and ρ_{Rob} are the length and the bending radius of the Robinson wiggler, $\bar{D}_x$ the average dispersion in the wiggler, ρ_0 the bending radius of the main bends, k the magnitude of the wiggler quadrupole gradient (in units of m^{-2}), and

$$F_\omega \equiv \frac{1}{2}N_w\frac{\rho_0}{\rho_w}\theta_w\,. \tag{4.107}$$

Unfortunately, the Robinson wiggler not only increases the damping but it can also blow up the equilibrium emittance, since it is preferably placed at a location with large dispersion.

Other wigglers. Further applications of wigglers include polarization wigglers for electron storage rings. These decrease the polarization time at low beam energies [23] or invert the spin direction [24].

4.4 Linac Emittance Control

4.4.1 Introduction

Preservation of the beam emittances in a linear accelerator differs from that in a lepton storage ring since in a linac there is no or little radiation damping. The normalized beam emittance therefore may increase due to many different effects.

In an electron linac, emittance dilutions may occur due to alignment errors of the accelerator components, which arise from steering the beam through misaligned structures and quadrupole magnets using beam-position monitors with residual offset errors [29]. The resulting transverse wake fields and dispersive effects increase the beam emittance. Some countermeasures that have been developed to minimize linac emittance growth are BNS damping, trajectory oscillations or 'wake field bumps', and dispersion-free steering, which we have already encountered in the previous chapter. In this section we discuss the BNS damping in more detail. We also present a further example for the application of 'wake field bumps' in the SLC, and briefly recall the underlying concept of dispersion-free steering.

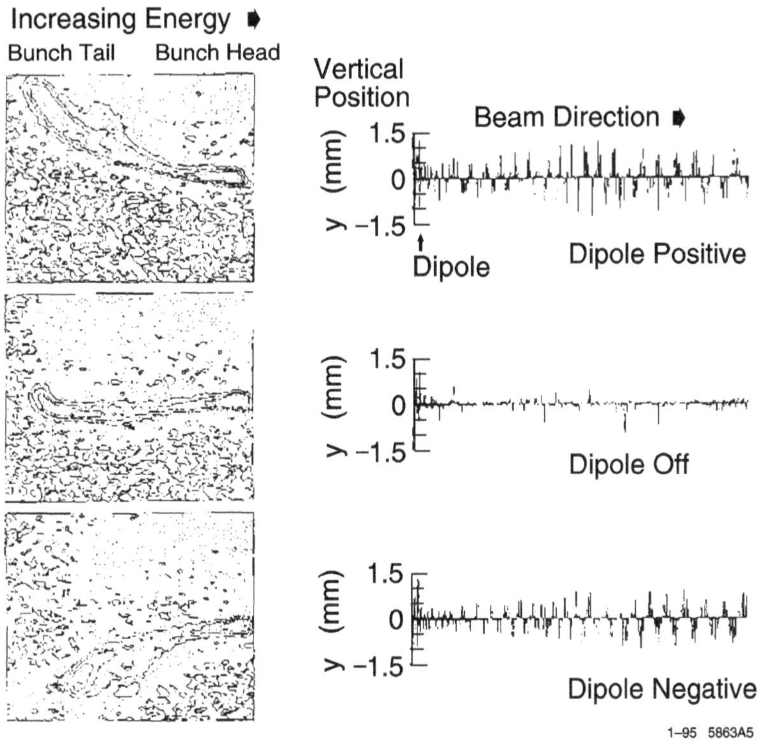

Fig. 4.12. Profile monitor measurements in a region of nonzero dispersion after the end of the SLC linac and vertical centroid trajectories with a positive purturbation to the bunch orbit (*top*), under nominal conditions (*middle*), and with a negative perturbation (*bottom*) (Courtesy J. Seeman, 2000)

As an example, Fig. 4.12 shows profile monitor measurements and trajectories for three different initial vertical displacements [25] of the linac beam in the SLC. The middle plots correspond to an optimized orbit. In the top plots the beam was kicked in one direction and in the bottom plot in the other direction. The increase in vertical amplitude towards the tail of the bunch shows the intrabunch particle displacements due to the transverse wakefields. Also evident from this measurement is a position-energy correlation[4]. The observed decrease in energy along the bunch depends on the cancellation between the rf slope due to the induced field and the rf slope of the accelerating rf (see Chap. 8).

Viewed independently, shown in Fig. 4.13 are the transverse beam profiles measured at the end of the linac for various initial beam displacements [26]. These measurements were made [27] by deflecting the beam using a fast kicker magnets located within the linac so that the true transverse profile $y(x)$ is represented. Notice that while the slice emittances of Fig. 4.12 are almost constant, the projected emittance is significantly increased as shown in Fig. 4.13.

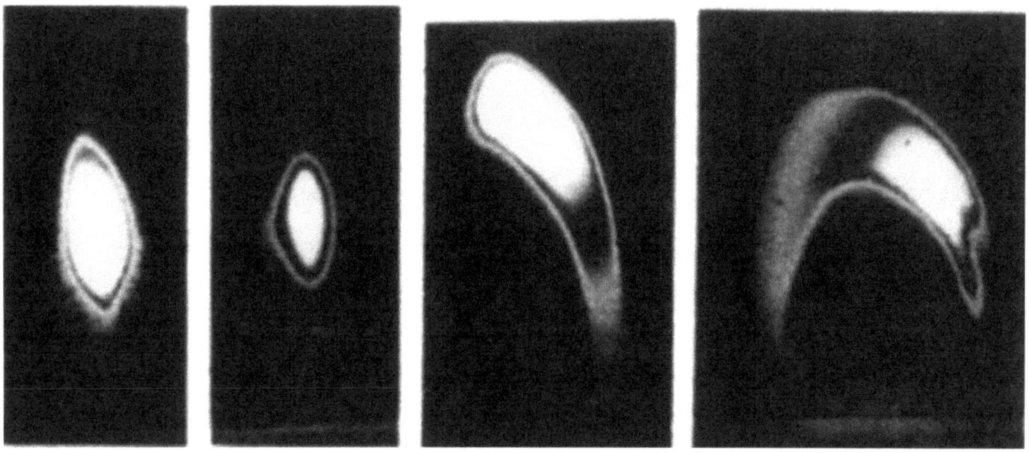

Fig. 4.13. Measured beam profiles demonstrating emittance growth due to wakefields as a function of increasing oscillation amplitude. From *left* to *right* the amplitudes in the applied horizontal trajectory displacement are 0 mm, 0.2 mm, 0.5 mm, and 1.0 mm. The single-bunch charge was 2×10^{10} electrons (Courtesy J. Seemann, 2000)

4.4.2 BNS Damping

The wake field effect can be reduced by proper adjustment of the rf phase profile along the linac. By passing the rf wave off-crest a position-energy

[4] these measurements were obtained by deflecting the beam onto a fluorescent screen using a kicker magnet located in a dispersive region (in the collider arcs) so that the measured horizontal position indicates an energy deviation; i.e., the profile monitor shows $y(E)$

correlation is generated along each linac bunch, such that the tail particles have lower energy than the particles in the bunch head. This results in so-called BNS damping [28], where the defocusing due to the wake fields is compensated by the stronger focusing for lower-energy tail particles.

To illustrate the mechanism, consider a 2-particle model, each with half the total bunch charge and a distance z apart. Let the first (head) particle be at the design energy and assume that the bunch head performs a pure betatron oscillation,

$$y_1(s) = \hat{y} \cos s/\beta \,, \tag{4.108}$$

with β the average vertical beta function. For simplicity. we here employ a smooth approximation for the betatron oscillation and the lattice focusing, i.e., the beta function is constant. Then the equation of motion for the second particle with a relative momentum deviation $\delta \equiv \Delta p/p_0$ (p_0 denotes the design momentum) is [30]

$$\frac{\mathrm{d}y_2(s)}{\mathrm{d}s^2} + \frac{1}{\beta^2(\delta)} y_2(s) = \frac{N r_e W_1(z)}{2\gamma L} \hat{y} \cos s/\beta \,, \tag{4.109}$$

where $W_1(z)$ denotes the value of the transverse wake function per cavity (in units of m^{-2}), N is the bunch population, and L the cavity period. We have ignored the effect of acceleration.

Equation (4.109) shows that there exists a value of δ for which, in first order, the bunch tail exactly follows the bunch head. Writing $\beta(\delta) = \beta(0) + \Delta\beta(\delta)$ this value corresponds to [30]

$$\frac{\Delta\beta(\delta)}{\beta(0)} = -\frac{N_b r_e \beta^2 W_1(z)}{4\gamma L} \,, \tag{4.110}$$

a condition which is also known as 'autophasing'. The relative change in beta function as a function of energy can easily be expressed using the linac chromaticity ξ using the relation

$$\frac{\Delta\beta}{\beta} = -\xi\delta \,. \tag{4.111}$$

For an optical FODO cell, we have

$$\xi = -\frac{2}{\mu} \tan\frac{\mu}{2} \,, \tag{4.112}$$

where μ is the betatron phase advance per cell. In case of an accelerated beam, the autophasing condition is still given by (4.110), if we simply replace the factor $1/\gamma$ by $\ln(\gamma_f/\gamma_i)/\gamma_f$ where γ_i and γ_f characterize the initial and final energies in units of the rest mass.

In practice, BNS damping can only partially be realized, since the energy spread introduced at low energies must be restored later in the linac to fit inside the energy acceptance of the downstream beam delivery system.

4.4.3 Trajectory Oscillations

In addition to BNS damping, empirically distributing a set of short-range oscillations, or wake-field bumps (see Sect. 3.6), along the accelerator proved indispensable for SLC operation [29]. Examples of betatron oscillations intentionally induced in the SLAC linac are shown in Fig. 4.14 [29].

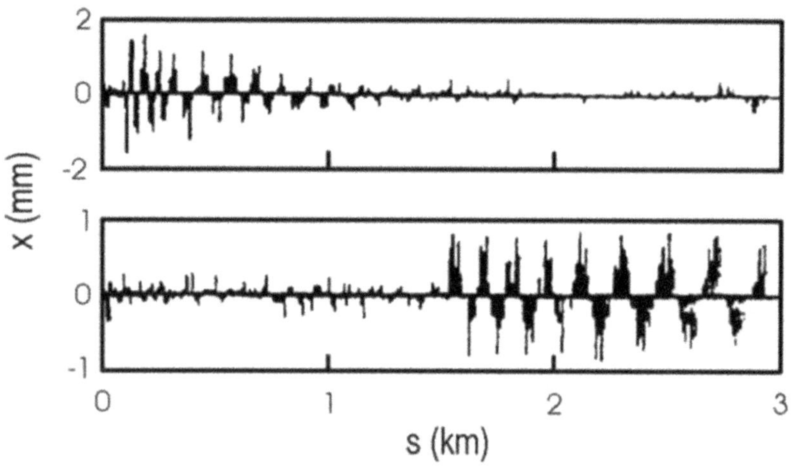

Fig. 4.14. Two trajectory ocillations in the SLAC linac, which were used to study the effect on the downstream emittance (Courtesy F.J. Decker, 1999)

Wakefields and dispersive effects in the linac generate both emittance growth and a mismatch. The mismatch induced early in the linac has completely filamented when the beam reaches its end, while perturbations near the linac end also result in residual unfilamented tails and in a phase-space mismatch, which is conventionally characterized by the parameter B_{mag}, (4.59). This factor specifies the emittance growth after filamentation. A matched beam fulfills $B_{mag} = 1$.

Similarily, trajectory oscillations induced in the early parts of the linac only change the beam emittance $\gamma\epsilon$, while those in the latter sections also affect the measured betatron mismatch. This is illustrated in Fig. 4.15, which displays the measured normalized emittance versus the amplitude of the two trajectory bumps in Fig. 4.14.

The SLC employed a series of more than 10 orbit feedback loops with roughly equidistant spacing along the SLAC linac. These feedbacks continually maintained constant values of offset and slope at certain pairs of beam-position monitors, by adjusting the strengths of a few steering correctors. The feedback set points for position and slope were set to empirically determined target values.

A closed trajectory oscillation was most easily generated by changing a feedback set point (for either slope or position). The induced trajectory oscillation was then automatically taken out by the next feedback downstream, because the latter attempted to restore the original orbit.

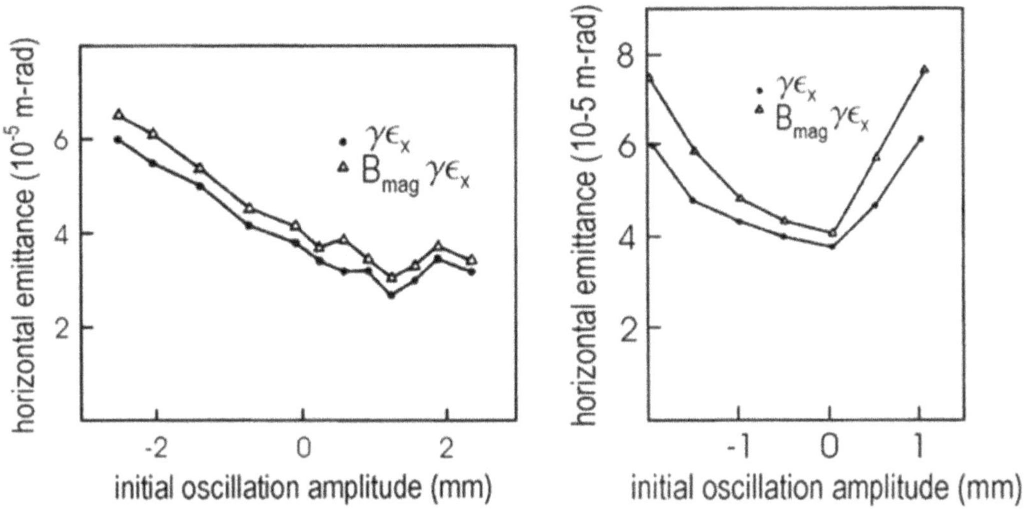

Fig. 4.15. Change in the normalized emittance as a function of the amplitude of a trajectory oscillation induced early in the linac (*left*) and towards the end of the linac (*right*). In the first case, the betatron mismatch is constant, and the normalized emittance decreased by 25% for an oscillation amplitude of about 1.5 mm. In the lower plot, the trajectory oscillation applied later in the linac did not reduce the normalized emittance. Instead it enhances the observed betatron mismatch, which is evident by the separation of the two curves representing $\gamma\epsilon$ and $B_{mag}\gamma\epsilon$ (Courtesy F.J. Decker, 1999)

In the later years of SLC operation, typical oscillation amplitudes were of the order of 100 or 200 µm, comparable to the presumed misalignments of the accelerating structures.

4.4.4 Dispersion-Free Steering

A very efficient steering algorithm has been developed in order to minimize the dispersive emittance growth in a linac. By its effect, this method is known as 'dispersion-free steering' [31, 32]. The detailed algorithm was already presented in Sect. 3.7, in the context of orbit correction schemes.

The basic idea of this method is to steer the orbit such that the particle trajectories become independent of the particle energy. In practice this can be achieved, for example, by exciting the steering coils (orbit correctors), so as to minimize the orbit response to a constant relative change of all quadrupole strengths.

During initial studies of this algorithm, the quadrupoles and correctors were so scaled to mimic the change in beam energy. In later years, instead, advantage was taken of the fact that both electron and positron bunches traversed the same linac. As far as dispersion is concerned, a change in the sign of the charge is equivalent to a 200% energy variation.

The so-called two-beam dispersion-free steering then consisted in measuring the orbit of both electron and positron beams, and correcting the absolute orbit offset of one beam as well the difference to the orbit of the other beam. At the same time, the overall excitation strength of the steering correctors was also constrained. This steering method was implemented in the SLC control system by means of an SVD algorithm (SVD or *singular value decomposition* was discussed in Sect. 3.5), where weighting factors could be assigned to the different minimization constraints. Two-beam dispersion-free steering was also applied with great success at the circular collider LEP [33].

Exercises

4.1 Beta Mismatch

Suppose a beam is injected with a distribution characterized by optical functions β, α and γ different from the matched values β_0, α_0 and γ_0. Show that the beam emittance after filamentation is given by $\epsilon = \mathrm{B}_{\mathrm{mag}}\ \epsilon_0$, where ϵ_0 is the initial emittance of the injected beam, and $\mathrm{B}_{\mathrm{mag}}$ was defined in (4.65). Hint: filamentation corresponds to a randomization of the betatron phase and $\epsilon = \langle I \rangle$.

4.2 Propagation of Twiss Parameters

In Fig. 4.11 the Twiss parameters were measured at a single location yet the 'measured' values were shown as a function of position along the transport line. Derive the matrix for propagation of the Twiss parameters from a known location to an arbitrary location along the transport line. Hint: use the equation for a general phase space ellipse of area ϵ (not to be confused with the rms emittance)

$$\gamma x^2 + 2\alpha x x' + \beta x'^2 = \epsilon\,, \tag{4.113}$$

the relation $\beta\gamma - \alpha^2 = 1$, and the 2×2 transport matrix of the form

$$\begin{pmatrix} x \\ x' \end{pmatrix} = \begin{pmatrix} C(s) & S(s) \\ C(s)' & S(s)' \end{pmatrix} \begin{pmatrix} x_0 \\ x_0' \end{pmatrix}. \tag{4.114}$$

4.3 Static and Dynamic change of Partition Numbers

Assume parameters typical for the SLC damping rings: 40 quadrupoles, $k_q \approx 15\ \mathrm{m}^{-2}$, $D_{x,q} \approx 0.15$ m, $L_q \approx 18$ cm, $\rho \approx 1/2\ C/(2\pi)$, $C = 35$ m, harmonic number $h = 84$, rf frequency $f_{\mathrm{rf}} = 714$ MHz, and momentum compaction $\alpha = 0.0147$.

a) What is the change in $\mathcal{D}$ for an outward shift of all magnets by $\Delta x = 1.5\,\mathrm{mm}$?

b) What would be the equivalent change in the rf frequency?

4.4 Effect of Wiggler on Equilibrium Emittance

Consider a ring which consists mainly of wiggler magnets, with a peak magnetic field B_w of 40 kG, and a wiggler oscillation period of λ_p of 20 cm. Calculate the equilibrium emittance and the damping time in such a ring, assuming a beta function β_x of 5 m, and beam energies of 1 GeV and 5 GeV. Compare this with a typical damping-ring design for a future linear collider, where $\gamma\epsilon_x \approx 3$ μm, and $\tau_x \approx 3$ ms.

4.5 BNS Damping at the SLC

For the SLAC linac $\beta \approx 20$ m, $W_\perp(1\text{ mm}) \approx 1\text{ cm}^{-2}$, $L = 3.5$ cm, $N = 4 \times 10^{10}$, $\mu \approx \pi/2$, with an injected beam energy of 1.2 GeV and a final energy of 47 GeV. How large is the BNS energy chirp δ over the bunch length of 1 mm?

5 Beam Manipulations in Photoinjectors

The design of an electron source is a challenging task. The designer must reconcile the contradictory requirements for a small emittances, a high charge, a high repetition rate, and, possibly, a high degree of beam polarization.

Electron beams can be generated in a variety of ways. Accordingly a number of different devices exist which can serve as electron sources for linear colliders: thermionic guns, dc guns with laser photocathodes (used at the SLC), or rf guns. In the future, also polarized rf guns may become available.

In this chapter, we first outline the general principle of an rf photoinjector, emphasizing the limits on the minimum emittance that it can produce. We then discuss two approaches for manipulating, shaping and preserving the transverse emittance of the beam generated by such a photoinjector, namely the compensation of space-charge induced emittance growth using a solenoid, and the flattening of the beam by the combined action of a solenoid and subsequent skew quadrupoles.

5.1 RF Photoinjector

In a laser-driven rf gun, or rf photoinjector, a high-power pulsed laser illuminates a photocathode placed on the end wall of an rf cavity. The emitted electrons are accelerated immediately in the rf field. The time structure of the electron beam is controlled by the laser pulse, and the rapid acceleration minimizes the effect of space-charge repulsion.

Several effects contribute to the normalized emittance attainable by such an rf gun [1]:

- The *thermal* emittance is determined by the initial transverse momenta of the electrons at the moment of their emission. It can be estimated as

$$\gamma\epsilon_{x,y}^{th}\ [\mathrm{mm\ mrad}] \approx \frac{1}{4}\sqrt{\frac{k_B T_\perp}{m_e c^2}}\ \sigma_{x,y}\ [\mathrm{mm}]\,, \tag{5.1}$$

  where $k_B T_e \approx 0.1$ eV represents the thermal emission temperature.
- An *rf emittance* arises from the time-dependent transverse focusing in the rf field. At the exit of the rf structure, it is approximately given by

M. G. Minty et al., *Measurement and Control of Charged Particle Beams*,
https://doi.org/10.1007/978-3-662-08581-3_5

$$\gamma\epsilon^{\mathrm{rf}}_{x,y}\ [\mathrm{mm\ mrad}] \approx \frac{eE_{\mathrm{rf}}}{\sqrt{8}m_E c^4}\ \sigma^2_{x,y}\sigma^2_z\omega^2_{\mathrm{rf}}\,, \tag{5.2}$$

where E_{rf} denotes the peak accelerating field.

- The space-charge emittance arises from the repelling force between the equally charged beam particles. Taking into account the focusing component of the rf field, the residual space-charge emittance is [2]

$$\gamma\epsilon^{\mathrm{sc}}_{x,y}\ [\mathrm{mm\ mrad}] \approx \frac{2N_b r_e}{7\sigma_{x,y}W}\ \exp\left(-3\sqrt{W\sigma_y}\right)\sqrt{\frac{\sigma_y}{\sigma_z}}\,, \tag{5.3}$$

  where $W = eE_{\mathrm{rf}}\sin\phi_0/(2m_e c^2)$ and ϕ_0 is the rf phase at the beam center. Since the transverse space-charge force depends on the local charge density of the bunch, it disorients in phase space the transverse slices located at different longitudinal positions along the bunch. For round beams this dilution can be almost fully inverted by properly placed solenoids [3], as described in the following section.

5.2 Space-Charge Compensation

Nowadays, photoinjectors, rather than thermionic injectors, are used for all applications requiring the combination of high-peak current and low emittance [3]. After the electron emission from the cathode, at low energies, space charge forces are very important. Here we follow closely the work of B. Carlsten [3].

We first consider the case without compensation and also neglect rf focusing effects. In this case, scaling arguments, supported by simulations, show that the transverse emittance of a 'slug' beam of length L and radius a with peak current I grows to a value [3, 4]

$$\epsilon_{xN} \approx \frac{eIs}{16\pi\epsilon_0 m_0 c^3\gamma^2\beta^2}\ G\,, \tag{5.4}$$

provided that the bunch does not strongly deform over the drift distance s. The geometric factor G depends on the beam aspect ratio in the beam frame, $\left(\frac{\gamma L}{a}\right)$, and on the longitudinal distribution. In the long-bunch limit and assuming that the radial distribution is uniform, G can be calculated to be 0.556 for a Gaussian longitudinal distribution and 0.214 for a parabolic distribution.

The radial space-charge force is a function of position within the bunch. Following [3] we introduce cylindrical coordinates ρ and ξ within the bunch, $\rho = 1$ defining the radial edge, and $\xi = \pm 1$ the longitudinal ends. There is no emittance growth if the radial force is linear in ρ and independent of ξ [3], or, equivalently, if

$$\Lambda(\rho, \xi, t) \equiv \frac{eE_r(\rho, \xi, t)}{m_0\gamma^3\beta^2c^2} = \rho_0\Lambda_0(t)\,, \tag{5.5}$$

where we have introduced the normalized force Λ, and E_r is the radial electric field in the laboratory frame.

If the longitudinal bunch density is not a constant, this condition is not fulfilled, and there will be a growth in the transverse emittance because different slices of the beam experience different radial space-charge forces. It is the projected emittance that increases, while the emittance of each short slice remains constant. In phase space the slices rotate against each other.

Now there exists an elegant method by use of a focusing solenoid to realign the different slices in the same phase space direction, and thus to recover the original emittance.

We consider again a slug beam. For simplicity, we assume that the space-charge force does not vary in time. If initially the beam at location $z = 0$ is non-divergent and has a radius r_0, a point in the slug at coordinates (ρ, ξ) will execute a non-relativistic transverse motion, so that after a distance z its radial coordinates will be

$$r(\rho, \xi, z) = \rho r_0 + \Lambda(\rho, \xi)\frac{z^2}{2} \tag{5.6}$$

and

$$r'(\rho, \xi, z) = \Lambda(\rho, \xi)z \tag{5.7}$$

after a distance z. We now place a lens (in practice, this lens is a solenoid) at the position $z = z_l$, and choose its focal length equal to [3]

$$f = \frac{z_d^2}{2(z_l + z_d)}\,, \tag{5.8}$$

where z_d denotes the distance from the lens to a point downstream. At this point, the ratio of the beam divergence to its radius becomes

$$\frac{r'(\rho, \xi)}{r(\rho, \xi)} = \frac{2(z_l + z_d)}{z_d(z_d + 2z_l)}\,, \tag{5.9}$$

which is independent of the particle's motion within the bunch. Thus the effect of the lens was to back-rotate the slices along the bunch with respect to each other so that they are again re-aligned after the total distance $(z_l + z_d)$. The normalized emittance can be written as [3]

$$\epsilon_{x,y} = \frac{1}{2}\beta\gamma\sqrt{\langle\Lambda^2\rangle\langle\rho^2\rangle - \langle\Lambda\rho\rangle^2}\left(2r_0(z_l + z_d) - \frac{z_d^2 r_0}{f}\right)\,, \tag{5.10}$$

where r_0 is the initial beam radius and the angular brackets indicate an average over the beam distribution. Equation (5.10) confirms that the emittance vanishes with the proper choice of lens (focal length f). The compensation recipe is illustrated schematically in Fig. 5.1.

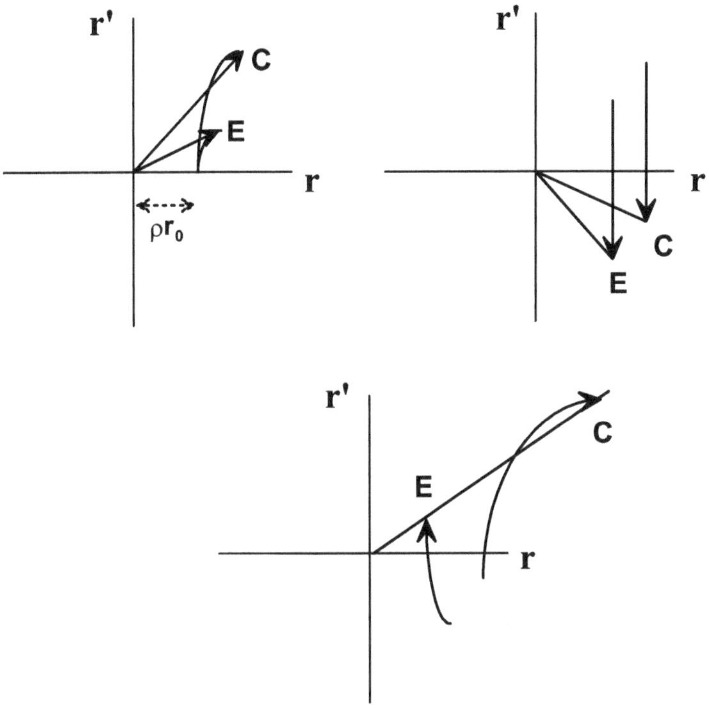

Fig. 5.1. Space charge compensation in photoinjectors. The two arrows illustrate the particle motion at the center (C) and at the end (E) of the bunch: (1) after initial drift, (2) after solenoid focusing, (3) after final drift until slice emittances are realigned [3]

In reality the physics is not quite so simple. In particular, the space-charge force is not constant in time. This complication results in a residual nonzero emittance. Nevertheless, already in the first beam experiments performed at Los Alamos [3] the above compensation scheme was shown to reduce the normalized rms emittance by up to an order of magnitude.

Let us assume the beam is focused to a beam-radius minimum. If the space-charge forces are weak compared with the external focusing, all particles cross through the beam's center. This can be called a *crossover* [3]. On the other hand, for strong space-charge forces, the particles will be deflected away from the center. This may be called a *waist* [3], but be careful not to confuse this with the notion of beam waist used to describe a generic position of minimum beam radius. In general, parts of the bunch will have a high density and particles there will experience a waist, while particles in the other parts will crossover. Indeed there exist particles at the border between these two regions, which are initially extremely close together and later on will be a finite distance apart. This is called a phase-space *bifurcation* [3].

The space-charge induced emittance growth can only be compensated for those particles which do not cross over, and only for those do the above approximations apply. Therefore, one of the most important design criteria for photoinjectors is to minimize the fraction of the beam crossing over.

The technique described here may be generalized to other situations where one wants to correct a correlated growth in the projected emittances that is induced by a nonlinear force.

5.3 Flat-Beam Transformation

Linear colliders require flat electron beams at the collision point, in order to maximize the luminosity while limiting the amount of synchrotron radiation emitted during the collision in the field of the opposing beam (this radiation is called *beamstrahlung*). Unfortunately, electron guns usually produce round beams.

A scheme by which one can transform a round beam ($\epsilon_x = \epsilon_y$) into a flat beam ($\epsilon_x \gg \epsilon_y$) was proposed by Y. Derbenev, R. Brinkmann, and K. Flottmann in 1999 [5, 6, 7]. We describe the idea following Edwards [8]. The basic scheme consists of two parts:

- the beam from a cathode immersed in a solenoidal field develops an angular momentum at exit from the solenoid;
- subsequently this beam is passed through a quadrupole (or skew quadrupole) channel with 90° phase advance difference between the two planes, and length scale defined by the solenoid field.

Consider electrons moving parallel to a solenoid field whose axis is oriented in the z direction. Maxwell's equations imply the presence of a radial magnetic field at the exit of the solenoid. This radial field gives rise to a transverse deflection, which depends on the distance from the solenoid axis. For example, the vertical deflection at the solenoid exit is

$$\Delta y' = \frac{1}{B\rho} \int B_x \, \mathrm{d}z = \frac{1}{B\rho} \frac{x_0}{2} B_z \,, \tag{5.11}$$

where B_z is the longitudinal field inside the solenoid and x_0 the horizontal offset. A similar expression holds for $\Delta x'$. Abbreviating, we write $\Delta y' = kx_0$, $\Delta x' = -ky_0$ with $k = B_z/(2B\rho)$. After leaving the solenoid, the beam takes on a clock-wise rotation

$$\begin{pmatrix} x \\ x' \\ y \\ y' \end{pmatrix}_0 = \begin{pmatrix} x_0 \\ -ky_0 \\ y_0 \\ kx_0 \end{pmatrix} . \tag{5.12}$$

We have neglected any initial uncorrelated momenta, assuming that these are much smaller than kx_0 or ky_0. However, actually these terms are important, as they determine the final flat-beam emittance. We will see this below.

Suppose now that the quadrupole channel behind the solenoid produces an I matrix in x and an additional 90° phase advance in y:

$$\begin{pmatrix} x \\ x' \\ y \\ y' \end{pmatrix}_1 = \begin{pmatrix} 1 & 0 & 0 & 0 \\ 0 & 1 & 0 & 0 \\ 0 & 0 & 0 & \beta \\ 0 & 0 & -1/\beta & 0 \end{pmatrix} \begin{pmatrix} x_0 \\ -ky_0 \\ y_0 \\ kx_0 \end{pmatrix} = \begin{pmatrix} x_0 \\ -ky_0 \\ k\beta x_0 \\ -\frac{1}{\beta} y_0 \end{pmatrix} . \tag{5.13}$$

If we choose $\beta = 1/k$, the final phase-space vector becomes

$$\begin{pmatrix} x \\ x' \\ y \\ y' \end{pmatrix}_1 = \begin{pmatrix} x_0 \\ -ky_0 \\ x_0 \\ -ky_0 \end{pmatrix} . \tag{5.14}$$

This is a flat beam inclined at 45°. If one uses a skew quadrupole channel instead of quadrupole channel, the beam can be made flat in the vertical plane, as shown next.

The general 4×4 transport matrix from the end of the solenoid through the skew quadrupole channel can be written as

$$M = R^{-1} T R \tag{5.15}$$

with

$$R = \frac{1}{\sqrt{2}} \begin{pmatrix} I_2 & I_2 \\ -I_2 & I_2 \end{pmatrix} , \tag{5.16}$$

where I_2 is 2×2 identity, and the matrix T represents a normal quadrupole channel:

$$T = \begin{pmatrix} A & 0 \\ 0 & B \end{pmatrix} . \tag{5.17}$$

Combining the above, we write M as

$$M = \frac{1}{2} \begin{pmatrix} A + B & A - B \\ A - B & A + B \end{pmatrix} . \tag{5.18}$$

The initial state after the solenoid exit is

$$X \equiv \begin{pmatrix} x_0 \\ -ky_0 \end{pmatrix} \quad \text{and} \quad Y \equiv \begin{pmatrix} y_0 \\ kx_0 \end{pmatrix} , \tag{5.19}$$

which we write more elegantly as

$$Y = SX \quad \text{using} \quad S \equiv \begin{pmatrix} 0 & -1/k \\ k & 0 \end{pmatrix} . \tag{5.20}$$

The final state is then

$$\begin{pmatrix} X \\ Y \end{pmatrix}_1 = \frac{1}{2} \begin{pmatrix} \{A + B + (A - B)S\}X \\ \{A - B + (A + B)S\}X \end{pmatrix} , \tag{5.21}$$

and the condition for a flat beam is $Y_1 = 0$, or $I = -(A - B)^{-1}(A + B)S$.

Using the Courant–Snyder parametrization [9] $A = \exp(J\mu)$, $B = \exp(J(\mu + \Delta))$, where J denotes the matrix

$$J = \begin{pmatrix} \alpha & \beta \\ -\gamma & -\alpha \end{pmatrix} , \tag{5.22}$$

the flat-beam condition becomes

$$I = -\frac{\cos(\Delta/2)}{\sin(\Delta/2)} \begin{pmatrix} k\beta & \alpha/k \\ -k\alpha & \gamma/k \end{pmatrix} . \tag{5.23}$$

This is fulfilled for $\Delta = -\pi/2$, $\alpha = 0$ and $\beta = 1/k$.

Finally, adding a random component to the slope of the initial vector, so that (5.12) is replaced by

$$\begin{pmatrix} x \\ x' \\ y \\ y' \end{pmatrix}_0 = \begin{pmatrix} x_0 \\ -ky_0 + x_0' \\ y_0 \\ kx_0 + y_0' \end{pmatrix} , \tag{5.24}$$

we can apply the same transformation M as above, (5.18), and, assuming that the beam at the source is round with $\sigma_{x0} = \sigma_{y0}$, $\sigma'_{x0} = \sigma'_{y0}$, and no initial correlation between the two transverse planes (e.g., $\langle x_0' y_0' \rangle = 0$), we find [6]

$$\epsilon_{y,1} = \frac{1}{2} \frac{{\sigma'_{y0}}^2}{k} \tag{5.25}$$

and

$$\epsilon_{x,1}/\epsilon_{y,1} = 1 + 4k^2 \frac{\sigma_{x0}^2}{{\sigma'_{x0}}^2} . \tag{5.26}$$

The larger the value of k, i.e., the stronger the solenoid field, the flatter the beam becomes.

First experimental tests of a flat beam electron source at Fermilab have demonstrated the viability of this scheme [10]. A similar application, which employs the inverse (flat-to-round) transformation, is the matching of a flat electron beam to a round proton beam, e.g., for electron cooling [5].

Exercises

5.1 Solenoidal Focusing

Verify that the ratio of the beam divergence to its radius in an rf photoinjector is given by (5.9).

5.2 Flat-Beam Transformer

a) Calculate the explicit form of the matrix M, (5.18), for $\mu = 2\pi$, $\Delta = -\pi/2$, $\alpha = 0$ and $\beta = 1/k$. See also the definition of A and B above (5.22).

b) Using the result, verify (5.25) and (5.26).

6 Collimation

Particles at large betatron amplitudes or with a large momentum error constitute what is generally referred to as a *beam halo*. Such particles are undesirable since they produce a background in the particle-physics detector. The background arises either when the halo particles are lost at aperture restrictions in the vicinity of the detector, producing electro-magentic shower or muons, or when they emit synchrotron radiation that is not shielded and may hit sensitive detector components. In superconducting hadron storage rings, a further concern is localized particle loss near one of the superconducting magnets, which may result in the *quench* of the magnet, i.e., in its transition to the normalconducting state.

In order to remove the unwanted halo particles, multi-stage collimation systems are frequently employed. Aside from the collimation efficiency, the collimators must also serve to protect the accelerator from damage due to a mis-steered beam. Especially for linear colliders, the collimator survival for different (linac) failure modes is of interest.

In this chapter we discuss a few sources of beam halo and then discuss several collimation issues, first for linear colliders and then for storage rings.

6.1 Linear Collider

In general, the beam entering the beam delivery system of a linear collider is not of the ideal shape, but it can have a significant halo extending to large amplitudes, both in the transverse and in the longitudinal direction. There are many sources of beam halo:

- beam-gas Coulomb scattering,
- beam-gas bremsstrahlung,
- Compton scattering off thermal photons [1],
- linac wakefields,
- the source or the damping ring.

The halo generation due to beam-gas Coulomb scattering can be reduced by using a higher accelerating gradient, while the halo formation due to beam-gas bremsstrahlung and thermal-photon scattering scales with the length of the

M. G. Minty et al., *Measurement and Control of Charged Particle Beams*,
https://doi.org/10.1007/978-3-662-08581-3_6

accelerator. The contributions of linac wake fields and of the injector complex to the size of the halo depend on many parameters; as a rough approximation, if measured as a fraction of the bunch population, such contributions could be considered as constant, independent of energy. A Monte-Carlo simulation study of beam loss in the next linear collider (NLC) beam-delivery system due to the first three processes given above and the positive effect of additional collimators is described in reference [2].

If the halo particles once generated strike the beam pipe or magnet apertures close to the interaction point, or if they traverse the final quadrupole magnets at a large transverse amplitude, they may cause unacceptable background. This background can be due to muons, electromagnetic showers, or synchrotron radiation. In particular, muons, with a large mean free path length, are difficult to prevent from penetrating into the physics detector. The muon generation occurs by a variety of mechanisms, the most important one being the Bethe–Heitler pair production [3]: $\gamma Z \to Z\mu^+\mu^-$. On average about one muon is produced for every 2500 lost electrons. Differential cross sections for muon production were derived by Tsai [4], and are used in simulations of the background induced by muons [3, 5]. In the Stanford Large Detector (SLD) at the Stanford Linear Collider (SLC) 1 muon per pulse entering the detector corresponded to a marginally acceptable background. Muons are produced when electrons and positrons impinge on apertures.

At the SLC, collimation upstream of the final focus was found to be essential for smooth operation and for obtaining clean physics events. In addition, large magnetized toroids had to be placed between the location of the collimators and the collision point to reduce the number of muons reaching the detector. When a muon passes through such a toroid it scatters, loses energy, and its trajectory is bent. A complex collimation system and muon toroids, whose length scales at least linearly with energy [6], will also be indispensable for future linear colliders [7, 8].

A conventional collimation system proposed for future linear colliders consists of a series of spoilers and absorbers. This collimation system serves two different functions: removing particles from the beam halo to reduce the background in the detector, and also protecting downstream beamline elements against missteered or off-energy beam pulses. The spoilers increase the angular divergence of an incident beam so that the absorbers located downstream can withstand the thermal loading of an entire bunch train [7]. A schematic is shown in Fig. 6.1.

Collimator shape (surface angle) and material should be chosen to minimize the fraction of re-scattered particles [9]. A further design criterion concerns wake fields generated by the collimators themselves [10]. An important criterion, which has influence on the length of the collimations system, requires that the collimators survive the impact of an entire bunch train. This implies that the collimators are located at positions where the β function is large. The correspondingly large area of the beam should ensure that the col-

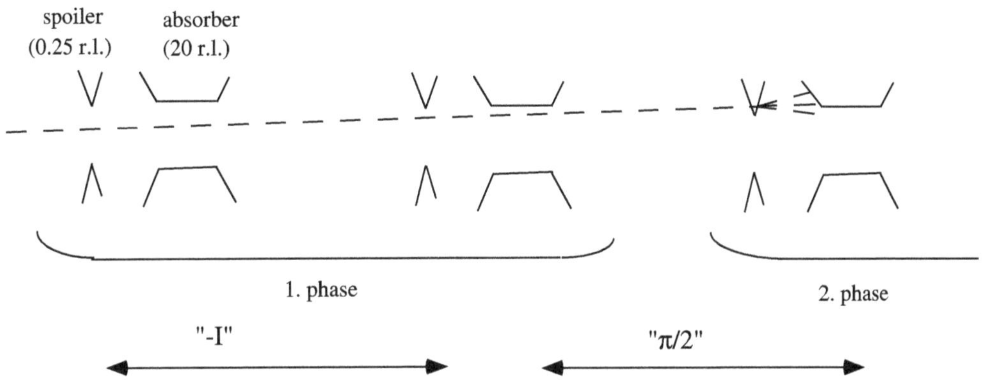

Fig. 6.1. Schematic of a conventional collimation system, consisting of a series of spoilers and absorbers. The lengths of the spoilers and absorbers are approximately 1/4 and 20 radiation lengths (r.l.), respectively

limator surface does not fracture or melt somewhere inside its volume in case it is hit by a mis-steered beam. For the NLC parameters, fracture and melting conditions give rise to about the same beam density limit (roughly 10^5 e^- per μm^2 for a copper absorber at 500 GeV [7]). While the surface fracture does not depend on the beam energy, the melting limit does, since the energy of an electromagnetic shower deposited per unit length increases in proportion to the beam energy. Therefore, for energies above a few hundred GeV, the beam area at the absorbers must increase linearly with energy. Since, in addition, the emittances decrease inversely proportional to the energy, the beta functions must increase not linearly but quadratically. Assuming that the system length l scales in proportion to the maximum beta function at the absorbers, this results in a quadratic dependence of the system length on energy: $l \propto \gamma^2$. Including both sides of the interaction point, the NLC collimation system is 5 km long. At 5 TeV the length of a conventional collimation system could approach 50 km.

Therefore, ideas for shorter and indestructible collimation schemes have been pursued, such as nonlinear collimation [11], laser collimation [12], plasma collimation [13], or nonlinear resonant collimation [14].

6.2 Storage Rings

Also the performance of storage rings can be limited by beam halo. At electron or positron rings the halo arises from beam-gas Coulomb scattering, beam-gas bremsstrahlung, beam-beam resonances, small tune drifts, and at high energies also from Compton scattering off thermal photons. In the case of proton or ion rings, halo may be caused by space-charge forces, injection errors, intrabeam scattering (multiple collisions of beam particles with each other), Touschek effect (single collision of particles within a bunch), diffusion driven by magnet nonlinearities or by the beam-beam interaction.

A collimation system proved invaluable at the HERA proton ring [15], and an advanced two-stage collimation system is contemplated for the LHC [16]. Here, the collimation also protects the superconducting magnets against local particle losses. The halo normally extends in both transverse and in the longitudinal direction, and collimation may be needed in all three planes.

The performance of LEP1 at 45.6 GeV (Z production) was limited by unstable transverse tails generated by the beam-beam interaction. Associated with these tails were a drop in the beam lifetime and background spikes (involving electromagnetic showers and hard synchrotron radiation from low-β quadrupoles), which frequently tripped the experiments. The partial cure consisted in changing the betatron tunes and the chromaticity, increasing the emittance (via a shift in the rf frequency), and opening the collimators. A lesson learned was that scraping into the beam halo close to the experiments had to be avoided.

For the higher energies and shorter damping times at LEP2 (80–100 GeV), background spikes were no longer observed. Stationary tails due to beam-gas scattering and thermal-photon scattering however were still present. Figure 6.2 compares a measurement of the beam tails in LEP using movable scrapers and the result of a Monte-Carlo simulation.

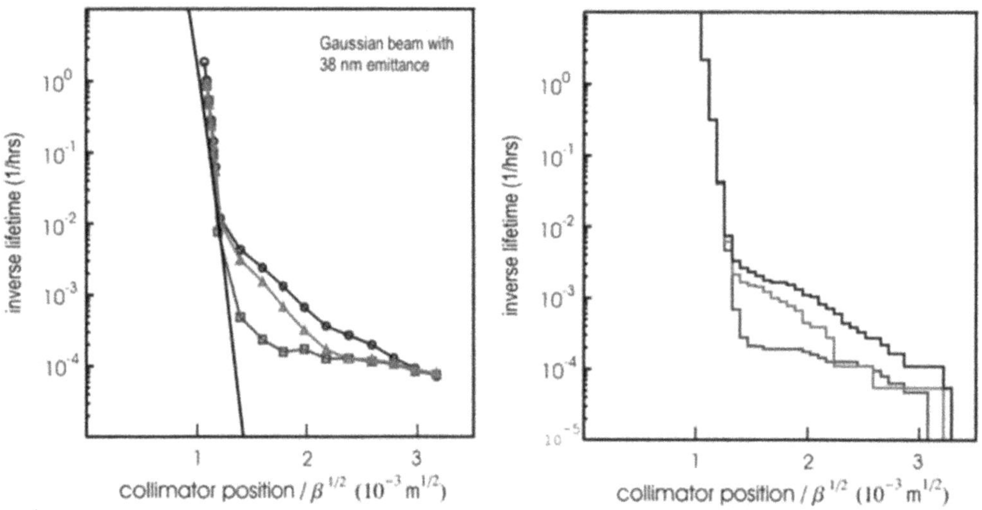

Fig. 6.2. Beam tails in LEP2 at 80.5 GeV: (*left*) measurement with collimation retracted (*circles*) and using movable scrapers at dispersive (*squares*) and nondispersive (*triangles*) locations, and (*right*) result of Monte-Carlo simulation [17, 18] (Courtesy H. Burkhardt, 1999)

An important scattering process for electron beams is beam-gas bremsstrahlung. The differential cross section for this process is

$$\frac{\mathrm{d}\sigma}{\mathrm{d}k} = \frac{A}{N_A X_0} \frac{1}{k} \left(\frac{4}{3} - \frac{4}{3} k + k^2 \right) , \tag{6.1}$$

where k denotes the ratio of the energy of the emitted photon and the beam energy: $k = E_\gamma/E_b$, X_0 is the radiation length ($X_0 \propto A/(Z(Z+1))$ or roughly $\sigma \propto Z^2$). For carbon monoxide molecules: $A/(N_A X_0) = 1.22$ barn, and the total cross section for an energy loss larger than 1% amounts to 6.5 barn (2.9% barn for an energy loss larger than 10%) [9]. For a gas pressure of 1 nTorr at a temperature of 300 K, the scattering probability is 2×10^{-14} m^{-1}.

The effect of elastic Coulomb collisions can also be significant. Here, the incident particles can scatter off residual nuclei or atomic electrons. In the first case, the energy change of the incident particle is relatively small and the primary effect is an angular deflection that may cause the particle to exceed the beam-pipe aperture. On the other hand, the energy change can be comparatively more important when scattering off the atomic electrons. The differential cross-section for Coulomb scattering off atomic nuclei can be written:

$$\frac{d\sigma_{en}}{d\Omega} = \frac{4F^2(q)Z^2r_e^2}{\gamma^2}\frac{1}{(\theta^2+\theta_{\min}^2)^2}, \tag{6.2}$$

where $\theta_{\min}$ is a function of the screening due to the atomic electrons, equal to $\theta_{\min} \approx (\hbar/pa)$ where p is the incident particle momentum and a is the atomic radius: $a \approx 1.4\lambda_e/\alpha Z^{1/3}$. In addition, $F(q)$ is the nuclear form factor which for relatively small scattering angles can be approximated by 1 and we have neglected the recoil of the nucleus; both of these later effects reduces the large angle scattering thus causing a slight overestimate of the scattering effect.

The second type of Coulomb collision is the elastic scattering off the atomic electrons. Here, the angular deflection can be roughly accounted for by replacing Z^2 with $Z(Z+1)$ in (6.2); again this will over-estimate the scattering, but the correction is small. In this case, however, the recoil of the electron cannot be neglected as it can result in a significant change in the energy of the incident particle. The differential cross-section for a relative energy change of δ is [19]:

$$\frac{d\sigma_{ee}}{d\delta} = \frac{2\pi Z r_e^2}{\gamma}\frac{1}{\delta^2} \tag{6.3}$$

and the cross section for scattering beyond a limiting energy aperture $\delta_{\min}$ is:

$$\sigma_{\delta_{\min}} = \frac{2\pi Z r_e^2}{\gamma}\frac{1}{\delta_{\min}}. \tag{6.4}$$

At energies higher than a few 10s of GeV, also the Compton scattering off thermal photons becomes significant [1, 20]. The photon density from Planck black-body radiation is

$$\rho_\gamma = \frac{2.4(k_BT)^3}{\pi^2(c\hbar)^3} \approx 20.2\left[\frac{T}{\mathrm{K}}\right]^3 \frac{1}{\mathrm{cm}^3}, \tag{6.5}$$

or, at room temperature,

$$\rho_\gamma(T = 300\mathrm{K}) \approx 5 \times 10^{14}\ \mathrm{m}^{-3}\,. \tag{6.6}$$

The scattering cross section is of the order of the Thomson cross section, $\sigma_T \approx 0.67$ barn. If all scattered particles are lost, the beam lifetime would be

$$\tau_{\mathrm{beam}} \approx \frac{1}{\rho_\gamma c \sigma_T}\,. \tag{6.7}$$

Another important source of backgrounds is synchrotron radiation generated in the focusing optics near the interaction point in lepton accelerators. At both LEP and the SLC the synchrotron radiation was minimized by weakening the last bending magnets closest to the interaction point by a factor ~10, which reduced the critical energy of the emitted photons as well as the number of photons emitted per unit length. In addition, radiation masks were installed to absorb the synchrotron radiation from the weak bend and from the upstream strong bending magnets. The layout of bends and synchrotron masks for LEP is illustrated in Fig. 6.3.

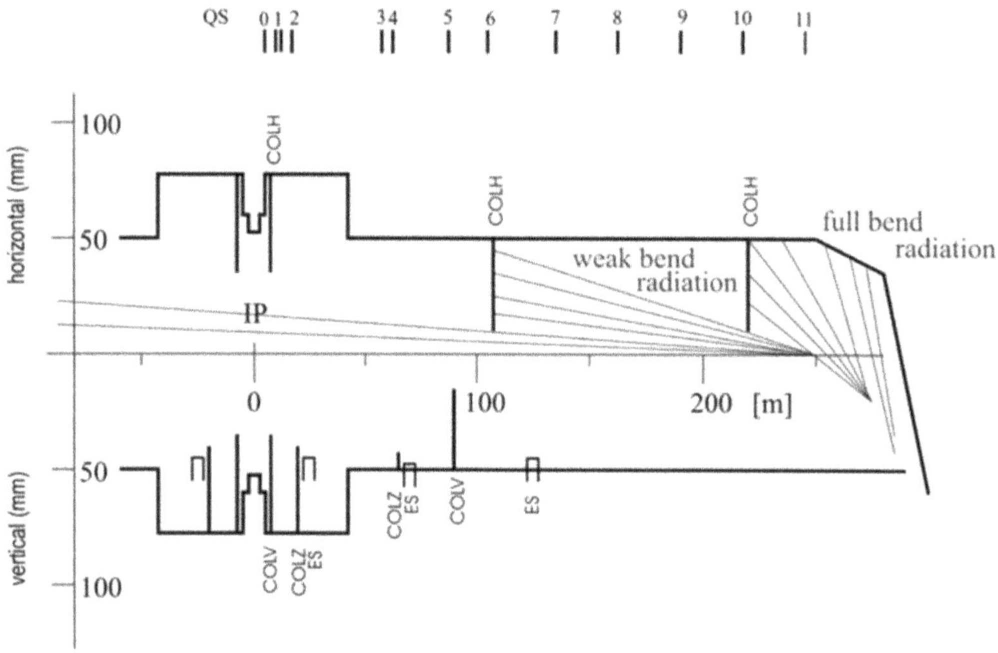

Fig. 6.3. Layout of the straight section around IP4 or IP8 in the horizontal and vertical planes. Shown are the quadrupoles (QS), electrostatic separators (ES), and collimators/masks (COLH, COLV, COLZ). The solid lines mark the inner vacuum chamber radii for the LEP1 layout [9, 21] (Courtesy H. Burkhardt, 1999)

Radiation collimators and masks around each LEP experiment provided complete shielding against direct photons and also against singly scattered synchrotron radiation, as illustrated in Fig. 6.4. For this reason, residual background at LEP arose mainly from multiply scattered radiation. Specular

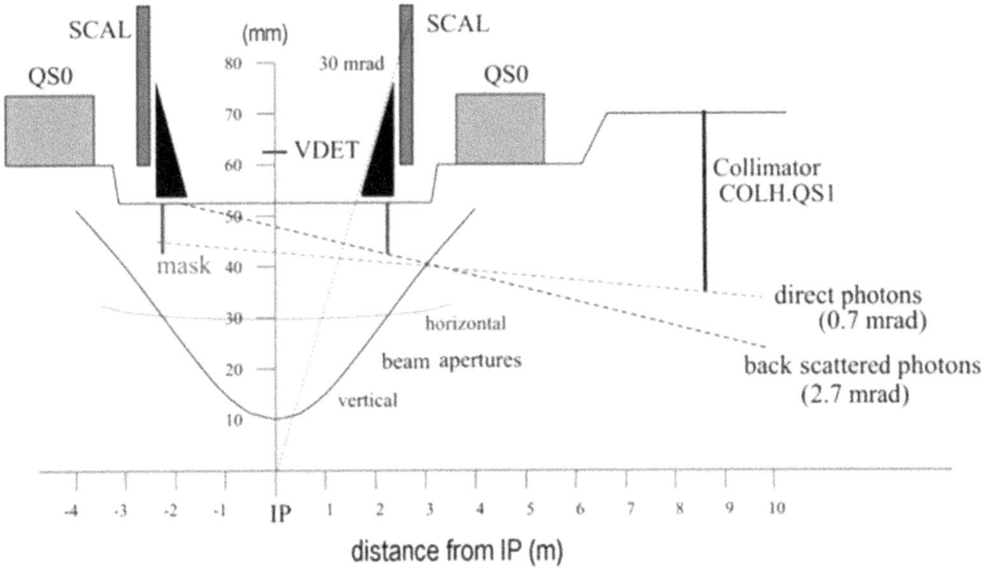

Fig. 6.4. Schematic of the synchrotron radiation masks around a LEP IP, indicating the constraints for mask solutions: (1) to stay outside the required LEP aperture (*solid lines*), (2) to cast a shadow over the entire unshielded IP beampipe length for small angle backscattered photons (*dashed*), (3) to stay outside of the very intense beam of direct photons collimated by the (8.5 m) synchrotron radiation collimator, when closed to 12 σ of the transverse beam distribution [9, 21] (Courtesy H. Burkhardt, 1999)

reflection of soft X rays is close to 100% at angles of incidence smaller than a so-called critical angle θ_c, where the angle is measured between the photon direction and the plane of impact. The critical angle is roughly

$$\theta_c \approx 30 \text{ mrad} \, \frac{\text{keV}}{E_\gamma} \, . \tag{6.8}$$

For a photon energy of 30 keV it is equal to about 1 mrad. Photons of this energy would still have a 95% chance of penetrating through a 1-mm Be layer. Multiple photon reflection can be reduced by coating or roughening of the vacuum chamber surface. In the LHC arcs, photon reflection will be reduced by impressing a sawtooth pattern on the beam screen which is installed inside the cold magnets [22].

Exercises

6.1 Scattering off Thermal Photons

Estimate the beam lifetime due to scattering off thermal photons:

a) in LEP at $T = 300\,\text{K}$, and

b) for a storage ring with a vacuum chamber cooled to $4\,\text{K}$.

c) Consider an NLC-like beam with 10^{12} electrons per bunch train. How many particles per train are scattered on thermal photons ($T = 300\,\text{K}$) over a linac length of 10 km?

7 Longitudinal Optics Measurement and Correction

Longitudinal focusing for a bunched beam is provided by both the change in path length with particle energy and by the time-dependent accelerating voltage. Usually one employs a smooth approximation, i.e., one ignores the discrete locations of the rf cavities, in describing the particle motion. The longitudinal motion can then be modelled by second order differential equations. For small oscillation amplitudes these equations simplify to those of harmonic oscillators.

The longitudinal optics and dynamics are closely linked to the transverse plane, so that 'cross-diagnostics' are possible. For example, an energy oscillation of the beam will result in a transverse oscillation at all places with nonzero dispersion; the signals recorded by a transverse pick-up therefore provide valuable information about the longitudinal dynamics. In analogy to the transverse motion (see Sect. 1.4), the oscillation frequency of the longitudinal or *synchrotron* motion is characterized by the number of oscillations per turn, which, in the longitudinal plane, is called the *synchrotron* tune.

In contrast to the transverse motion, there is an inherent strong nonlinearity which arises from the sinusoidal shape of the accelerating voltage. As a consequence the synchrotron tune decreases with oscillation amplitude and is zero at the separatrix. The separatrix describes the outer boundary of the *rf bucket* in phase space, beyond which particles are no longer contained longitudinally. In electron storage rings, particles outside the rf bucket are quickly lost, since they emit synchrotron radiation and lose energy. In proton storage rings such particles lose the time structure of the bunched beam forming a dc current. This phenomenon, seen at the Tevatron and HERA, has come to be called 'coasting beam' [1][2].

Observables which have a strong influence on the beam dynamics include the dispersion function, the momentum compaction factor, and the chromaticity. The values of the dispersion function together with the local bending radius determine the change in path length with beam energy, i.e., the *momentum compaction factor*. The change in the betatron tune with the beam energy, the *chromaticity*, may also couple the transverse and longitudinal degrees of freedom.

In this chapter, we discuss various techniques for measuring the synchrotron tune, the dispersion, the momentum compaction factor, and the chromaticity.

M. G. Minty et al., *Measurement and Control of Charged Particle Beams*,
https://doi.org/10.1007/978-3-662-08581-3_7

7.1 Synchronous Phase and Synchrotron Frequency

In a storage ring the equations of motion for small deviations from the ideal case can be expressed in terms of the phase difference of a particle within the bunch from the synchronous phase ϕ_s and the relative momentum error $\delta = \Delta p/p$. The synchrotron phase is defined such that a particle without momentum error launched at this phase, with respect to the accelerating rf wave, will arrive at the same phase on all subsequent turns. In linear approximation, the equations of motion are

$$\frac{\mathrm{d}\phi}{\mathrm{d}t} = \left(\alpha_c - \frac{1}{\gamma^2}\right)\omega_{\mathrm{rf}}\delta\,, \tag{7.1}$$

where ω_{rf} is the angular accelerating frequency, α_c the momentum compaction factor (compare (1.37)), γ the particle energy in units of the rest mass, and

$$\frac{\mathrm{d}\delta}{\mathrm{d}t} \approx \frac{e\dot{V}}{\omega_{\mathrm{rf}}\beta^2 E T_{\mathrm{rev}}}\,, \tag{7.2}$$

where the dot denotes a derivative with respect to time, E is the beam energy, and $\beta = v/c$ is the particle velocity in units of the speed of light. The momentum compaction factor α_c is a property of the magnetic focussing lattice of the accelerator (which is typically in the range of 10^{-4}, for high brightness accelerators and future storage rings, and is about 10^{-2} in existing storage rings). Note that the phase variation $\mathrm{d}\phi/\mathrm{d}t$ vanishes for $\gamma = \gamma_{\mathrm{t}} \equiv 1/\sqrt{\alpha_c}$, which is referred to as the transition energy. At this energy the beam is particularly susceptible to perturbations and instabilities. In addition, particular rf manipulations are required during 'transition crossing', such as a π phase jump of the accelerating rf wave. For these reasons, several accelerator complexes avoid this point of operation, by transferring beam from one ring with a small momentum compaction factor, operating below transition, to a second ring, where the energy of the injected beam is already above transition.

The quantity $\dot{V}$ represents the restoring force supplied by the cavity. Specifically, for a low current particle beam, $\dot{V}$ is the slope of the accelerating rf voltage evaluated at the particle position. For $V_{\mathrm{rf}}(t) = \hat{V}\cos(\omega_{\mathrm{rf}}t + \psi)$, where ψ is the phase with respect to an arbitrary reference point, the slope is given by

$$\dot{V} = -\omega_{\mathrm{rf}}\hat{V}\sin(\omega_{\mathrm{rf}}t + \psi)|_{\omega_{\mathrm{rf}}t+\psi=\phi_b(t)} = -\omega_{\mathrm{rf}}\hat{V}\sin\phi_b(t)\,, \tag{7.3}$$

where $\phi_b(t)$ the phase of a particle with respect to the crest of the cavity voltage $\hat{V}$. In the steady state, the phase of the beam centroid is approximately equal to the synchronous phase, i.e., $\langle\phi_b(t)\rangle = \phi_s$, where the angular brackets indicate an average over the beam distribution. In general, since both the external rf wave and the beam-nduced ('wake fields') are nonlinear, this relation is not exactly fulfilled.

The synchronous phase corresponds to the phase at which the energy gain from the accelerating cavities exactly compensates the energy lost per turn.

The total energy loss is $\sum U = U_0 + U_{\mathrm{hom}} + U_{\mathrm{par}}$, where U_0 is the energy loss per turn per particle due to synchrotron radiation, U_{hom} is the loss due to higher order modes in the cavities, and U_{para} represents all other losses arising, for example, from the interaction of the beam with components of the vacuum system. Taking as a zero phase reference the crest of the accelerating voltage, we have

$$e\hat{V}\cos\phi_s = \sum U \tag{7.4}$$

or

$$\phi_s = \cos^{-1}\left(\frac{\sum U}{e\hat{V}}\right) . \tag{7.5}$$

With this definition of ϕ_s (note that often the synchronous phase is defined with respect to the zero crossing instead), a synchronous phase of $\phi_s = \pi/2$ corresponds to zero energy loss.

A further word of caution may be in order. Equation (7.2) is based on a smooth approximation, which does not take into account the discrete location of the rf cavities used for restoring the particle energy. A more exact treatment would employ difference equations instead. For large high-energy storage rings, such as LEP, where the synchrotron tune may approach the half integer, the exact calculation can become necessary [3, 4].

The overvoltage factor $q = e\hat{V}/\sum U$ is useful for parametrizing the energy acceptance of an electron storage ring. As shown in reference [5], the low current energy aperture for a sinusoidal accelerating voltage is given by

$$\delta_{\mathrm{max}} = \sqrt{\frac{U_0}{\pi(\alpha_c - 1/\gamma^2)hE}F(q)} , \tag{7.6}$$

where $F(q)$ is the aperture function

$$F(q) = 2\left[\sqrt{q^2-1} - \cos^{-1}\left(\frac{1}{q}\right)\right] \tag{7.7}$$

and h the harmonic number. The harmonic number h follows from the accelerator circumference C, the rf frequency f_{rf}, and nominal particle velocity v, via

$$h = \frac{f_{\mathrm{rf}}C}{v} = \frac{f_{\mathrm{rf}}}{f_{\mathrm{rev}}} , \tag{7.8}$$

where f_{rev} is the revolution frequency.

Equations (7.1) and (7.2) can be combined to give a second order, uncoupled equation, which for small amplitudes ϕ_b or δ further simplifies to the equation of a harmonic oscillator (see (1.43)). The harmonic solutions to this equation are represented by the small amplitude contours in the phase space ($\phi - \delta$) plot. For small oscillation amplitudes, the constant energy trajectories are ellipses centered about the synchronous phase and energy. The oscillation frequency is called the synchrotron frequency and, at small amplitudes, is given by

$$f_s = \frac{\omega_s}{2\pi} = \sqrt{\frac{(\alpha_c - 1/\gamma^2) he\hat{V} {f_{\rm rev}}^2 \sin\phi_s}{2\pi\beta^2 E}} \, . \tag{7.9}$$

It is convenient to define the synchrotron tune Q_s by normalizing the measurable synchrotron frequency f_s to the beam revolution frequency $f_{\rm rev}$ as

$$Q_s = \frac{f_s}{f_{\rm rev}} = \sqrt{\frac{(\alpha_c - 1/\gamma^2) he\hat{V} \sin\phi_s}{2\pi\beta^2 E}} \, . \tag{7.10}$$

For small amplitude oscillations a particle or bunch returns to the same place in phase space every $1/Q_s$ turns.

If the beam centroid performs synchrotron oscillations, the arrival time of the particle at a beam position monitor (BPM) is modulated at the synchrotron frequency and synchrotron sidebands will appear around every harmonic of the revolution frequency. Stronger sidebands are observed with BPMs located at places with nonzero dispersion, due to the modulation of the transverse beam position at the synchrotron frequency. An example of measured synchrotron sidebands about a revolution harmonic is shown in Fig. 7.1. Note that a nonzero chromaticity gives rise to additional synchrotron sidebands around the betatron tune (not shown).

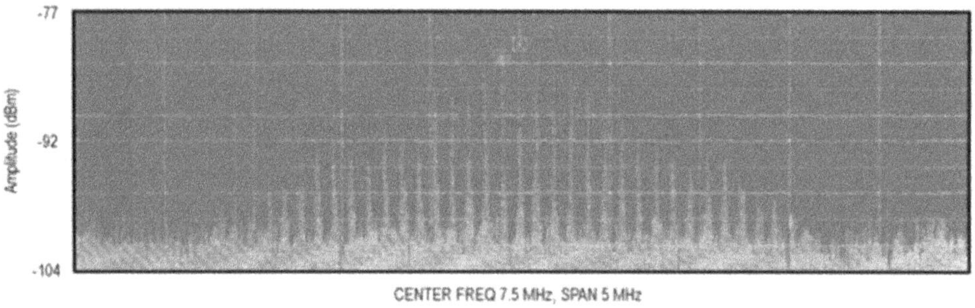

Fig. 7.1. Measurement of multiple synchrotron sidebands at injection in the SLAC electron damping ring. The synchrotron frequency is given by the difference in frequency between the fundamental and the nearest synchrotron sideband

In a manner quite similar to the measurement of the betatron tunes, turn-by-turn BPM measurements may be recorded and Fourier analyzed to detect the modulation of the beam signal due to synchrotron oscillations, e.g., if the selected BPM is in a region of nonzero dispersion. A typical example was shown in Fig. 2.5.

7.2 Dispersion and Dispersion Matching

The horizontal dispersion D_x describes the transverse displacement $x(s)$ of a particle as a function of its relative momentum deviation δ, via

$$x(s) = D_x(s)\delta \, . \tag{7.11}$$

The vertical dispersion is defined analogously. Superimposed on the energy-dependent displacement of (7.11) could be a fast betatron oscillation (cf., (1.6)).

7.2.1 RF Frequency Shift

In most storage rings the dispersion function is inferred from the orbit change induced by a shift in the rf frequency. This measurements makes use of the fact that a frequency shift Δf_{rf} changes the relative momentum deviation of the beam centroid by an amount

$$\delta = -\frac{1}{\alpha_c - \gamma^{-2}} \frac{\Delta f_{\mathrm{rf}}}{f_{\mathrm{rf}}} \approx -\frac{1}{\alpha_c} \frac{\Delta f_{\mathrm{rf}}}{f_{\mathrm{rf}}} . \tag{7.12}$$

The last approximation, which ignores the change in particle velocity, is usually applicable for electron rings, where $\gamma \gg 1$.

Combining (7.11) and (7.12), we can relate the dispersion to the measured orbit change:

$$D(s) = \left(\gamma^{-2} - \alpha_c\right) \frac{\Delta x(s)}{\Delta f_{\mathrm{rf}}/f_{\mathrm{rf}}} . \tag{7.13}$$

This 'static' dispersion measurement is quite simple. It requires the capability of being able to smoothly change the ring rf frequency (e.g., to 'unlock' it from the injector rf using an independent voltage controlled oscillator, for example) and a reasonably large energy aperture. By energy aperture we here refer to the range over which f_{rf} can be changed without beam loss. The

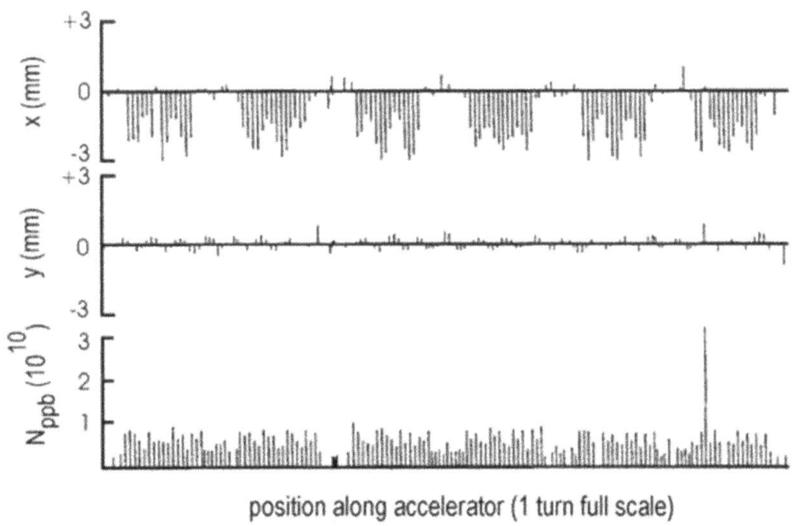

Fig. 7.2. Static dispersion measurement on the PEP-II HER: the orbit change induced by a 2 kHz shift in rf frequency. The nominal rf frequency is 476 MHz, the harmonic number $h = 3492$, and the momentum compaction factor $\alpha_c \approx 0.0024$ (Courtesy U. Wienands, J. Seeman et al., 1998)

residual vertical dispersion is obtained from the vertical orbit shift, in quite the same way as for the horizontal plane.

As an illustration, Fig. 7.2 shows a static dispersion measurement at the PEP-II HER, and Fig. 7.3 a dispersion measurement at the KEK/ATF Damping Ring before and after applying a correction based on exciting steering magnets.

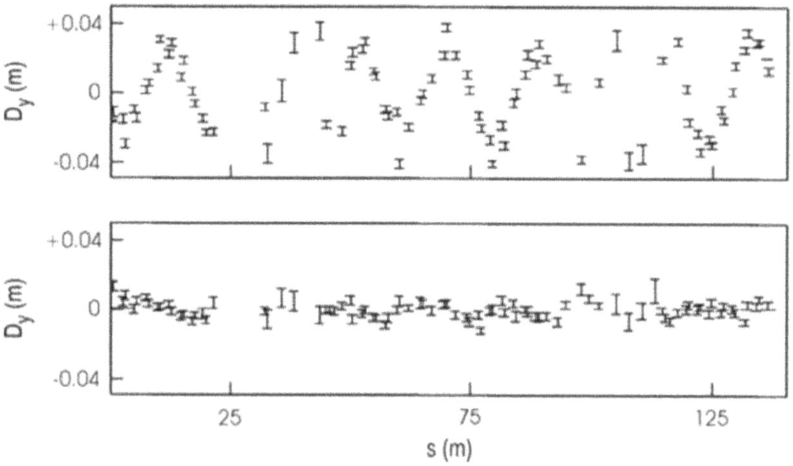

Fig. 7.3. Vertical dispersion measured via a ±5 kHz rf frequency ramp at the KEK ATF Damping Ring before [*top*] and after correction [*bottom*] [6]; a dispersion of $D_y = 5$ mm corresponded to an orbit change of $\Delta x \approx 15$ μm (Courtesy J. Urakawa, 2000)

7.2.2 RF Modulation

In very large rings, operating at high energy, the above method may not be applicable because of a finite energy aperture and the lack of orbit reproducibility. Therefore, at LEP, a dynamic measurement was occasionally applied [7], where the phase of the rf voltage was harmonically modulated at a frequency close to the synchrotron frequency. Fourier-analysing turn-by-turn BPM data, the frequency component of the induced (resonant) orbit variation at the synchrotron frequency could be inferred, which is proportional to the dispersion at the BPM.

The result of such a dynamic dispersion measurement is displayed in Fig. 7.4. If the dispersion at the cavities is nonzero, the dynamic measurement will give a result different from the static measurement [8]. The reason is that, for nonzero (horizontal) dispersion or nonzero slope of dispersion, an energy change ($\Delta\delta$) due to the rf modulation induces a horizontal betatron motion via $\Delta x_\beta = -D_x\, \Delta\delta$ and $\Delta x'_\beta = -D'_x\, \Delta\delta$. This additional or 'spurious' component of the measured response in amplitude propagates around the ring like a betatron oscillation. In principle, a precise phase measurement from BPM to BPM can be used to correct for this effect.

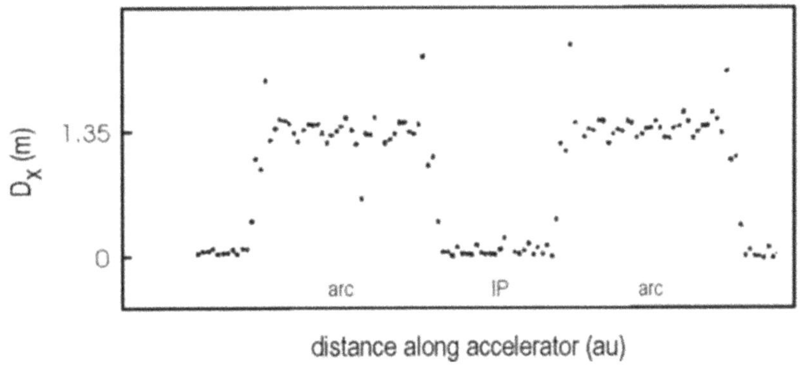

Fig. 7.4. Dynamic dispersion measurement at LEP [7]: the rf voltage was modulated at the synchrotron frequency and the dispersion was deduced by a harmonic analysis of the beam response at each BPM (Courtesy C. Bovet, 1998, G. Morpurgo, 2002)

In the arcs, the maximum value of the spurious dispersion arising from nonzero dispersion at the rf cavities is given by [8]:

$$|\Delta D|_{x,y;\max} = \sqrt{\beta_{x,y}\mathcal{H}_{x,y}} \left| \frac{\sin(2\pi Q_s)\ \sin(\pi Q_{x,y})}{\cos(2\pi Q_s) - \cos(2\pi Q_{x,y})} \right| , \tag{7.14}$$

where $Q_{x,y}$ and Q_s are the betatron and synchrotron tune, respectively, $\mathcal{H}_{x,y} = (D_{x,y}^2 + (\alpha_{x,y}D_{x,y} + \beta_{x,y}D'_{x,y})^2)/\beta_{x,y}$ is the dispersion invariant [5] in the straight section, and $\beta_{x,y}$ the beta function in the collider arcs.

7.2.3 RF Amplitude or Phase Jump

Similar dynamic schemes have been tested at the SLC and at the ATF damping rings. In both these rings, a longitudinal oscillation can be induced by a shock excitation: either a sudden step-change to the rf voltage (at the SLC [9]) or a fast phase jump (at the ATF [10]). Like the rf modulation technique, these methods can also give spurious results, if there is residual dispersion at the rf cavities. On the other hand, the dynamic schemes may be used to correct the synchrobetatron coupling, which arises from nonzero dispersion at the rf cavities, by empirically minimizing the spurious dispersion.

7.2.4 Resonant Correction of Residual Dispersion

Since betatron oscillations and residual vertical dispersion propagate around the ring at the same (betatron) frequency, a closed-orbit distortion can resonantly generate a significant dispersion. For this reason, the Fourier spectrum, with respect to the azimuthal position around the ring, of the residual dispersion (normalized to the square root of the local beta function),

$$\tilde{D}_y(k) = \sum_l \frac{D_y(s_l)}{\sqrt{\beta_y(s_l)}} e^{-ik\frac{\mu(s_l)}{Q_y}} , \tag{7.15}$$

usually contains a large peak at k equal to the integer part of the betatron tune Q_y. In (7.15), the sum is over the ring BPMs, which are located at s_l, and $\mu(s_l)$ denotes the betatron phase at the lth BPM. This 'resonant' dispersion, corresponding to the peak of the Fourier spectrum, can be compensated by special orbit bumps across arcs, as also demonstrated in LEP [11]. Consider an orbit bump of normalized amplitude Y extending over one arc

$$y_{\rm co} = Y\sqrt{\beta(s)}\sin\left[\mu(s) - \mu(s_i)\right] , \tag{7.16}$$

where s_i denotes the start of the arc. The associated dispersion is

$$\begin{aligned} D_y(s) = & -y_{\rm co} - \frac{Y\sqrt{\beta(s)}}{2\sin(\pi Q_y)}\int_{\rm bump} {\rm d}s' \left[\beta(k_1 - k_2 D_{x,0})\right]_{s'} \cos\left[\pi Q_y\right. \\ & \left. - \left|\mu(s) - \mu(s')\right|\right] \sin\left[\mu(s') - \mu(s_i)\right] . \end{aligned} \tag{7.17}$$

Here, k_1 is the quadrupole strength, k_2 the sextupole strength, and $D_{x,0}$ the nominal horizontal dispersion. With a phase advance of 90° per cell, and two families of sextupoles placed near the focusing and defocusing quadrupoles, respectively, from (7.17) the generated dispersion is [11]

$$D_y(s)/\sqrt{\beta(s)} \approx AY\sin\left[\pi Q_y - \mu(s) + \mu(s_i)\right] , \tag{7.18}$$

where A denotes the amplification factor for a single arc, and is given by

$$A = \frac{\pi N_{\rm cell} Q'_{\rm cell}}{\sin(\pi Q_y)} , \tag{7.19}$$

where $N_{\rm cell}$ is the (even) number of regular arc cells covered by the bump, and $Q'_{\rm cell}$ the chromaticity of a single FODO cell.

Bumps across various arcs can be combined in a symmetric or asymmetric manner, so as to cancel and control either the dispersion or the slope of the dispersion at the collision points [11].

7.2.5 Higher-Order Dispersion in a Transport Line or Linac

In a manner similar to that applied in circular accelerators, the dispersion or, more precisely, the R_{16} matrix element in a transport line can be inferred from the measured variation of the beam orbit as a function of the incoming beam energy.

We note that in a transport line the dispersion is not uniquely defined, but depends on the location at which the beam energy is varied. The dispersion or R_{16} (R_{36}) measured in this way does not necessarily correspond to the energy-position correlation within the beam, i.e., for a transport line or linac, in general, we expect that $R_{16} \neq \langle x\delta\rangle/\langle\delta^2\rangle$, where the angular brackets denote an average over the beam. A procedure which directly measures the correlation $\langle x\delta\rangle$ in the beam distribution is described in [12]. It employs two wire scanners at dispersive locations, separated by an optical $-I$ (minus identity) transformation.

We can extend the concept of dispersion or R_{16} matrix element by including higher-order nonlinear terms[1], of the form:

$$\Delta x(s) = R_{16}(s)\delta + T_{166}(s)\delta^2 + U_{1666}(s)\delta^3 \,, \tag{7.20}$$

$$\Delta x'(s) = R_{26}(s)\delta + T_{266}(s)\delta^2 + U_{2666}(s)\delta^3 \,. \tag{7.21}$$

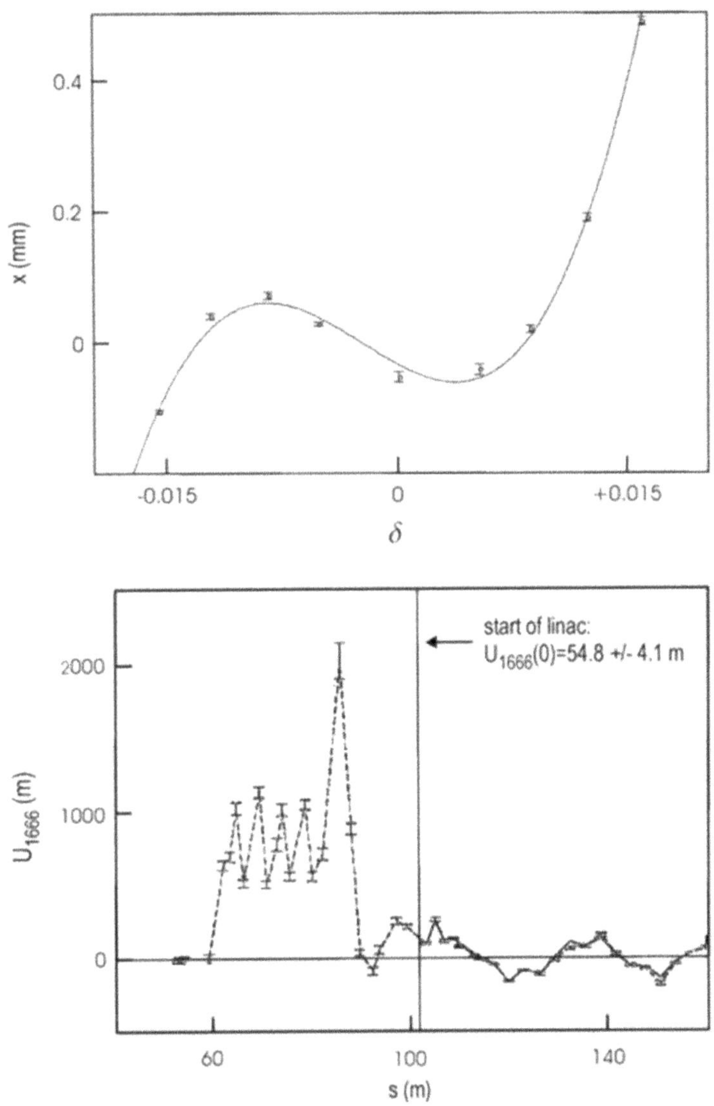

Fig. 7.5. Evidence of 3rd order dispersion in the SLC ring-to-linac transfer line (RTL) [14]: (*top*) Horizontal beam position versus beam energy; (*bottom*) 3rd order dispersion inferred for all BPMs in the RTL and in the early linac. The 3rd order dispersion in the linac is fitted to calculate the magnitude of the U_{1666} and U_{2666} matrix elements (Courtesy P. Emma, 1998)

1 Here we adopt the notation from TRANSPORT [13].

Sufficiently large energy changes allow a measurement not only of the first-order dispersion matrix element, R_{16}, but also of the 2nd and 3rd order contributions, $T_{166}(s)$ and $U_{1666}(s)$.

Such measurements have been performed at the North ring-to-linac transfer line (NRTL) of the SLC. Under normal operation, the phase of the rf compressor at the entrance to the RTL is set so that the beam center passes at the zero crossing of the rf wave. For a dispersion measurement, the phase is shifted such that the beam center is positioned at the top of the rf crest, and the beam energy is varied by changing the amplitude of the rf voltage.

Figure 7.5 (top) shows the beam position on one of the RTL BPMs as a function of the beam energy. Clearly visible is a nonlinear dependence, which indicates the presence of 3rd order dispersion. The value of the 3rd order dispersion at this BPM can be obtained by fitting a 3rd order polynomial to the measurement. Plotted in the bottom figure is the 3rd order dispersion function obtained from multiple BPMs as a function of position along the RTL and in the early part of the SLAC linac.

The large 3rd order dispersion led to undesired and irrecoverable emittance growth. To correct this, in 1991 two octupole magnets were installed which cancelled the U_{1666} and U_{2666} terms of (7.20) and (7.21). The optimum octupole strength was found by minimizing the linac emittance as a function of the octupole excitation. Such a measurement is shown in Fig. 7.6. The

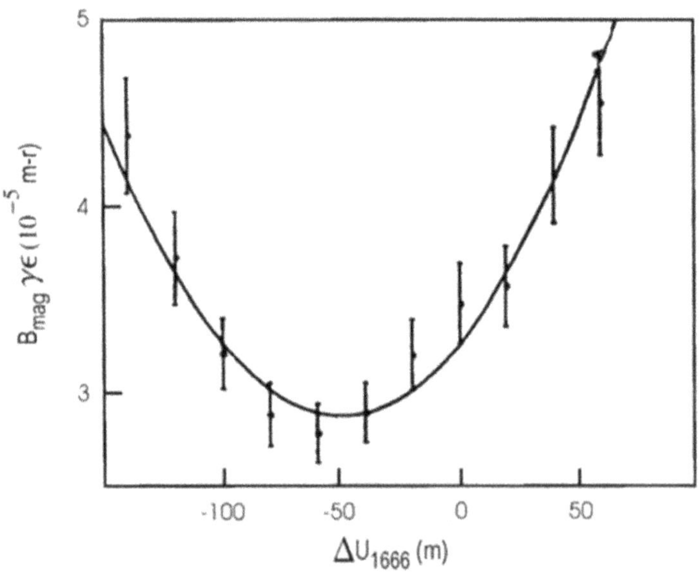

Fig. 7.6. RTL emittance minimization using an octupole for correction of the 3rd order dispersion. Shown on the horizontal axis is change in the octupole strength in units of the generated 3rd order dispersion (ΔU_{1666}). The vertical axis represents the product of B_{mag} (cf. (4.65)) and normalized emittance. The octupole strength for which the emittance is minimum agrees with the magnitude of U_{1666} estimated from the fit in Fig. 7.5 (*bottom*) (Courtesy P. Emma, 1998)

octupole setting for minimum emittance and the corresponding U_{1666} value are in good agreement with the 3rd order dispersion inferred from the BPM readings, which was depicted in Fig. 7.5 (bottom).

7.3 Momentum Compaction Factor

It is sometimes of interest to measure the momentum compaction factor α_c,

$$\alpha_c = \frac{\Delta C/C}{\delta} = \frac{1}{C}\oint \frac{D_x(s)}{\rho(s)}\,\mathrm{d}s\,, \tag{7.22}$$

for example, in storage rings operating near $\alpha_c = 0$ or as a basic check of the accelerator optics. We next present several methods with which to do so.

7.3.1 Synchrotron Tune

If the rf voltage is well calibrated, one can invoke the definition of the synchrotron tune, (7.10), to infer the momentum compaction factor from the measured dependence of the synchrotron tune on $\hat{V}$, taking into account that the synchronous phase angle ϕ_s is also a function of the rf voltage $\hat{V}$. However, often the rf voltage calibration is not very accurate. In addition, if the ring accommodates several rf cavities, these may not be optimally phased with respect to each other, complicating the calculation of the total rf voltage. It is then advantageous to confirm the momentum compaction without having to assume a value for the rf voltage.

Despite the aforementioned difficulties, the synchrotron tune contains information on a variety of parameters, which may be extracted by a judicious choice of measurements and proper fitting strategies. For example, the longitudinal loss factor U may be obtained by measuring the synchrotron tune Q_s as a function of rf voltage $\hat{V}$ for various beam currents, and fitting the result to

$$Q_s^2 = \frac{(\alpha_c - 1/\gamma^2)h}{2\pi}\left(\frac{g^2 e^2 \hat{V}^2}{E^2} + M g^4 \hat{V}^4 - \frac{1}{E^2}U^2\right)^{1/2}\,, \tag{7.23}$$

where the $\hat{V}^4$ term is included to account for the discrete locations of the rf cavities. The coefficient M can be computed from the ring optics [3, 15]. An example of a loss-factor measurement using (7.23) is given in Fig. 7.7. From the fit also α_c was determined with a precision better than 10^{-3} (see [15]). In addition, if the beam energy is known at one point, e.g., on a spin resonance, the Q_s vs. $\hat{V}$ curve can be used to calibrate the rf voltage (Fig. 7.8) [15].

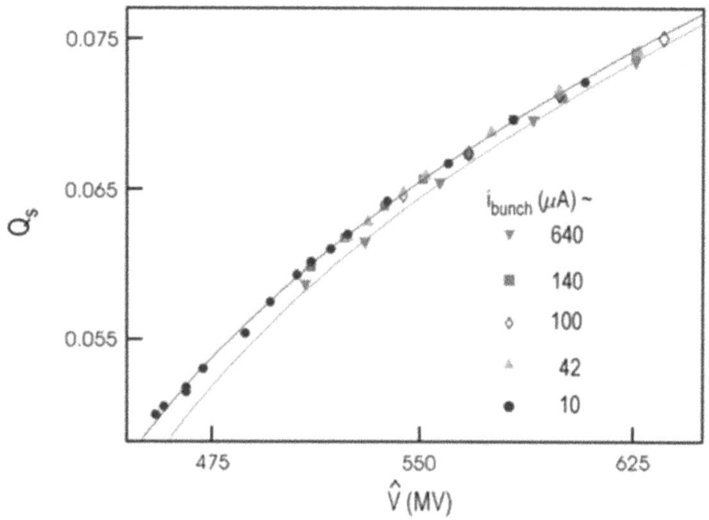

Fig. 7.7. Synchrotron tune Q_s as a function of total rf voltage in LEP; the two curves are fits to the 640 μA and 10 μA data. The difference due to current-dependent parasitic mode losses is clearly visible. Here the nominal beam energy E_{nom} was 60.589 GeV [15] (Courtesy A.-S. Müller, 2000)

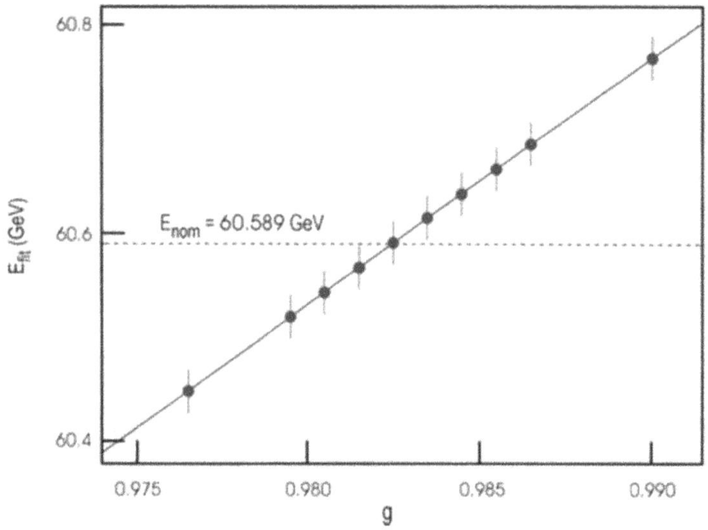

Fig. 7.8. Beam energy fitted from Q_s vs. $\hat{V}$ measurement as a function of rf voltage calibration factor g; the *dotted line* indicates the known energy value, determined independently by resonant depolarization [15] (Courtesy A.-S. Müller, 2000)

7.3.2 Bunch Length

The rms bunch length in an electron or high-energy proton ring can be inferred from the duration of the synchrotron-light pulse using a streak camera. The rms bunch length σ_z is proportional to α_c and to the rms energy spread [5], namely

$$\sigma_z = \frac{c(\alpha_c - 1/\gamma^2)}{2\pi Q_s f_{\text{rev}}} \sigma_\delta \,, \tag{7.24}$$

where c is the speed of light and $f_{\rm rev}$ is the revolution frequency. The rms energy spread can either be deduced from the measured decoherence of a transverse oscillation due to nonzero chromaticity and its subsequent recoherence after one synchrotron period (see also (7.36)) [16], or it can be calculated from [5]

$$\sigma_\delta^2 = \frac{C_q \langle G^3 \rangle \gamma^2}{J_\epsilon \langle G^2 \rangle} , \tag{7.25}$$

where $C_q = 3.84 \times 10^{-13}$ m, $G = 1/\rho$ the inverse bending radius, $\langle \ldots \rangle$ indicates an average over the ring, γ is the beam energy in units of the particle rest mass, and J_ϵ the longitudinal damping partition number. The value for the latter can be verified by measuring either the horizontal emittance (which is inversely proportional to the horizontal partition number J_x, where $J_x = (3 - J_\epsilon)$ for a planar accelerator) or the longitudinal damping time (cf. (4.74)).

Plotting the measured bunch length as a function of the inverse synchrotron tune immediately gives the value of α_c as the slope [17] from (7.24). Note that the synchrotron frequency $f_s = \omega_s/2\pi$ can be measured very precisely. Figure 7.9 shows a measurement of bunch length vs. synchrotron tune in PEP-II.

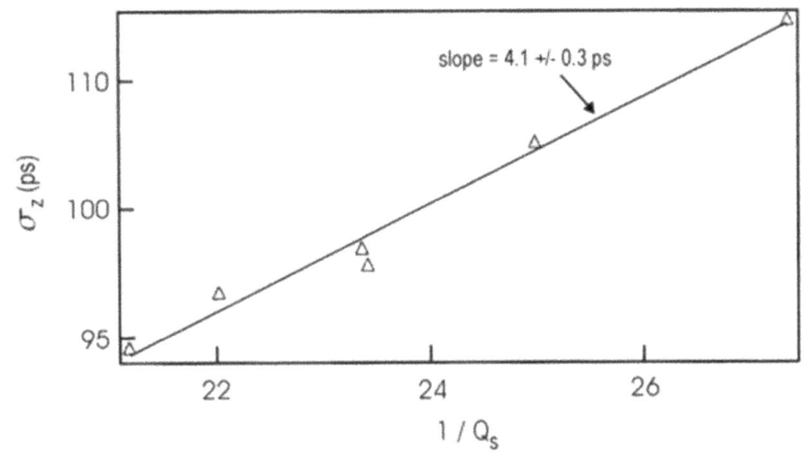

Fig. 7.9. Rms bunch length in the PEP-II HER as a function of the inverse synchrotron tune [18]. The fitted slope determines the momentum compaction factor α_c, if the rms energy spread is known (Courtesy A. Fisher, 1998)

7.3.3 Lifetime

A different approach [17] for determining α_c, applicable for electron rings, is to measure the quantum lifetime [5],

$$\tau_q = \frac{\tau_\delta}{2} \frac{e^{\xi_q}}{\xi_q} , \tag{7.26}$$

where τ_δ is the longitudinal damping time, and ξ the ratio of the energy aperture δ_{max} and the rms relative energy spread σ_δ:

$$\xi_q = \frac{\delta_{\mathrm{max}}^2}{2\sigma_\delta^2} \,. \tag{7.27}$$

A formula for the rms relative energy spread σ_δ was given in (7.25). At low current, the energy aperture, δ_{max}, if limited by the rf bucket size, is (see also (7.6) and (7.7)) [5]

$$\delta_{\mathrm{max}} \approx \left(\frac{U_0}{\pi(\alpha_c - 1/\gamma^2)hE} F(q) \right)^{1/2} , \tag{7.28}$$

with

$$F(q) = 2\left(\sqrt{q^2 - 1} - \cos^{-1}(1/q)\right) \tag{7.29}$$

and

$$q = \frac{e\hat{V}}{U_0} \,. \tag{7.30}$$

The term $U_0 = C_\gamma E^4 L\langle G^2\rangle/(2\pi)$ is the energy loss per turn, and $C_\gamma = 8.85 \times 10^{-5}$ m GeV^{-3}.

We may express σ_δ in terms of σ_z using (7.24), and in addition replace the rf voltage $\hat{V}$ in the definition of q by Q_s and α_c, making use of (7.10). We then arrive at an equation for the quantum lifetime τ_q in terms of the measurable quantities Q_s and σ_z, and the unknown parameter α_c. The latter can then be obtained from a fit to data taken at different rf voltages [17].

7.3.4 Path Length vs. Energy

The momentum compaction factor or, in a transport line, the R_{56} matrix element can also be measured directly by changing the beam energy at the entrance to the beamline of interest, and observing the shift in arrival time at the end of that section.

Such measurements were performed to fine-tune the optics in the nominally achromatic arc of the KEKB linac. The time of arrival at the exit of the arc was measured by a streak camera, which converts the time structure of a pulse of synchrotron radiation from a bend, or of optical transition radiation from a target, into a vertical deflection at the CCD camera.

During commissioning of the KEKB linac, the streak camera trigger signal was locked to the linac rf frequency upstream of the arc. The beam energy was varied by adjusting the voltage of the last klystrons prior to the arc. Figure 7.10 shows two measurements of the R_{56}, performed before and after the strengths of a few quadrupoles were adjusted to match the dispersion, as inferred from the energy dependence of the orbit. Figure 7.10 demonstrates that the dispersion match also eliminated the linear component of R_{56}; the

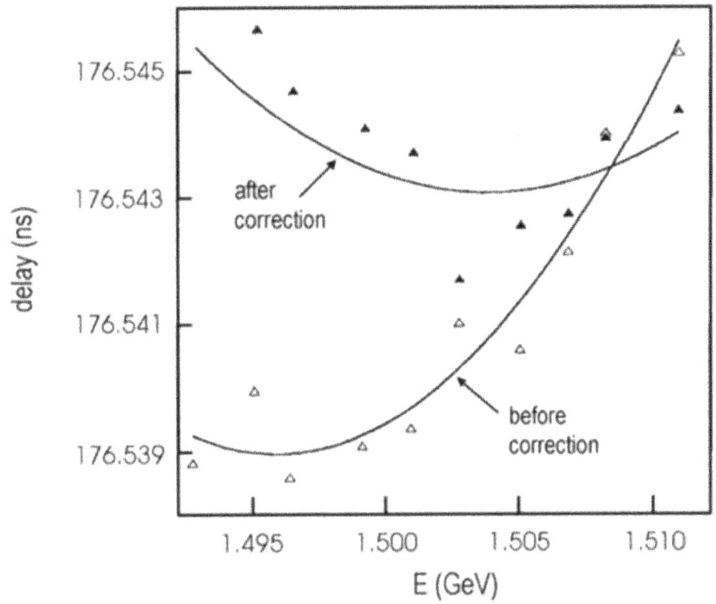

Fig. 7.10. R_{56} measurement for the asynchronous arc of the KEKB linac, before and after dispersion correction. A streak camera was used to measure the arrival time (*vertical axis*) as a function of the beam energy (Courtesy H. Koiso and K. Oide, 1998)

remaining path length dependence on energy is purely quadratic. In the future, it is planned to reduce this quadratic component, as well as the second order dispersion, by adjusting sextupole magnets.

7.3.5 Beam Energy via Resonant Depolarization

In storage rings with polarized beams the beam energy can be determined with a very high precision, using a resonant depolarization technique. The spin tune (see Chap. 10) for an electron is given by

$$\nu_0 = a_e \gamma = \frac{E \text{ [MeV]}}{440.6486 \text{ [MeV]}} , \tag{7.31}$$

where a_e is the electron anomalous magnetic moment. If a radially oscillating field generated by a coil is in resonance with the fractional part of the spin tune, the effect of the field adds up over many turns and the nominally vertical spin vector may precess towards the horizontal plane. The exact value of the resonance frequency determines the beam energy via (7.31).

With this technique, it is possible to precisely measure the energy variation induced by a change in the rf frequency. The slope of this measurement gives the momentum compaction factor:

$$\frac{\Delta p}{p} = \frac{1}{\gamma^{-2} - \alpha_c} \frac{\Delta f_{\rm rf}}{f_{\rm rf}} \approx -\frac{1}{\alpha_c} \frac{\Delta f_{\rm rf}}{f_{\rm rf}} . \tag{7.32}$$

An application of this technique at LEP is shown in Fig. 7.11.

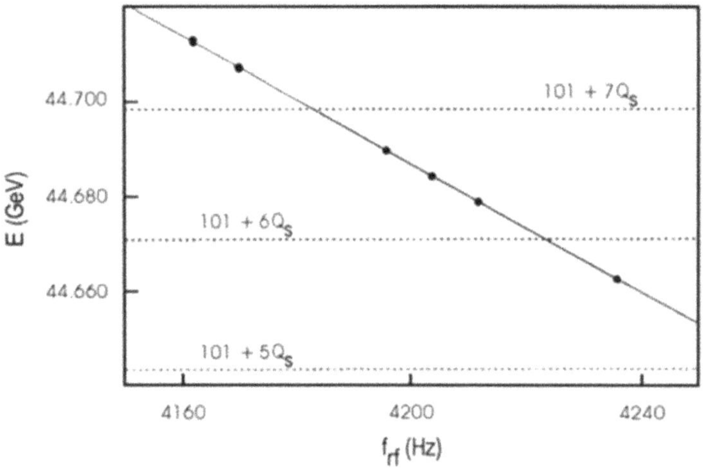

Fig. 7.11. Change of beam energy E as a function of the rf frequency f_{rf} in LEP [19]. Only the last four digits of the rf frequency are shown (the nominal value is $f_{\mathrm{rf}} = 352\ 254\ 170$ Hz). Several strong spin resonances are indicated by the dotted lines. From this measurement the momentum compaction factor was determined to be $(1.86\pm0.02)\times10^{-4}$, which compared well with the calculated value of 1.859×10^{-4} (Courtesy R. Assmann, 1998)

7.3.6 Change in Field Strength for Unbunched Proton Beam

The energy of an unbunched proton beam is constant (neglecting energy losses due to synchrotron radiation for ultrarelativistic beams). If the strength of all magnets (dipoles and quadrupoles) is increased by a factor $\Delta B/B$, the orbit moves inwards and the revolution time is reduced. This change in revolution period Δt can be detected with a Schottky monitor [20]. The momentum compaction factor α_c then simply follows from the relation

$$\frac{\Delta T}{T} = -\left(\alpha_c - 1/\gamma^2\right)\frac{\Delta B}{B}, \tag{7.33}$$

where T denotes the revolution period.

7.4 Chromaticity

The dependence of the focusing force on beam energy is generally referred to as chromaticity. In a storage ring this is characterized by the energy dependence of the betatron tunes, which is denoted as $\xi = (\Delta Q/Q)/(\Delta p/p)$ or $Q' = \Delta Q/(\Delta p/p)$. Note that ξ is the normalized chromaticity, related to Q' via $\xi = Q'/Q$.

The natural chromaticity due to the energy dependence of the quadrupole focusing is usually compensated by means of two or more sextupole families at locations with nonzero dispersion. Often a total chromaticity close to zero is desired as this minimizes the tune spread induced by a finite energy spread, and also the amount of synchrobetatron coupling. The chromaticity should be slightly positive to avoid the head-tail instability. Since a positive chromaticity gives head-tail damping, sometimes Q' is intentionally increased in order to counteract beam instabilities.

7.4.1 RF Frequency Shift

The total chromaticity can easily be determined by measuring the tune shift as a function of the rf frequency f_{rf} using

$$Q'_{x,y} = \frac{\Delta Q_{x,y}}{\Delta p/p} = -\left(\alpha_c - \frac{1}{\gamma^2}\right)\frac{\Delta Q_{x,y}}{\Delta f_{\mathrm{rf}}/f_{\mathrm{rf}}}\,, \tag{7.34}$$

where α_c is the momentum compaction factor. As an example, Fig. 7.12 shows a chromaticity measurement performed at LEP.

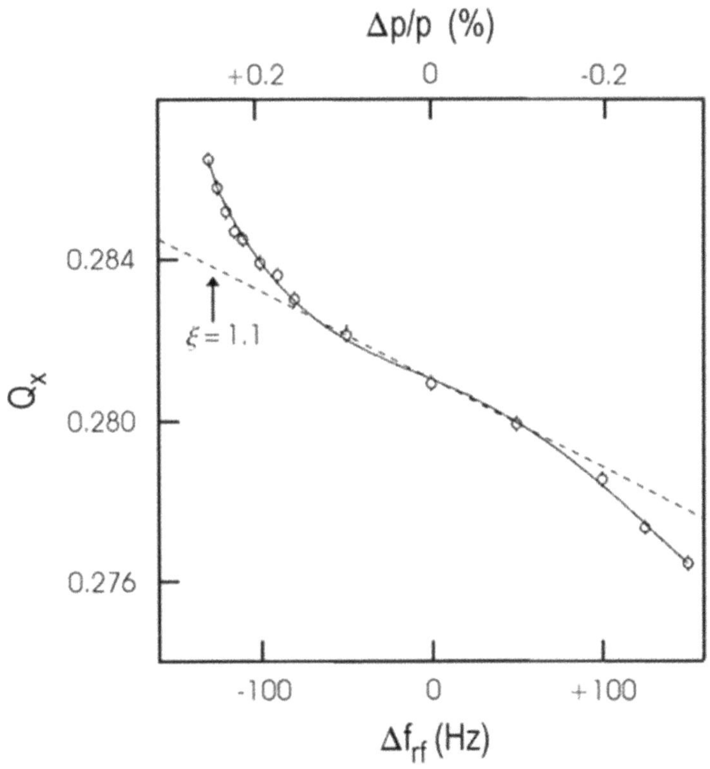

Fig. 7.12. Horizontal tune as a function of the change in rf frequency measured at LEP [21]. The *dashed line* shows the linear chromaticity as determined by the measurements at $\Delta f_{\mathrm{rf}} = \pm 50$ kHz (Courtesy H. Burkhardt, 1998)

7.4.2 Head-Tail Phase Shift

Deflecting the beam transversely and measuring the oscillation of the bunch head and bunch tail separately over a few hundred successive turns also allows the determination of the chromaticity. The underlying relation is [22]

$$\xi_{x,y} = \frac{(\alpha_c - 1/\gamma^2)\Delta\phi(n)}{Q_{x,y}\omega_{\mathrm{rf}}\ \Delta\tau[\cos(2\pi n Q_s) - 1]}\,, \tag{7.35}$$

where $Q_{x,y}$ is the betatron tune, ω_{rf} the angular revolution frequency, $\Delta\tau$ the sample time delay between head and tail signal, and $\Delta\phi(n)$ the phase difference at turn n. After every full synchrotron period, $n = n_s$, the head and tail are again in phase and $\Delta\phi(n_s) = 0$.

The phase at each turn is obtained by a sweeping harmonic analysis, i.e., fitting an oscillation to sets of consecutive data points. The phase difference is determined by simply subtracting the phases of head and tail measured using a wide-band pick-up on each turn. Figure 7.13 shows an example measurement from the CERN SPS, where for technical reasons the phase shift between the head and center of the bunch was detected. The advantage of this method is that it is fast. It will be used to correct rapid changes in the chromaticity at the start of acceleration in the LHC, due to persistent currents in the superconducting magnets.

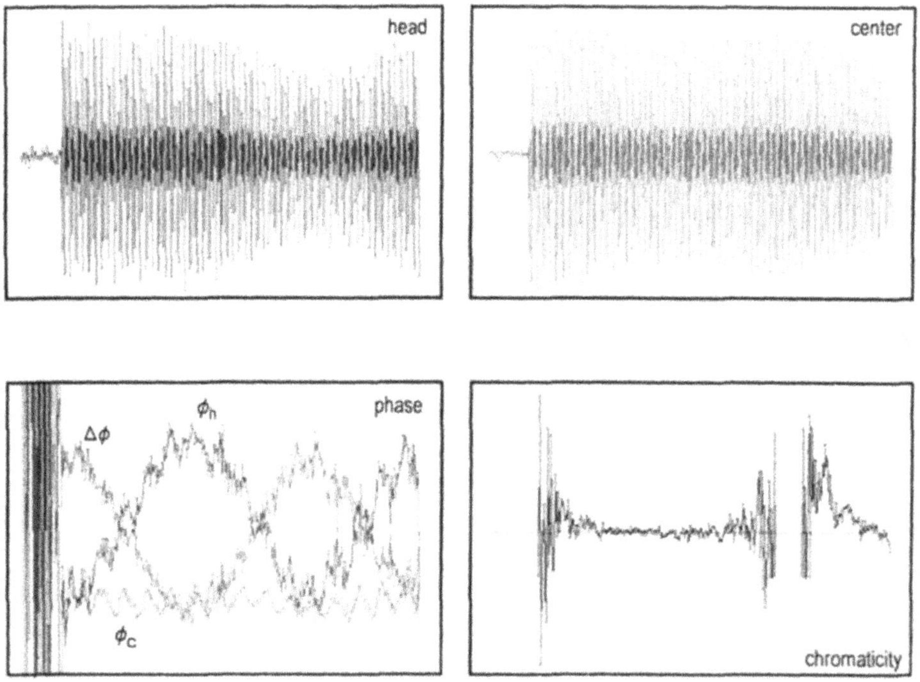

Fig. 7.13. Chromaticity inferred from measurements of the phase shift difference between the head and center of a bunch at the CERN SPS [22]. *Top left*: head oscillation after applying a kick (a.u.); *top right*: center oscillation after the kick (a.u.); *bottom left*: phase of head ϕ_h and center ϕ_c as well as their difference $\Delta\phi$ (with 1.34 radians full scale); *bottom right*: the chromaticity inferred for each turn (with 16.3 units full scale). In all plots the full scale of the horizontal axis is 409 turns (Courtesy R. Jones, 2000)

7.4.3 Alternative Chromaticity Measurements

Other possibilities of measuring the chromaticity include the detection and analysis of synchrobetatron sidebands in the tune spectrum [23] or, as studied at Novosibirsk [24], of the impulse excitation of the beam and the ensuing reversible decoherence due to the chromatic phase.

Due to chromatic beating, the centroid of the 'beam arc' (referring to the beam shape in the transverse phase space) oscillates at the synchrotron frequency. After a vertical impulse excitation, the dipole signal at the betatron frequency is of the form [16]

$$A(t) \propto e^{-\frac{t^2}{2\tau^2}} e^{-\left(\frac{Q'_y}{Q_s}\sigma_\delta\right)^2 (1-\cos\Omega_s t)} , \tag{7.36}$$

where τ is the decoherence time due to the betatron tune shift with amplitude (anharmonicity), and σ_δ is the rms energy spread. For $Q'_y \geq \sigma_\delta Q_s$ separate peaks were seen in the experiment [16]. The width of the peaks is $\tau_E^{-1} = Q'_y \sigma_\delta \omega_0$, which allows measuring either Q'_y or the energy spread σ_δ [16]. We note that in the Novisibirsk experiment, the beam oscillations and decoherence were measured by detecting the synchrotron-radiation photons passing through a limiting half-aperture.

7.4.4 Natural Chromaticity

The natural chromaticity is the chromaticity that derives from the energy dependence of the quadrupole focusing. In other words it is the chromaticity the ring would have without sextupole magnets. Fortunately, to measure the natural chromaticity, it is not necessary to turn off the sextupoles, which might be impossible. Rather the latter can be obtained by detecting the variation of the betatron tune as a function of the main dipole field strength. For an electron beam since the rf frequency is constant, the total path length is constant and the orbit at the sextupoles remains unchanged. The sextupoles therefore do not contribute to a change in tune. (This is a good assumption for FODO lattices, however, it is conceivable that for certain low-emittance lattices the orbit in the sextupoles might change when the dipole field strength is varied. This effect can be estimated with computer codes. One can also monitor the orbit stability at the sextupoles when the dipole field is varied.)

In this case, the absolute beam energy E varies in proportion to the field change: $\Delta E/E = \Delta B/B$. Thus, the natural chromaticity ${Q'}^{\mathrm{nat}}_{x,y}$ is given by

$${Q'}^{\mathrm{nat}}_{x,y} \approx \frac{\Delta Q_{x,y}}{\Delta B/B} . \tag{7.37}$$

A typical measurement, from the PEP-II High Energy Ring (HER), is depicted in Fig. 7.14.

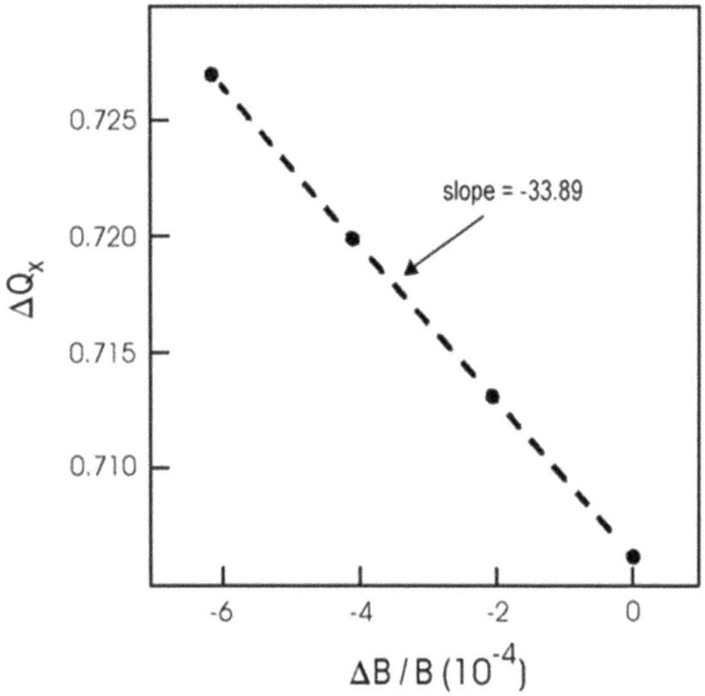

Fig. 7.14. Measurement of the natural chromaticity in the PEP-II HER. Shown is the horizontal tune as a function of relative variation in the main dipole field. The slope of -33.9 is the natural chromaticity inferred from this measurement (Courtesy U. Wienands, J. Seeman et al., 1998)

Similarly, the natural chromaticity can be measured for a proton beam. However, for protons, the rf frequency must be changed in proportion to the dipole field as $\Delta\omega_{,\mathrm{rf}}/\omega_{\mathrm{rf}} = (1/\gamma^2)\Delta B/B$, in order to maintain a constant orbit in the sextupoles.

7.4.5 Local Chromaticity: $\mathrm{d}\beta/\mathrm{d}\delta$

Measuring the beta functions (e.g., with the tune shift method of (2.29)), and using

$$\beta_{x,y} \approx \pm 4\pi \, \frac{\Delta Q_{x,y}}{\Delta k} \, , \tag{7.38}$$

or its more precise equivalent, (2.28), for different values of the rf frequency yields informations on the local chromaticity. This can help to identify the origin of chromatic errors or to find sources of chromatic nonlinearities.

7.4.6 Chromaticity Control in Superconducting Proton Rings

In superconducting proton rings the natural chromaticity is small compared with the chromaticity arising from the persistent-current sextupole components in the dipole magnets. For example, in the HERA superconducting proton ring the sextupole component in the dipoles contributes a chromaticity that is 5 times larger than the natural chromaticity. At injection energy, a significant part of the persistent current decays in time, causing a large variation in chromaticity. This is illustrated in Fig. 7.15, which also shows the effect of an automatic correction system used at HERA. The correction is done locally by exciting sextupole correction coils mounted inside all bending magnets. The excitation for these correction magnets is determined from the instantaneous sextupole field measured using rotating coils in two reference magnets, which are connected in series with the main superconducting magnet circuit.

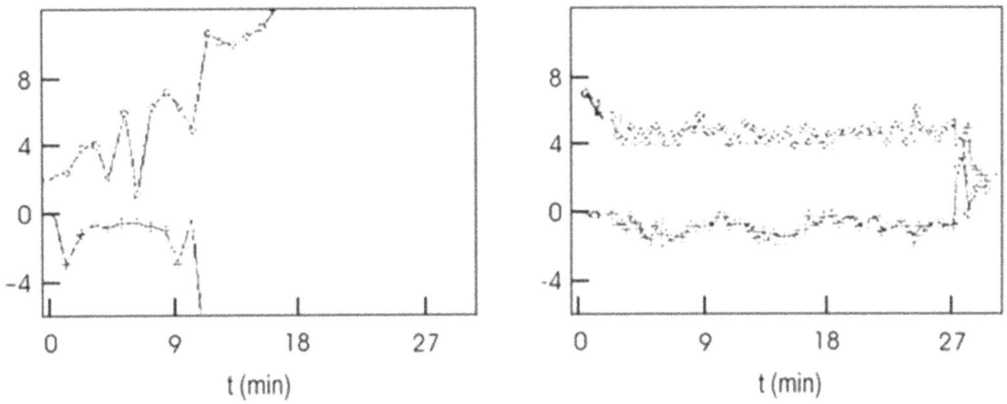

Fig. 7.15. Variation of chromaticity in time, due to persistent-current decay without (*left*) and with (*right*) chromaticity control based on continuous measurements of the sextupole fields in two dipole reference magnets at the HERA proton ring [25]. The horizontal axis is the time in units of 3 minutes per division. The vertical axis refers to the horizontal (*upper trace*) and vertical chromaticity (*lower trace*) in dimensionless units (Courtesy B. Holzer, 1998)

At HERA, the persistent-current sextupole field in the dipole magnets decays during injection at 40 GeV. It is reinduced by the change in dipole field at the start of acceleration, resulting in large variations of the chromaticity. Figure 7.16 shows and example of the change in chromaticity during acceleration from 40 GeV to 70 GeV. The figure compares the actual chromaticity, i.e., the change in tune detected per relative rf frequency change, (7.34), measured without continuous correction; the chromaticity predicted by the reference magnets; and the chromaticity measured with a correction derived from the reference magnets [26].

Another noteworthy feature of the persistent-current sextupole field is that it is not very reproducible from cycle to cycle. At HERA, after each magnet cycle, with the ring magnets set for the injection energy, the chromaticity is first corrected manually by means of a direct measurement (tune shift versus rf frequency). Subsequently, the chromaticity is held constant using the automatic control based on the reference magnets.

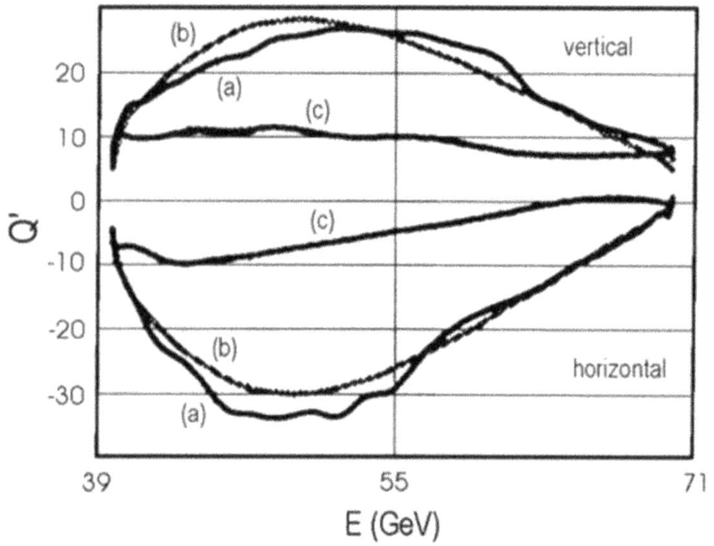

Fig. 7.16. Variation of the chromaticity in the HERA proton ring during acceleration from 40 GeV to 70 GeV [26]: (*a*) measured chromaticity without correction; (*b*) change in chromaticity expected from the reference-magnet measurements; (*c*) measured chromaticity with correction (Courtesy O. Meincke, 1998)

7.4.7 Application: Measuring the Central Frequency

Measuring the chromaticity for different sextupole strengths determines the 'central frequency'. This is the rf frequency for which the orbit on average passes through the center of all sextupoles [27, 28]. An example of such a measurement is shown in Fig. 7.17. Usually adjacent sextupoles and quadrupoles are well aligned with respect to each other, so that one can ex-

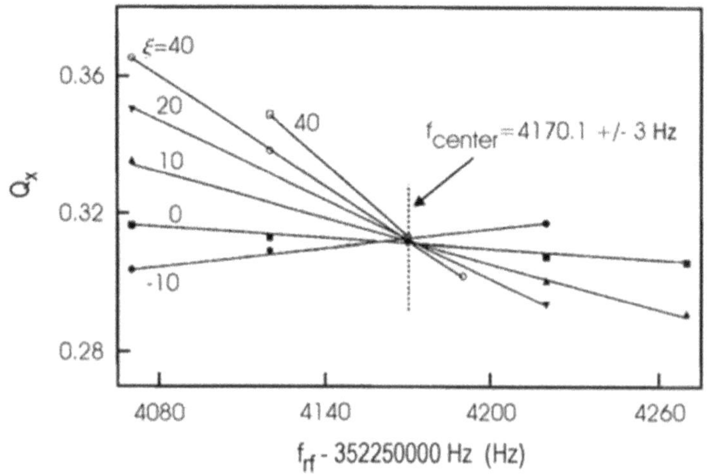

Fig. 7.17. LEP chromaticity measurements for different sextupole excitation patterns with net chromaticities in the range $Q' = -10$ to $+40$ [21]. The intersection of the different lines determines the central frequency, where the orbit is on average centered in the sextupoles (Courtesy H. Burkhardt, 1998)

pect that at the central frequency the beam also passes (on average) through the center of the quadrupoles.

More specifically, four different center frequencies can be measured by changing the strength of the horizontal or vertical sextupole families and by measuring the resulting change in the horizontal or vertical tune, respectively. In most cases, the four central frequencies so obtained are found to be the same, supporting the hypothesis that the magnets are usually well aligned on short length scales.

One can then monitor changes of the beam energy, using the relation

$$\frac{\Delta p}{p} = \left(\frac{1}{\gamma^{-2} - \alpha_c} \right) \frac{\Delta f_{\mathrm{rf}}}{f_{\mathrm{rf}}} . \tag{7.39}$$

This energy-monitoring technique was applied at BEPC [29] and LEP [30].

Exercises

7.1 Review of Fourier Transformations and an Application

a) Calculate the Fourier spectrum of the current signal for two bunches of equal charge and asymmetric spacing in a storage ring. Assume that the particle distribution of each bunch can be represented as a δ-function, and that the bunches have equal synchrotron oscillation amplitudes. That is, in the time domain, take the current distribution to be

$$\begin{aligned} i(t) = Q \sum_{n=-\infty}^{\infty} & \delta\left(t - nT - \tau_a \cos(\omega_s nT)\right) \\ & + \delta\left(t - nT - \frac{T}{2} - \Delta t - \tau_a \cos(\omega_s nT + \phi)\right) , \end{aligned} \tag{7.40}$$

where Q is the individual bunch charge, T is the revolution period, τ_a is the synchrotron oscillation amplitude, ω_s is the angular synchrotron frequency, Δt is the relative timing offset between the bunches, and ϕ is the relative phase between the two bunches.

Some useful properties of Fourier transformations are:

The Poisson sum rule

$$\frac{1}{2\pi} \sum_{n=-\infty}^{\infty} e^{-2\pi i n(\omega/\omega_r)} = \omega_r \sum_{n=-\infty}^{\infty} \delta(\omega - n\omega_r) , \tag{7.41}$$

A Property of Delta-functions

$$\int_{-\infty}^{\infty} f(t)\delta(t - x)\,\mathrm{d}t = f(x) , \text{ and} \tag{7.42}$$

The Bessel function sum rule

$$e^{iz\cos\alpha} = \sum_{k=-\infty}^{\infty} i^k J_k(z) e^{ik\alpha} . \tag{7.43}$$

b) Considering dipole mode oscillations only, sketch the frequency spectrum assuming $\Delta t = 0$ for "0-mode" oscillations (bunches oscillate in phase, that is $\phi = 0$) and for "π-mode" oscillations (for which the bunches oscillate out of phase; i.e., $\phi = \pi$). Show that these two normal modes of oscillation can be distinguished from one another by measuring the frequency spectrum.

c) (Optional) Suppose you wanted to build a passive cavity to damp the π-mode oscillations. (At high beam currents the 0-mode oscillations are Robinson damped.) What would be suitable choices[2] of resonant frequency for the passive cavity assuming (i) and equal bunch spacing ($\Delta t = 0$) or (ii) an unequal bunch spacing. Comment on how the optimum frequency depends on the bunch spacing.

7.2 Adjusting the Incoming Beam Energy

An off-energy beam will have orbit contributions $\Delta x = D_x\delta$ similar to that shown in Fig. 7.2. For a proton beam, the dispersive offset will change on successive turns according to the net bending field and the rf frequency. For electrons the beam will slowly be damped to the on-energy equilibrium orbit due to synchrotron radiation. In either case, describe a procedure using difference orbits at fixed rf frequency for correcting the energy of the incoming beam. Hint: consider a beam injected perfectly into the center of an rf bucket and determine, at which turn one is maximally sensitive to beam energy deviations.

7.3 Resonant Depolarization

Resonant depolarization at the IUCF cooler ring was initially observed at a driving frequency slightly different from expectation assuming the beam energy. Using well-known kinematic reactions with an internal target, the beam energy was precisely determined. Show that the apparent discrepancy can be explained by a small adjustment to the assumed orbit circumference.

7.4 Approximate Expression for the Momentum Compaction Factor

a) Using the approximate formula for the average dispersion function

$$\langle D_x \rangle \approx \frac{\langle \beta_x \rangle}{Q_x} , \tag{7.44}$$

and (7.22), show that a good approximation for the momentum compaction factor is given by $\alpha \approx 1/Q_x^2$. Give a numerical example.

b) Find an analogous expression for the transition energy γ_t using these approximations.

---

[2] we recognize F. Pederson for pointing out the important consequences of unequal bunch spacing in application at the SLC damping rings

7.5 Achieving Design Parameters in the Presence of Unknowns

Suppose upon initial commissioning of an electron storage ring, the beam energy E in the ring, the ring circumference C, and main dipole field B are known to only about $\pm(1-2)\%$. Describe a strategy for setup that ensures the desired beam energy, a dipole field strength matched to this energy, and that dictates the required rf frequency to center the beam in the quadrupoles. Assume that the energy of the injected beam can be determined with a good resolution. Which additional observables could one use to simplify this problem?

7.6 Chromatic Phase Advance[3]

Maintaining the second order driving terms in Hill's equations for the particle motion, we have

$$\begin{aligned} x'' + kx &= kx\delta - \frac{m}{2}(x^2 - y^2)\,, \\ y'' - ky &= -ky\delta + mxy\,, \end{aligned} \tag{7.45}$$

where m is the sextupole strength in units of m^{-3} introduced in Chap. 1 (behind (1.11)).

a) Assuming horizontal dispersion (D_x) only show that

$$\begin{aligned} x_\beta'' + kx_\beta &= (k - mD_x)x_\beta\delta\,, \\ y_\beta'' - ky_\beta &= -(k - mD_x)y_\beta\delta\,, \end{aligned} \tag{7.46}$$

where the higher-order, so-called geometric aberrations, have been set to zero.

b) Noting that the perturbation in betatron tune may be expressed as

$$\Delta Q = -\frac{1}{4\pi}\int \beta(s)\Delta k(s)\,\mathrm{d}s\,, \tag{7.47}$$

where $\Delta k(s)$ represents the focussing error in units of m^{-2} and $\beta(s)$ the beta function at the location of the error, show that the chromatic phase shifts $\Delta\phi_{x,y}$ over one revolution period for a relative momentum offset δ are given by

$$\begin{aligned} \Delta\phi_x &= -\frac{\delta}{2}\int_s^{s+C} \beta_x(k - mD_x)\,\mathrm{d}s\,, \\ \Delta\phi_y &= \frac{\delta}{2}\int_s^{s+C} \beta_y(k - mD_x)\,\mathrm{d}s\,, \end{aligned} \tag{7.48}$$

where C is the ring circumference.

[3] from text in [31]

c) Using the definition of tune and chromaticity, show that

$$
\begin{aligned}
Q'_x &= -\frac{1}{4\pi}\int \beta_x (k - mD_x)\,\mathrm{d}s\,,\\
Q'_y &= \frac{1}{4\pi}\int \beta_y (k - mD_x)\,\mathrm{d}s\,.
\end{aligned}
\tag{7.49}
$$

The first term on the right-hand side of the last two equations represents the natural chromaticity while the second term shows the additive contributions arising from the sextupoles.

Taking into account the second order expansion of the dispersive trajectories (see (1.6)):

$$
\begin{pmatrix} x \\ x' \\ y \\ y' \end{pmatrix} = \delta \begin{pmatrix} D_1 \\ D_2 \\ D_3 \\ D_4 \end{pmatrix} .
\tag{7.50}
$$

the chromatic phase shifts for an arbitrarily coupled lattice may be derived. As shown in [32] the chromatic phase advances are given by

$$
\begin{aligned}
\Delta\phi_x &= -\frac{\delta}{2}\int_s^{s+C} \left[\beta_x ({K_x}^2 + k - \lambda D_1) + 2\alpha_x K_x D_2 - \gamma_x (K_x D_1 + K_y D_3)\right] \mathrm{d}s\,,\\
\Delta\phi_y &= -\frac{\delta}{2}\int_s^{s+C} \left[\beta_y ({K_y}^2 + k - \lambda D_1) + 2\alpha_y K_y D_4 - \gamma_y (K_x D_1 + K_y D_3)\right] \mathrm{d}s\,,
\end{aligned}
$$

where λ is an eigenvalue of the one-turn transfer matrix and $\gamma_{x,y}$ is the usual Twiss parameter defined in the plane (x or y) of interest.

8 Longitudinal Phase Space Manipulation

In this chapter we describe various techniques used to control the longitudinal properties of particle beams. We concentrate on the manipulation of the second moments of the longitudinal distribution; that is, on the bunch length and energy spread. As will be shown, the bunch length can be varied using accelerating cavities to compress, coalesce, split, and lengthen stored bunches. The energy spread of the beam can also be adjusted (usually to be a minimum) by proper phasing of the rf, by invoking cancellations between the applied and beam-induced rf, and by more sophisticated techniques for the case of long bunch trains. A practical application of the use of rf systems to affect the beam's transverse emittance is presented lastly.

8.1 Bunch Length Compression

Bunch length compression using dedicated accelerating structures and beamlines is common to all linear collider designs [1, 2, 3] and FEL linac drivers [4, 5]. Compression is usually performed in two or more steps. First an rf section (for example an accelerating structure) is used to introduce a correlation between the particle energy and position within a bunch. In the second step the beam passes through a transport line with nonzero dispersion (i.e., bends) where the actual compression occurs due to the energy dependence of the particle trajectory.

If the bunch is made to arrive near a zero crossing of the rf wave with the voltage decreasing from positive to negative values, the longitudinal phase space through the compressor evolves then as follows. Let (z_1, δ_1), (z_2, δ_2), and (z_3, δ_3) denote the longitudinal position z and relative energy δ at the entrance of the compressor, downstream of the compressor rf structure, and at the end of the compressor, respectively. A particle within the bunch is transported through the compressor cavity as

$$\begin{aligned} z_2 &= z_1 \\ \delta_2 &= \delta_1 - \frac{eV_{\rm rf}}{E} \sin\phi \ , \end{aligned} \tag{8.1}$$

where e is the particle charge, $V_{\rm rf}$ is the compressor voltage, E is the beam energy, and ϕ is the relative phase of the particle with respect to the zero cross-

M. G. Minty et al., *Measurement and Control of Charged Particle Beams*,
https://doi.org/10.1007/978-3-662-08581-3_8

ing of the compressor voltage. A particle arriving earlier ($\phi < 0$), will acquire a higher energy ($\delta_2 > \delta_1$). The phase ϕ can also be written as $\phi \equiv \phi_1 - \phi_c$, where ϕ_1 is the phase of the particle and ϕ_c the phase of the compressor rf voltage, both with respect to a common reference.

For simplicity, in this section we consider an ultrarelativistic particle, travelling at the speed of light, and thus, we do not distinguish between relative momentum deviation and relative energy deviation. Under this assumption, after passing through the dispersive downstream arc, the longitudinal position of a particle is

$$\begin{aligned} z_3 &= z_2 + R_{56}\delta_2 \\ \delta_3 &= \delta_2 \,. \end{aligned} \tag{8.2}$$

A particle, which after the compressor cavity has a higher energy than nominal ($\delta_2 > 0$), will be slowed down, i.e., $z_3 < z_2$, provided R_{56} is less than zero, as is the case for a regular arc. For a non-relativistic particle, (8.2) must be modified to include an additional factor $\beta = v/c$ (velocity divided by the speed of light) and a term representing the effect of the change in velocity.

Combining (8.1) and (8.2), the particle's longitudinal position and energy at the exit of the compressor are given in terms of its initial values by

$$\begin{aligned} z_3 &= z_1 + R_{56}\left(\delta_1 - \frac{eV_{\mathrm{rf}}}{E}\sin\phi\right) \\ \delta_3 &= \delta_1 - \frac{eV_{\mathrm{rf}}}{E}\sin\phi \,. \end{aligned} \tag{8.3}$$

The phase ϕ is related to the initial longitudinal position via $\phi = -\omega_{\mathrm{rf}} z_1/c$ (a positive z indicates a position ahead of the bunch center). If the phases are small compared with π, we may linearly expand the sine function, and (8.3) can be approximated by

$$\begin{aligned} z_3 &\approx \left(1 + R_{56}\frac{e\omega_{\mathrm{rf}}V_{\mathrm{rf}}}{cE}\right) z_1 + R_{56}\delta_1 \\ \delta_3 &\approx \delta_1 - \frac{eV_{\mathrm{rf}}\omega_{\mathrm{rf}}}{cE} z_1 \,. \end{aligned} \tag{8.4}$$

Then the final bunch length is

$$\sigma_{z,f} = \langle {z_3}^2 \rangle^{1/2} \approx \sqrt{\left(1 + R_{56}\frac{eV_{\mathrm{rf}}\omega_{\mathrm{rf}}}{Ec}\right)^2 \sigma_{z,0}^2 + R_{56}^2\sigma_{\delta,0}^2}\,, \tag{8.5}$$

where σ_{z0} is the initial bunch length and $\sigma_{\delta,0}$ is the initial beam energy spread. In the approximation it has been assumed that the incoming beam distribution has no energy-position correlation ($\langle \delta_1 z_1 \rangle = 0$). If the rf voltage is adjusted to give

$$-R_{56}\frac{eV_{\mathrm{rf}}\omega_{\mathrm{rf}}}{Ec} = 1\,, \tag{8.6}$$

then the final bunch length is minimum, equal to $R_{56}\sigma_{\delta,0}$ and is independent of the initial bunch length, σ_{z0}. This is called the condition for full compression. If the voltage is smaller, one operates with undercompression, and for higher rf voltage one has overcompression.

According to (8.5), a larger compressor voltage and a smaller value of R_{56} may provide shorter bunches. However, as can easily be inferred from (8.3), for large rf voltages the final energy spread increases roughly in proportion to the rf voltage. Given a limited momentum acceptance in the downstream system, a compromise has to be made, which introduces a lower limit on R_{56}.

Phase errors and phase fluctuations may be critical in a compressor, particularly if the compression takes place upstream of a linear accelerator with strict tolerances on the injection phase[1]. For the (single-stage) compressor scheme described above with $\phi = -\omega z/c$, the resulting beam phase ϕ_3 in terms of the initial beam phase ϕ_1 is

$$\begin{aligned}\phi_3 &= \phi_1 - R_{56}\frac{\omega}{c}\left[\delta_1 - \frac{eV_{\mathrm{rf}}}{E}\sin(\phi_1 - \phi_c)\right] \\ &\approx -R_{56}\frac{\omega}{c}\delta_1 + \left[1 + R_{56}\frac{\omega_{\mathrm{rf}}}{c}\frac{eV}{E}\right]\phi_1 - R_{56}\frac{\omega_{\mathrm{rf}}}{c}\frac{eV_{\mathrm{rf}}}{E}\phi_c\,, \end{aligned} \tag{8.7}$$

where ω_{rf} is the angular accelerating frequency of the structure. Assuming that the errors in the injected beam phase $\Delta\phi_i$ and the compressor phase $\Delta\phi_c$ are independent, and that the initial momentum deviations are independent of these phases ($\mathrm{d}\delta_1/\mathrm{d}\phi_1 = \mathrm{d}\delta_1/\mathrm{d}\phi_c = 0$) we find

$$\frac{\mathrm{d}\phi_3}{\mathrm{d}\phi_c} = \eta \quad \text{and} \quad \frac{\mathrm{d}\phi_3}{\mathrm{d}\phi_1} = 1 + \eta\ , \tag{8.8}$$

where we have defined $\eta \equiv (-R_{56}\frac{\omega_{\mathrm{rf}}}{c}\frac{eV}{E})$. Combining the two contributions of (8.7) in quadrature gives for the final phase error

$$(\Delta\phi_3)^2 = \eta^2(\Delta\phi_c)^2 + (1+\eta)^2(\Delta\phi_1)^2\ . \tag{8.9}$$

In particular, for $\eta = -1$ (full compression) the final phase is independent of the initial phase error $\Delta\phi_1$.

An example of a bunch compressor designed for the Next Linear Collider [6] is shown in Fig. 8.1. This design comprises a two-fold compression scheme. The principle of the first bunch compressor (BC1) is as described above, but with an energy-dependent path length generated by a wiggler magnet with $R_{56} < 0$. At higher energy, the second bunch compression (BC2) is performed, which consists of an arc, a second rf section, and a magnetic chicane. In BC2 a net 360° phase space rotation is used to minimize the sensitivity to incoming energy errors, which might arise either from phase errors at the entrance to BC1 of from beam loading in the linac located between BC1 and BC2.

[1] at the SLC, the phase stability at injection into the main linac required for imperceptible influence on the luminosity was $< 0.1°$, or < 1 ps, at the linac frequency of 2856 MHz

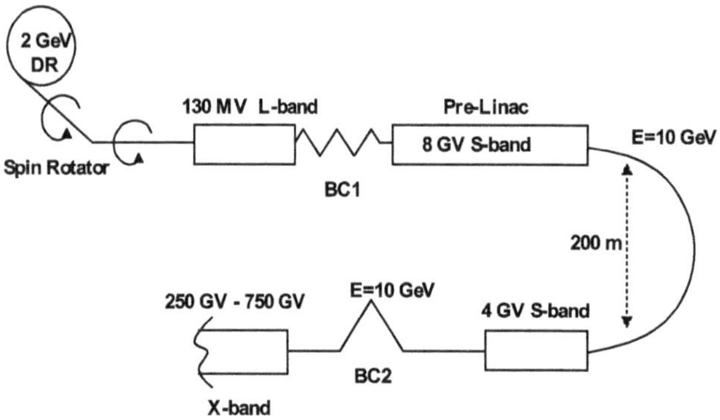

Fig. 8.1. Schematic of an NLC bunch compressor [6]

8.2 Bunch Length Precompression

Bunch precompression by means of rf systems is frequently used for reducing the bunch length (at the expense of an increased energy spread) for transferral of particle beams from one accelerator to a downstream accelerator operating at higher rf frequency. Bunch precompression has also been used to alleviate the consequences of current-dependent bunch lengthening (i.e., beam loss) in lepton accelerators.

For a relativistic beam ($\alpha_c - 1/\gamma^2 \approx \alpha_c$), the equation of motion for the bunch length, $\sigma_z = \sigma_\phi(c/w)$, is obtained as follows. Using a dot to denote a derivative with resepct to time, i.e.,

$$\dot{()} \equiv \frac{\mathrm{d}}{\mathrm{d}t}() \quad \text{and} \quad \ddot{()} \equiv \frac{\mathrm{d}^2}{\mathrm{d}t^2}() \,, \tag{8.10}$$

and the equalities

$$\begin{aligned} \langle \dot{\phi^2} \rangle &= 2\langle \phi\dot{\phi} \rangle \,, \\ \langle \dot{\phi\delta} \rangle &= \langle \phi\dot{\delta} + \dot{\phi}\delta \rangle \,, \\ \langle \dot{\delta^2} \rangle &= 2\langle \delta\dot{\delta} \rangle \,, \end{aligned} \tag{8.11}$$

one finds

$$\begin{aligned} \langle \ddot{\phi^2} \rangle &= \frac{\mathrm{d}}{\mathrm{d}t} 2\langle \phi\dot{\phi} \rangle = 2\alpha\omega \langle \dot{\phi\delta} \rangle \text{ using (7.1)} \\ &= -\frac{2\alpha e\dot{V}}{ET} \langle \phi^2 \rangle + 2(\alpha\omega)^2 \left[\frac{1}{\langle \phi^2 \rangle} \left(\epsilon^2 + \frac{\langle \dot{\phi^2} \rangle^2}{(2\alpha\omega)^2} \right) \right] \\ &= -\frac{2\alpha e\dot{V}}{ET} \langle \phi^2 \rangle + \frac{2(\alpha\omega)^2}{\langle \phi^2 \rangle} \epsilon^2 + \frac{1}{2} \frac{\langle \dot{\phi^2} \rangle^2}{\langle \phi^2 \rangle} \,, \end{aligned} \tag{8.12}$$

where we have used the definition of the longitudinal emittance ϵ

$$\epsilon^2 = \langle \phi^2 \rangle \langle \delta^2 \rangle - \langle \phi\delta \rangle^2 \,. \tag{8.13}$$

Using

$$\frac{\mathrm{d}^2}{\mathrm{d}t^2}\sigma_\phi^2 = 2(\sigma_\phi\ddot{\sigma}_\phi + \dot{\sigma}_\phi^2) \,, \tag{8.14}$$

the equation of motion for the bunch length (in phase units) is

$$\ddot{\sigma_\phi} = -\frac{\alpha_c e\dot{V}}{ET}\sigma_\phi + \frac{(\alpha_c\omega_{\mathrm{rf}})^2\epsilon^2}{{\sigma_\phi}^3} \,, \tag{8.15}$$

where $\dot{V}$ was given in (7.3). A similar analysis for the rms energy spread is likewise calculable. The results are summarized [7] as

$$\begin{aligned} \ddot{\sigma_\phi} - {\omega_s}^2\sigma_\phi &= (\alpha\omega_s)^2\frac{\epsilon^2}{{\sigma_\phi}^3} \,, \\ \ddot{\sigma_\delta} - {\omega_s}^2\sigma_\delta &= \frac{(e\dot{V})^2}{E\omega T}\frac{\epsilon^2}{{\sigma_\phi}^3} \,. \end{aligned} \tag{8.16}$$

Bunch rotations are used for better capture efficiency of the proton beam at HERA [8, 9]. There two schemes were tried to shorten the bunch at extraction from the upstream PETRA ring. Initially, the bunch rotation was made by introducing a 180° phase jump in the accelerating rf which places the beam distribution next to an unstable fixed point. The ensuing slow motion of particles along the separatrices translates into a mismatch. The phase was then restored to its original setting and the beam was extracted about a quarter of a synchrotron period later. Beam loading effects however caused bunch shape distortions during the phase jump [8].

Presently bunch precompression at HERA is achieved by amplitude modulating the rf system to induce a quadrupole mode oscillation. A similar scheme was used at the SLC damping ring primarily for reducing transmission losses in the downstream transport line [7]. Here bunch precompression was implemented in a two step process (see Ex. 8.2). The first step change in the requested cavity voltage resulted in a longitudinal phase space mismatch which elongated the bunch. The resulting beam phase oscillation was then eliminated while amplifying the bunch length oscillation by application of a second, appropriately timed, step change to the cavity voltage.

Figure 8.2 shows diagnostic measurements illustrating bunch precompression from the SLC. Plotted are the cavity voltage (measured using a diode detector), the bunch length (obtained from a peak current measurement, which is inversely proportional to the bunch length, using a single stripline of a position monitor), and the mean energy of the beam. The centroid energy was measured using a horizontal beam position monitor in a region of high dispersion in the damping ring. The peak-current signal at extraction was increased, corresponding to a shorter bunch.

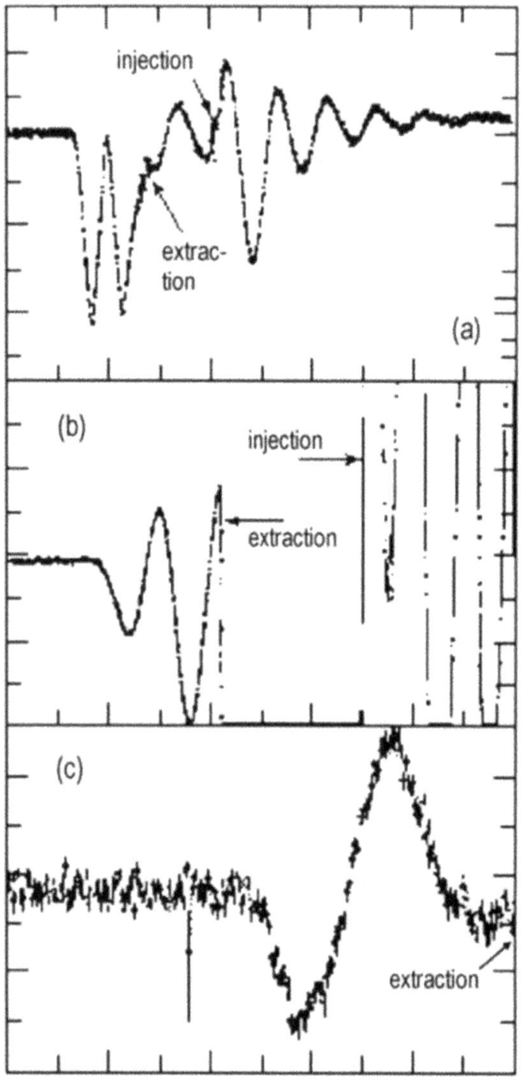

Fig. 8.2. Bunch precompression in the SLC damping rings: (**a**) measured cavity voltage [50 kV/dvsn, 10 μs/dvsn], (**b**) peak current [10%/dvsn, 5 μs/dvsn], and (**c**) centroid energy [50 μm or 0.77%/dvsn, 20 turns or 2.34 μs/dvsn]

8.3 Bunch Coalescing

Bunch coalescing, used primarily in hadron accelerators, consists of combining multiple bunches into a single bunch in order to achieve a high peak intensity. At Fermilab two types of coalescing are used [10]: standard coalescing for the low intensity antiprotons and 'snap' coalescing for high intensity proton beams.

Experimental data from the Fermilab Main Ring are shown in Fig. 8.3 which demonstrate the bunch coalescing concept. The different traces correspond to different times. Initially there were 11 bunches captured in 53 MHz rf buckets. The vector sum of the rf voltages was then adiabatically reduced, or 'paraphased' by shifting the relative phases between the accelerating cavities, which lengthens the bunch while preserving the longitudinal beam emittance.

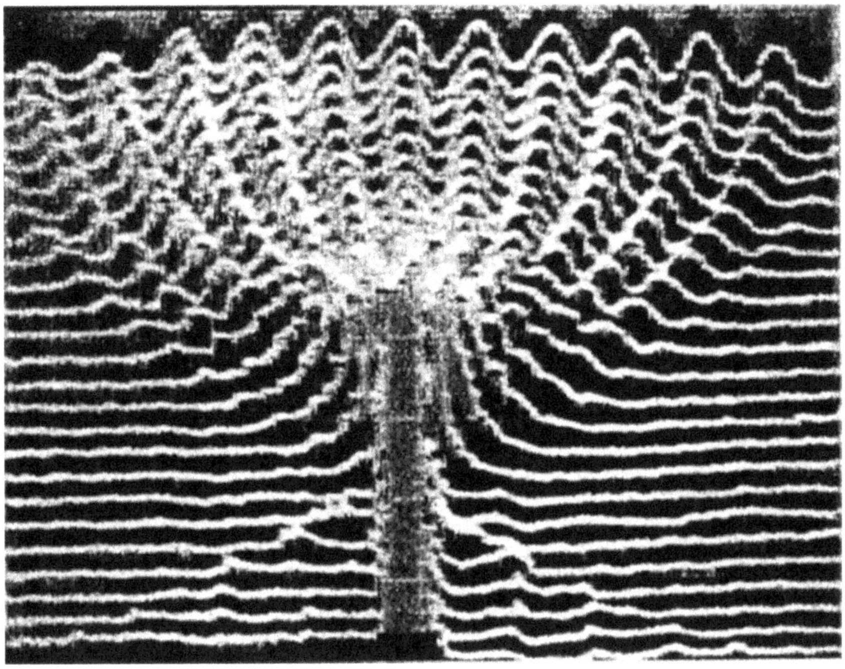

Fig. 8.3. Experimental data from the Fermilab main ring showing multiple bunches being coalesced into a single bunch. Successive traces are spaced by 6.8 ms intervals (Courtesy P. Martin, 1999)

The bunches were next subjected to a higher voltage 2.5 MHz rf system. The bunches rotate with the synchrotron frequency in this low frequency rf potential. In practice [11], a 5 MHz rf is also applied to help linearize the rotation. When the bunches were vertically aligned in the 2.5 MHz rf bucket, they are finally captured in a single 53 MHz rf bucket.

The *snap coalescing* scheme replaces the adiabatic voltage reduction from above with a phase space rotation. Here the coalescing procedure is initiated with a fast reduction of the primary rf amplitude. The beam is then longitudinally mismatched and shears in longitudinal phase space. After one quarter synchrotron oscillation the low frequency rf systems are turned on and the bunches are recaptured back into the primary rf bucket. Simulations [10] have shown that the capture efficiency of snap coalescing is about 10% less than with the adiabatic coalescing. However, at high currents, beam instabilities have been observed during adiabatic paraphasing of the 53 MHz rf systems. This is avoided with snap coalescing..

Two practical issues associated with bunch coalescing are the increased longitudinal emittance (minimized by adjusting the amplitude of the rf during initial bunch lengthening) and the production of satellite bunches (i.e., particles in neighboring rf buckets) which can arise during the recapturing process due to rf nonlinearities. The satellite bunches, which can cause detector backgrounds, may be eliminated using a longitudinal damper to discard the offending bunches [12]. Recently new 2.5 MHz cavities have been installed in the main ring allowing for a threefold increase in the vector voltage at this lower frequency. The coalescing efficiencies were increased by 14% for the antiproton beam and 10% for the proton beam and the satellite bunches were eliminated using the new cavities [13].

8.4 Bunch Splitting

The LHC beam may consist of bunch trains of 72 bunches each containing about 10^{11} protons with a 25 ns bunch-to-bunch spacing. The original concept for producing these bunch trains in the proton synchrotron (PS) was to debunch 6 or 7 highly intense bunches and then to recapture the beam using a higher frequency rf system. A longitudinal microwave instability arose during the debunching and rebunching process [14] however and resulted in non-uniform beam distributions.

An alternate solution was adapted consisting of splitting of bunches using additional harmonic cavities in the injection chain for the Large Hadron Collider (LHC) [15]. To produce as many bunches as possible, the bunches are split in the upstream PS accelerator in a 2 step process. First, at low proton momentum (3.57 GeV/c) each of the 6 bunches from the booster ring is split into 3 as shown in the simulation results of Fig. 8.4. Then, after ramping to high momentum (26 GeV/c) the bunches are further split into 4, as shown in Fig. 8.5. On the left of these figures is shown the relative amplitude of each of the different harmonic rf systems as a function of time during the bunch splitting process.

Such bunch splitting has first been experimentally demonstrated [17] in the PS booster at CERN in application to neutrino experiments. Shown in Fig. 8.6 is the measured evolution of the longitudinal distribution using tomographic measurement techniques [18, 19] from the CERN PS [15, 19]. Plotted

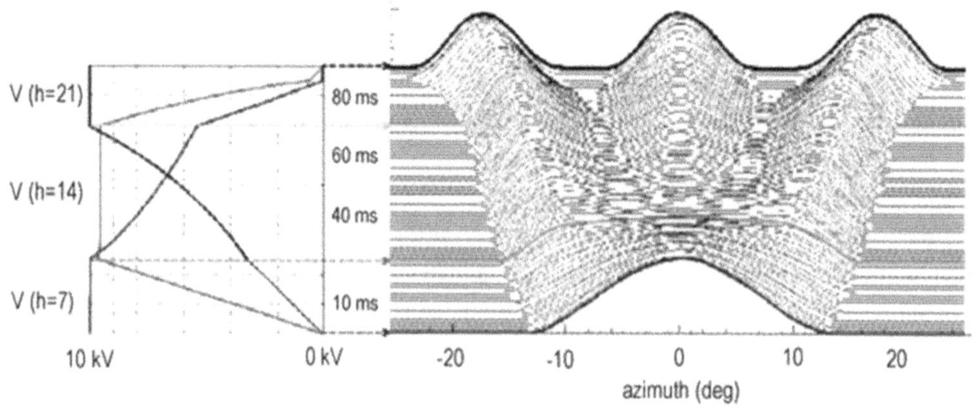

Fig. 8.4. Simulation of bunch triple splitting in the CERN PS at low energy (3.57 GeV/c) in preparation for injection into the LHC. Compare with the measurements shown in Fig. 8.7 (Courtesy R. Garoby, 1999)

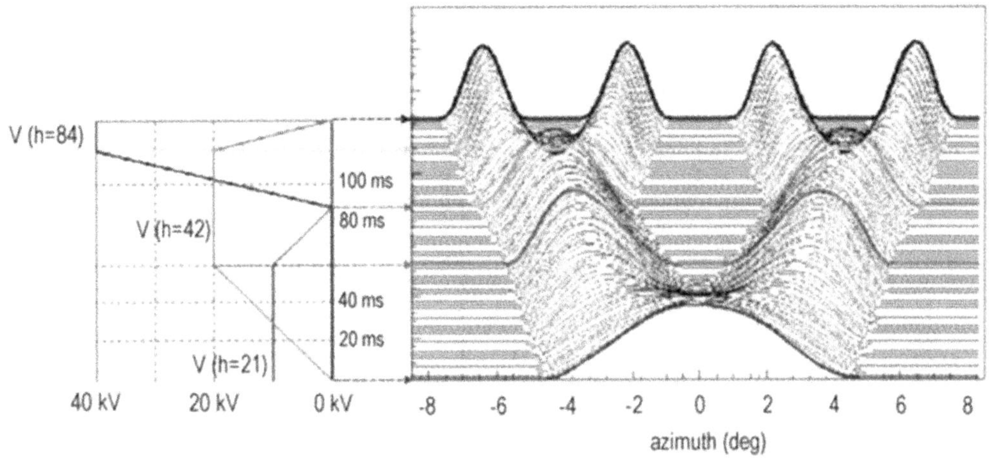

Fig. 8.5. Simulation of further bunch splitting in the CERN PS at high energy (26 GeV/c) in preparation for injection into the LHC. Compare with the measurements shown in Fig. 8.8 (Courtesy R. Garoby, 1999)

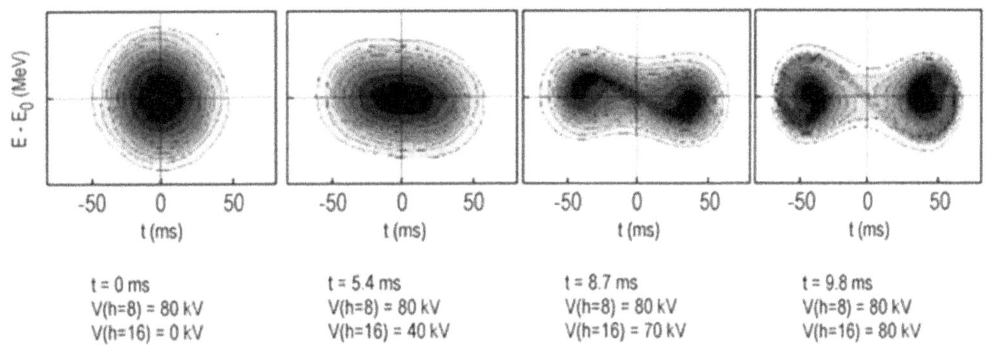

Fig. 8.6. Tomographically reconstructed phase space showing bunch splitting in the CERN PS booster ring after acceleration with 3×10^{12} protons (Courtesy R. Garoby, 1999)

is the phase space at the indicated times, and the corresponding amplitudes of the primary and harmonic cavity voltages are indicated below each picture.

After a series of hardware upgrades the feasibility of the bunch filling scheme required for the LHC has been demonstrated [14] in the CERN PS ring. First the triple-splitting scheme was verified experimentally [20], and then the newly installed harmonic cavities operating at 40 MHz ($h = 84$) and 80 MHz ($h = 168$) were used to accommodate also bunch quadruple-splitting and bunch rotation [16]–[21]. In the triple-splitting scheme, the 10 ferrite-loaded cavities of the PS were tuned before the splitting process as follows: 4 inactive, 2 at $h = 7$, 2 at $h = 14$, and 2 at $h = 21$. After splitting, as shown in Fig. 8.7, all cavities were tuned on $h = 21$ to allow full rf voltage for further acceleration.

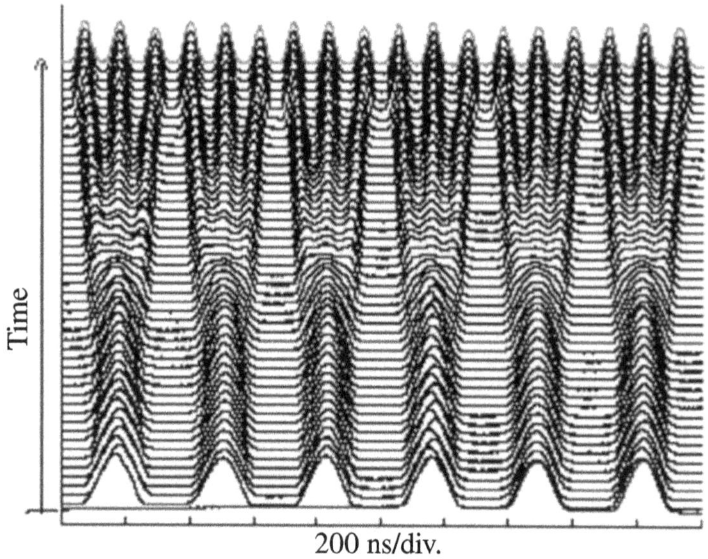

Fig. 8.7. Measured bunch triple-splitting at the CERN PS (E = 1.4 GeV, I_p = 8×10^{12}, 400 revolutions per trace). Compare with Fig. 8.4 (Courtesy R. Garoby, 2001)

The quadruple splitting was achieved in two steps as shown in the measured results of Fig. 8.8. The bunches were then rotated in longitudinal phase space using the 40 and 80 MHz cavities to match into the acceptance of the downstream Super Protron Synchrotron (SPS) accelerator. This bunch splitting technique is used in routine operation of the CERN PS for the pro-

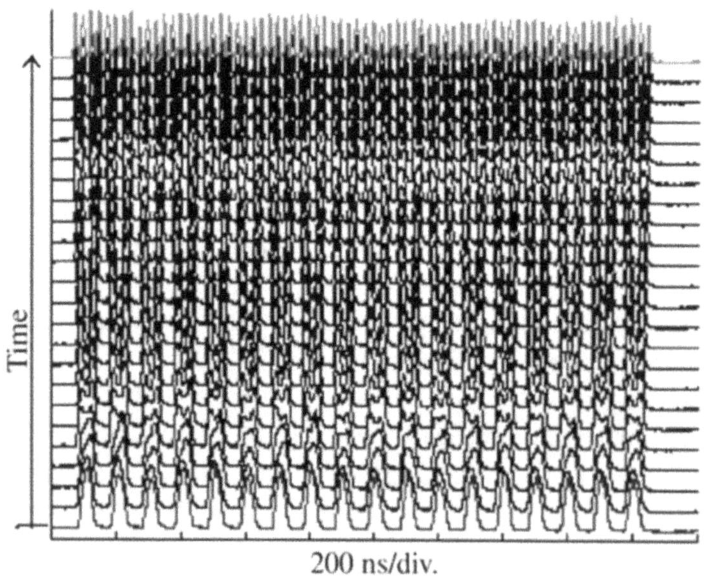

Fig. 8.8. Subsequent quadruple splitting at the CERN PS (E = 25 GeV, I_p = 8×10^{12}, 1250 revolutions per trace). Compare with Fig. 8.5 (Courtesy R. Garoby, 2001)

duction of an LHC type beam which is sent to the SPS, where it is further accelerated up to the LHC injection energy of 450 GeV.

At the IUCF Cooler Ring, using a different technique proton bunches were observed to split by application of either a phase [22] or an amplitude [23, 24] modulation of the rf signal driving the accelerating cavities. An example is shown in Fig. 8.9. In this case, longitudinal modulation resulted from application of a sinusoidal field variation close to the synchrotron frequency to a transverse dipole located in a region of high dispersion. Since the time these

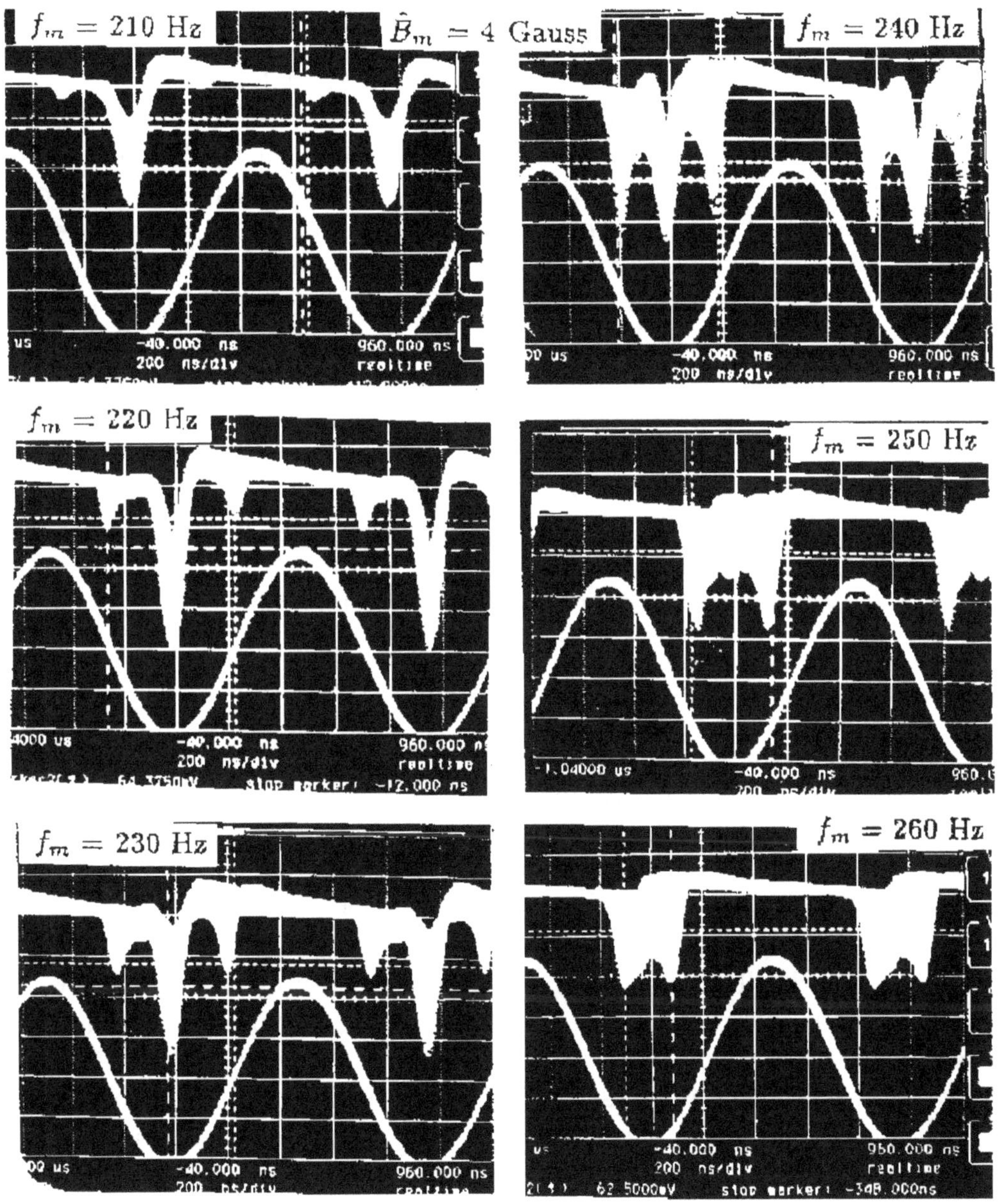

Fig. 8.9. Longitudinal beam profile observed using an oscilloscope (*top*) and the rf wave form (*bottom*) during rf modulation of proton bunches in the IUCF Cooler Ring (Courtesy S.Y. Lee, 1999)

data were taken, further experiments at the Advanced Light Source (ALS) with electron beams have included the use of streak camera for a more direct measurement of the evolution of the bunch length [25].

8.5 Harmonic Cavities

The use of additional rf cavities for longitudinal phase space manipulation has a variety of applications – we have already seen two such examples for bunch coalescing and splitting. Applications of harmonic cavities to reduce emittance growth when crossing transition (see Sect. 7.1) has been demonstrated at the Fermilab Main Ring [26, 27] and at BNL [28]. In this section we describe the use of harmonic cavities for lengthening the bunch. This approach has been adopted at various synchrotron light sources [29, 30] in order to either avoid intrabeam scattering effects and so increase the beam lifetime, or to increase the bunch length to avoid longitudinal beam instabilities, for example, at DAPHNE [31, 32] and as once proposed at LEP [33, 34].

Electron beam lifetimes at energies of the order 1 GeV or below may be dominated by large-angle intrabeam scattering otherwise known as Touschek [35] scattering. This process refers to single collisions between particles inside the same bunch, in which sufficient energy is transferred from the transverse to the longitudinal direction, such that particles are scattered outside of the ring energy acceptance. The energy acceptance is either equal to the height of the rf bucket, or given by a physical aperture at a position with large dispersion, whichever value is smaller.

In a somewhat simplified picture, the beam lifetime τ is approximately given by

$$\frac{1}{\tau} = \frac{\sigma c}{N} \int \rho^2 \mathrm{d}V \,, \tag{8.17}$$

where σ is the cross section for scattering beyond the energy acceptance, ρ is the bunch particle density, N the total number of particles in the bunch, and V represents the volume. For a constant beam energy and charge, the lifetime can be increased by increasing the bunch length and thus reducing the volume density ρ (nominally not coupled to the transverse emittances so that the transverse brightness remains unchanged). We note that a more accurate expression of the Touschek beam lifetime can be obtained by integrating the differential cross section over the momentum distribution of the beam, taking into account the correlations between position and momentum and between the different degrees of freedom as well as the variations of the optical functions around the ring, and always weighting with an appropriate convolution of two beam distribution functions [36, 37].

The increase of the bunch length using higher harmonic rf systems can be easily understood from Fig. 8.10. Here a third harmonic cavity is added to the primary rf such that the vector voltage seen by the beam is constant over the length of the bunch.

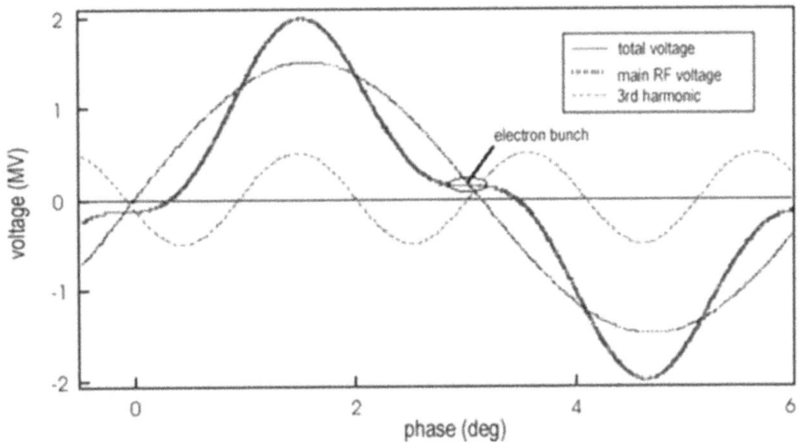

Fig. 8.10. Conceptual illustration of bunch lengthening with a higher harmonic cavity (Courtesy J. Byrd, 1999)

For electrons with nonzero synchronous phase, the total voltage is given by

$$V(t) = \hat{V}\left[\sin(\omega_{\mathrm{rf}}t + \phi_1) + k\sin\left(n(\omega_{\mathrm{rf}}t + \phi_h)\right)\right] , \tag{8.18}$$

where ω_{rf} and $\hat{V}$ are the angular rf frequency and voltage of the fundamental rf, n is the ratio of higher harmonic to main accelerating cavity frequencies, k is the desired net voltage ratio for the two rf amplitudes, ϕ_1 is the phase of the primary rf system measured with respect to the zero crossing, and ϕ_n the harmonic phase. We assume that the bunch center arrives at time $t = 0$. Note that in (7.5) we defined the synchronous phase ϕ_s with respect to the crest of the rf, so that $\phi_s = \phi_1 - \pi/2$.

Then, we must have $eV(0) = U$, where U is the energy loss per turn (taken here to be dominated by synchrotron radiation). For optimum bunch lengthening both the slope and the curvature of the net voltage at the position of the bunch are zero, i.e.,

$$\left.\frac{\mathrm{d}V}{\mathrm{d}t}\right|_{t=0} = 0 \quad \text{and} \quad \left.\frac{\mathrm{d}^2V}{\mathrm{d}t^2}\right|_{t=0} = 0\,. \tag{8.19}$$

The potential seen by the beam with and without a third harmonic rf system is shown in Fig. 8.11.

Equations (8.18) and (8.19) determine the optimum amplitude and phase of the harmonic cavity (see also [30]):

$$\begin{aligned} k^2 &= \frac{1}{n^2} - \frac{(U/(e\hat{V}))^2 n^2}{(n^2-1)^2}\,, \\ \tan\phi_h &= \frac{nU/(e\hat{V})}{\sqrt{(n^2-1)^2 - (n^2U/(e\hat{V}))^2}}\,, \\ \sin\phi_1 &= \frac{n^2}{n^2-1}\frac{U}{e\hat{V}}\,. \end{aligned} \tag{8.20}$$

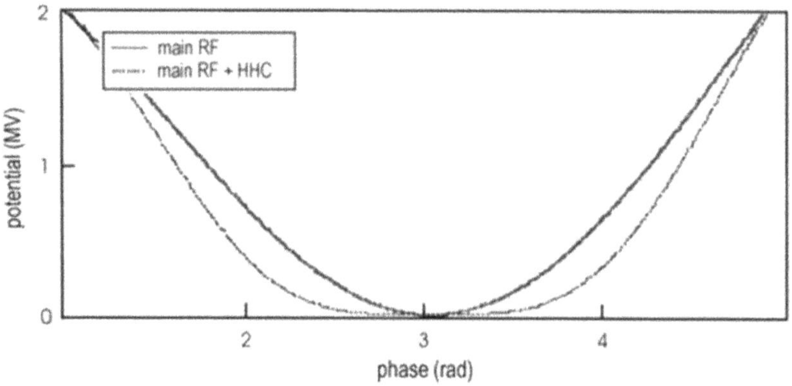

Fig. 8.11. Potential seen by circulating bunch with and without a harmonic cavity for the parameters of the ALS (Courtesy J. Byrd, 1999)

Simulations for the expected longitudinal density distribution are shown for the case of the ALS in Fig. 8.12.

The third harmonic RF system at the APS [29] consists of five single-cell copper 1.5 GHz cavities which are driven passively by the beam. With the resulting decreased peak charge density, an increase in beam lifetime exceeding a factor of two has resulted [38].

The beam current distribution, or fill pattern, has been shown to strongly affect the obtainable beam lifetimes [38] due to transient beam loading effects. Transient beam loading effects arise, because the beam is not a continuous current, but consists of bunches, and often these bunches are not uniformly distributed around the rings. A frequent beam pattern is one or several trains of equally spaced bunches, separated by 'gaps' without any bunches. Such gaps may be introduced, for example, to remove ions or photo-electrons, which otherwise are attracted and trapped by the beam electric field.

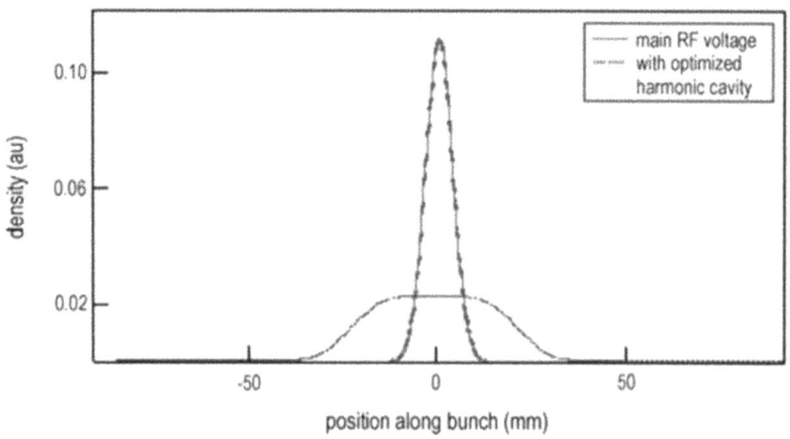

Fig. 8.12. Bunch length with and without a third harmonic cavity at the ALS (Courtesy J. Byrd, 1999)

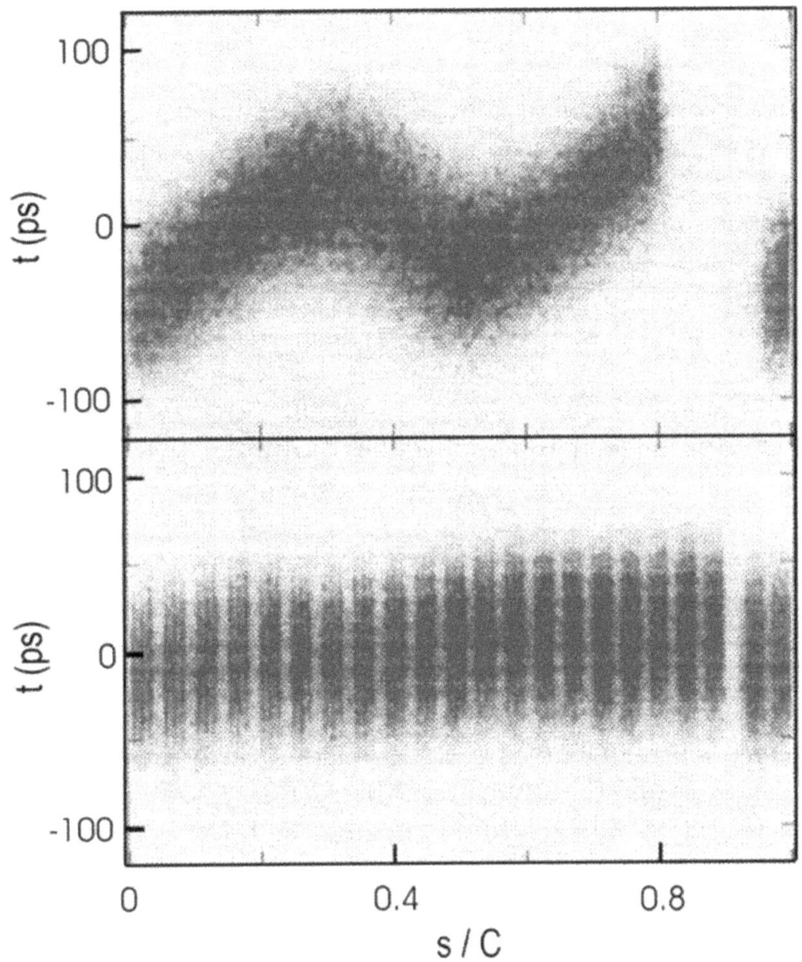

Fig. 8.13. Beam phase modulations measured at the ALS with a large (17% gap, *top*) and a small (2.4% gap, *bottom*) in the beam fill pattern (Courtesy J. Byrd, 2001)

Streak camera measurements of the longitudinal distribution versus bunch number are shown in Fig. 8.13 for the case of a large gap (17%) and for a small gap (2.4%) in the fill pattern. To date, simulation results of the variation in synchronous phase show good qualitative agreement with the measurements, yet do not explain the measured increases in the bunch length and beam lifetime [38].

The implications of phase variations along the fill pattern are two-fold. Most importantly, the gain in bunch-by-bunch feedback loops is reduced by such a phase variation. On the other hand, the increased interbunch or intrabunch phase spread contributes to increased Landau damping.

8.6 Energy Spread

As compared with the other 5 dimensions of a beam's phase space, the second moment corresponding to the beam energy spread is perhaps the most difficult to measure and control. In circular accelerators, lepton beams are naturally radiation damped to the limit of quantum fluctuations. Hadron beams on the other hand experience emittance dilutions particularly if subjected to internal targetry. For this reason various cooling mechanisms (see chapter 11) have been devised to combat large energy spreads.

Experience at both lepton and proton accelerators with high beam currents has shown that as the currents are increased, single-bunch instabilities due to longitudinal wake fields, dominated by the so-called microwave instability, can lead to an increase in the beam energy spread. Measurements at the SLC damping ring made with a wire scanner in a region of high dispersion in the extraction line are shown in Fig. 8.14. These measurements demonstrated a dramatic increase in the bunch energy spread beyond currents of about 1.5×10^{10} particles per bunch [39]. While relatively unimportant provided that the distribution remains stable from pulse-to-pulse, observations have shown that the increased energy spread is often associated with random turbulent bunch lengthening. Detailed analyses of this yet not fully understood phenomenon are beyond the scope of this book. Rather we will focus on methods for preserving, controlling, and minimizing the beam energy spread in linear accelerators and transport lines, assuming a constant incoming energy spread. We have already seen one example whereby the energy spread of a beam is increased as the bunch length is decreased using bunch precompression and bunch compression.

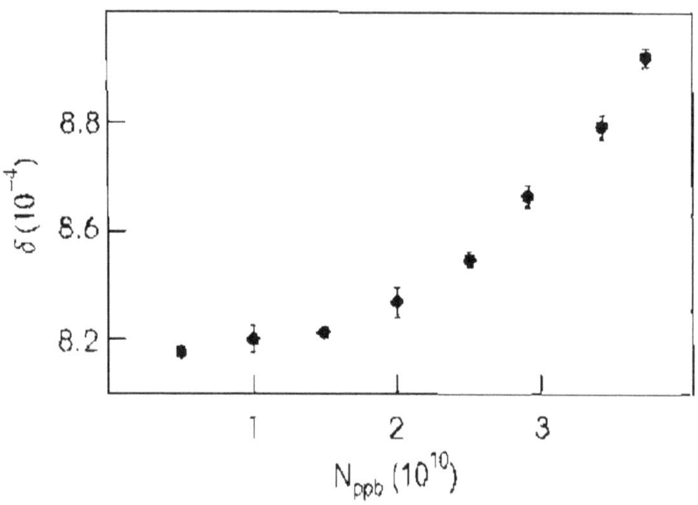

Fig. 8.14. Measured energy spread of beams exiting the SLAC electron damping ring as a function of beam current. The total cavity voltage in this measurement was 945 kV

In a linear accelerator, the energy of the particles with longitudinal density distribution $\rho(z)$ is a sum of the injected beam energy E_0, the energy gained in acceleration from each klystron ΔE_i, and the losses from longitudinal wake fields $W_{\parallel}$ [40, 41]:

$$E(z) = E_0 + \sum_{i=1}^{N_{\text{klys}}} \left[\Delta E_i \cos(\phi_i + \phi(z)) + \Delta s_i N_b e \int_z^{\infty} W^i{}_{\parallel}(z' - z)\rho(z')\,\mathrm{d}z' \right] , \tag{8.21}$$

where ϕ_i is the klystron phase (i.e., referring to the arrival time of the beam with respect to the crest of the rf in the ith acceleration section), $\phi(z) = 2\pi z/\lambda_{\text{rf}}$, $\lambda_{\text{rf}} = c/f_{\text{rf}}$ the rf wavelength at frequency f_{rf}, N_b the bunch population, $W_{\parallel}(z)$ is the longitudinal wake function per unit length in units of eV/(Cm) as a function of particle distance z, $\rho(z)$ is the normalized longitudinal beam density, and Δs_i gives the distance between successive accelerating sections powered by different klystrons.

The energy spread of the bunch σ_E is obtained by averaging over the particle distribution after subtracting out the mean energy $\langle E \rangle$ of the bunch. Normalized to the mean energy of the bunch

$$\langle E \rangle = \int_{-\infty}^{\infty} E(z)\rho(z)\,\mathrm{d}z , \tag{8.22}$$

the relative energy spread is

$$\frac{\sigma_E}{E} = \frac{1}{\langle E \rangle} \left[\int_{-\infty}^{\infty} \left(E(z) - \langle E \rangle\right)^2 \rho(z)\,\mathrm{d}z \right]^{\frac{1}{2}} . \tag{8.23}$$

In the low-current limit, the beam does not take away any energy from the accelerating structures and the beam is placed on the crest of the rf wave to achieve both maximum acceleration and minimum energy spread within the bunch. At higher beam currents while invoking BNS damping, the klystrons in the first part of the linac are phased to impart relatively higher energy to the head of the bunch while in the latter part of the linac the klystrons are phased to restore the energy spread to less than the energy acceptance of the downstream target area or beam-delivery system.

At even higher bunch currents, one must carefully balance the two terms in square brackets in (8.21), that is cancel the energy variation along the bunch arising from the slope of the rf and that from the longitudinal wake field. Shown in Fig. 8.15 are sketches illustrating such cancellation. The effective rf gain representing the vector sum of all accelerating stations is plotted versus time (left) together with the projection of the charge distribution which shows the resultant energy spread of the bunch (right). At low current, a bunch placed on crest has minimum energy spread. Off crest, a position-energy correlation is introduced and the energy spread is increased. At high current, due to longitudinal wake field, or *beam loading*, a bunch placed on crest has a large energy spread. For short, high intensity bunches, the energy spread may

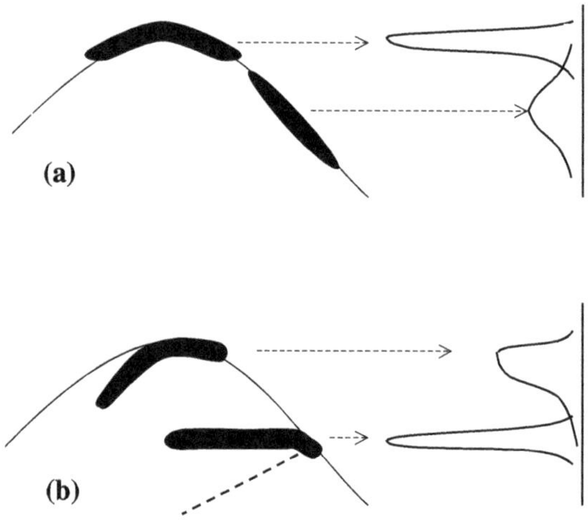

Fig. 8.15. Effective energy gain (*left*) and energy spread (*right*) for low (**a**) and high (**b**) current bunches illustrating optimum phasing of the rf structures for minimum energy spread

be minimized by placing the beam off-crest as shown. In this case the beam-induced wake field (dotted line) exactly cancels the slope of the rf across the bunch.

The energy spread can also be modified by changing the longitudinal beam distribution in the linac. A clever technique of adjusting the bunch shape in the SLC linac consisted in raising or lowering the compressor voltage, relative to the ideal case of maximum compression (see Sect. 8.1). These two modes of operation are called overcompression or undercompression, respectively. Although the same rms bunch length can be obtained with either mode, the shape of the bunch is quite different, which is partly due to an initial distortion caused by impedance effects in the upstream damping ring. Operation with overcompression produced the desired forward-peaked distribution [42] and was used to avoid undesirable energy tails at the end of the linac [43] (see also Fig. 8.20).

Simulation results using (8.3) are shown in Fig. 8.16. While perhaps nonintuitive, simply by 'overcompressing' the bunch, the tails in the energy spread distribution could be eliminated without diluting the longitudinal beam emittance. Measurements of the beam at the end of the SLC linac in a dispersive region are shown [43] in Fig. 8.17 with undercompression (left) and with overcompression (right). The absence of the low energy tails in the latter case justified the routine use of over-compressed beams at the SLC.

In linear accelerators with high current bunches the longitudinal density profile can be further adjusted (so-called bunch shaping) to minimize transverse emittance dilutions arising from short-range wake fields and/or dispersion. This is particularly useful since the outgoing energy spread, and especially energy 'tails', often cause even further emittance dilutions downstream due to chromatic aberrations in the final focus systems.

Minimization of the energy spread depends critically on the single-bunch charge. In the single particle approximation, the energy gained through an accelerating structure is

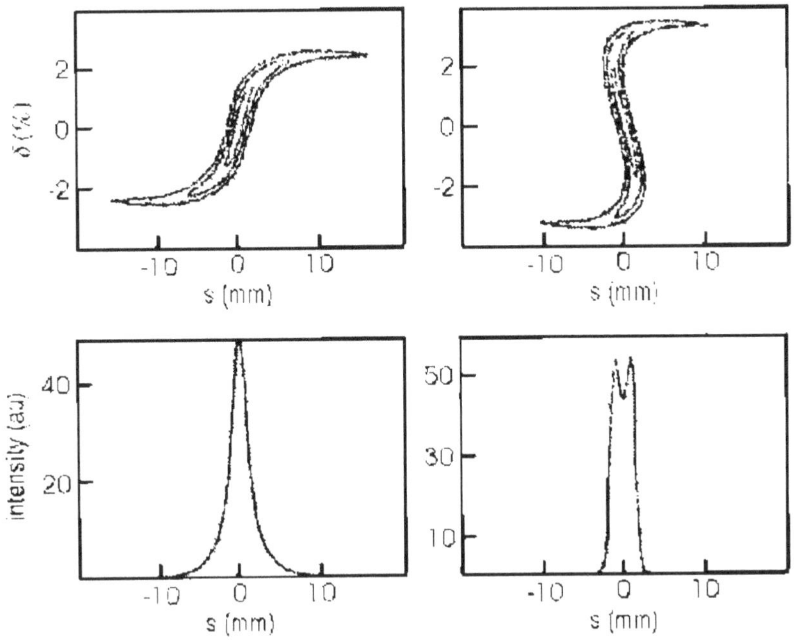

Fig. 8.16. Longitudinal phase space (*top*) and projections onto time axis (*bottom*) with undercompression (*left*) and with overcompression (*right*) (Courtesy F.-J. Decker, 1999)

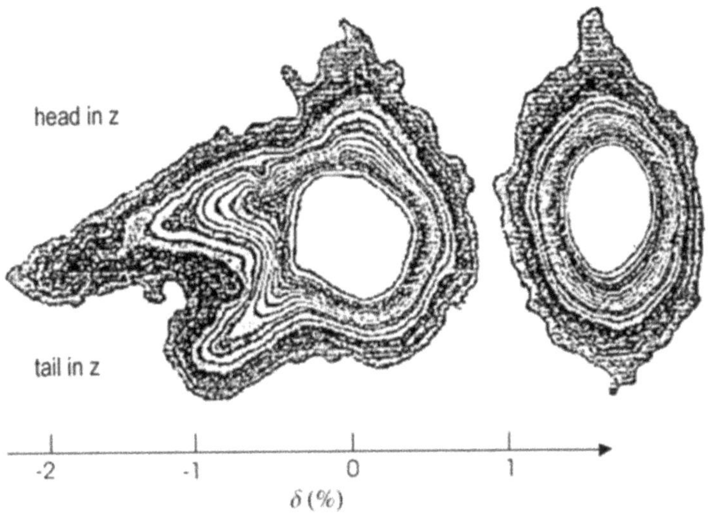

Fig. 8.17. Measurements of the beam profile at a dispersive location at low compressor voltage (*left*) and with bunch overcompression (*right*) (Courtesy F.J. Decker, 2000)

$$eV = E_k L_k \cos \phi_k \,, \tag{8.24}$$

where E_k is the accelerating gradient in units of MeV/m, L_k is the length of the accelerating region, and ϕ_k is the time-like variable representing the phase of the particle relative to the crest of the rf. The magnitude of E_k

and the definition of the phase ϕ_k include possible effects of klystron phase errors.

For a bunch of particles, the energy seen by a given particle is reduced due to loading of the accelerating structure by leading particles within the bunch. Letting ϕ_0 represent the phase at the head of the bunch with respect to the crest of the rf, the total energy gain at the end of the linac becomes

$$eV(\phi_k) = E_k L_k \cos\phi_k + N_b e \int_0^{\phi_0-\phi_k} \rho(\phi') W_{\|}(\phi_0 - \phi_k - \phi')\,\mathrm{d}\phi' \,, \qquad (8.25)$$

where $\rho(\phi')$ represents the bunch charge distribution, and $W_{\|}$ is the wake function for the entire accelerator and is given by the product of the single-bunch wake field times the number of accelerating structures.

Minimum energy spread within the bunch requires that $V(\phi_k)$ is independent of ϕ_k; that is

$$\frac{\partial V(\phi_k)}{\partial \phi_k} = 0 \,. \qquad (8.26)$$

It has been shown [42] that there exists a solution for the bunch charge distribution $\rho(\phi)$ which satisfies this criterion, and which can be obtained by numerically solving

$$\rho(x) = \frac{E_k L_k}{W_{\|}(0)} \sin\left(\phi_0 - \tilde{\phi}\right) + N_b e \int_0^{\tilde{\phi}} \frac{\frac{\partial W_{\|}}{\partial \tilde{\phi}} \left(\tilde{\phi} - \phi'\right) \rho(\phi')}{W_{\|}(0)} \mathrm{d}\phi' \,, \qquad (8.27)$$

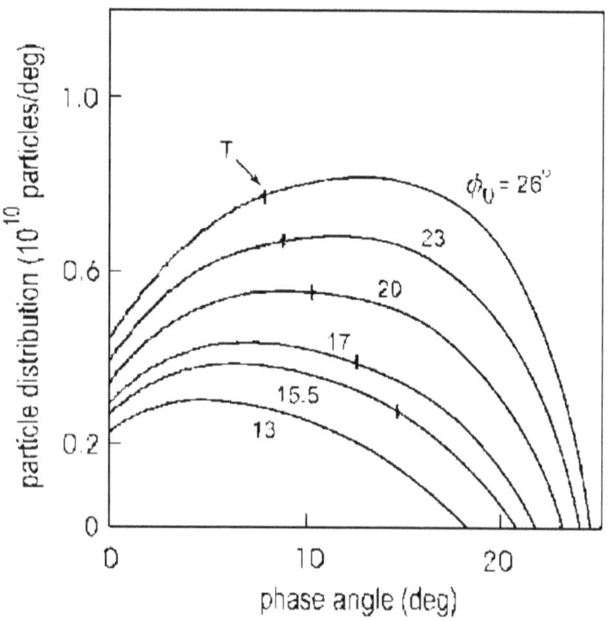

Fig. 8.18. Optimum bunch shape for beam loading compensation in the SLC linac. The bunch head is on the left (Courtesy G. Loew, 1999)

where $\tilde{\phi} \equiv (\phi_0 - \phi_k)$. The interpretation of (8.27) is quite analogous to that shown in Fig. 8.10. Only in this case, the voltage provided by the higher harmonic cavity is replaced by the decelerating voltage induced by beam loading of the leading particles within the bunch.

The solution given in (8.27) is shown in Fig. 8.18 for the SLC linac. The horizontal axis shows the phase angle of particles within the bunch measured with respect to ϕ_0 with the leading particles located at zero phase angle. The different curves labelled by ϕ_0 designate the position of the head of the bunch with respect to the crest of the rf in units of degrees, and the points marked 'T' indicate where the integrated bunch charge reaches the design single bunch population of $N_b = 5 \times 10^{10}$ particles per bunch. As can be seen, for minimum energy spread, the preferred charge distributions tend in general to have a steeply rising edge.

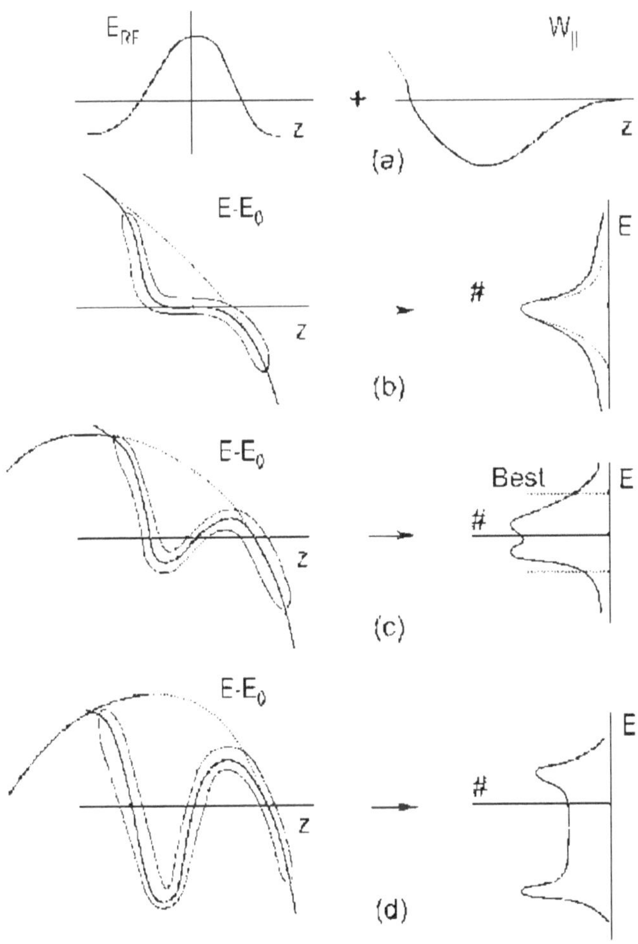

Fig. 8.19. Conceptual illustration for optimizing the relative phase of the beam for the case of very long bunches. The accelerating field E_{rf} and the longitudinal wake field $W_{\|}$ are shown in (**a**). The longitudinal profiles and their projections onto the energy axes are shown for different relative phases (between the beam and the rf wave) in (**b**)–(**d**) (Courtesy J. Seeman, 1999)

The tradeoff between bunch length and energy spread in a linear accelerator is illustrated in Fig. 8.19, which shows the effect of the combined voltages from the power source $\mathrm{E_{RF}}$ and the longitudinal wake field $W_{\|}$. On the left is depicted the longitudinal phase space of a long bunch while the projection onto the energy axis is given on the right. In the limit of long bunches one can see that a 'double-horned' distribution produces the minimum rms energy spread.

Measured energy spread profiles taken at the end of the SLAC linac are shown in Fig. 8.20. A wire scanner located in a dispersive region of a downstream transport line was used to measure the profile σ_w. The energy spread σ_δ was inferred by subtracting out, in quadrature, the contribution from the betatron beam size $\sigma_\beta = \sqrt{\epsilon_x \beta_x}$:

$$\sigma_\delta = \sqrt{{\sigma_w}^2 - {\sigma_\beta}^2} \, . \tag{8.28}$$

The angle ϕ denoted in the figure shows the BNS phase angle at the time of the measurement. These data show clearly the effects of not only misphasing the linac, but the additive contributions of the short-range longitudinal wake field and have been used together with simulation [44, 45] to determine the longitudinal bunch distribution at the SLC.

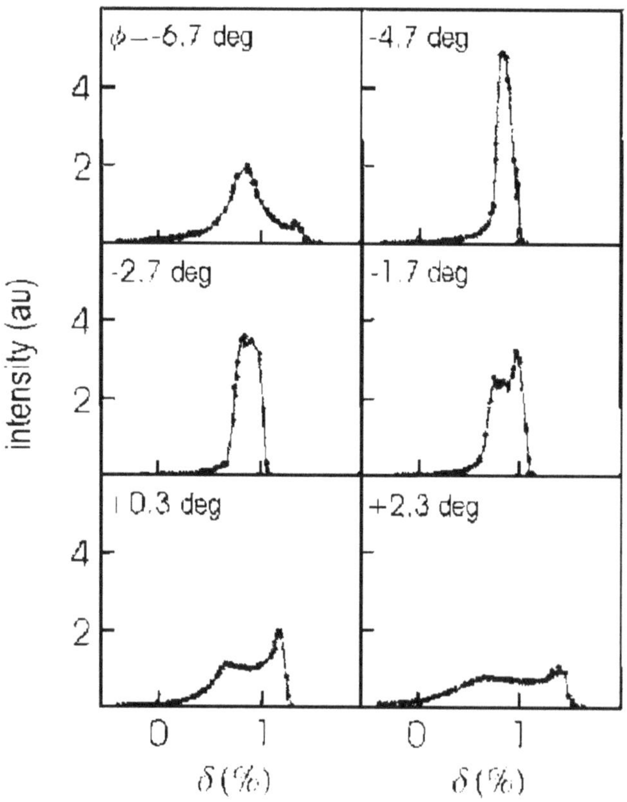

Fig. 8.20. Energy spread measurements taken at the end of the SLAC linac for different BNS phases (ϕ) (Courtesy K. Bane, 1999)

8.7 Energy Compression

The energy spread of a single bunch can be made smaller (at the expense of increased bunch length) using reverse application of a bunch length compressor. In this case, the beam first passes through a region with an energy-dependent path length, in which the particles of different energy are separated in time. Following this is an accelerating section which decelerates and accelerates the high and low energy particles, respectively. Energy compression for single bunches was designed and implemented [46] for the positron transport line into the SLAC positron damping ring. The net increase in particle yield was observed to be about 10%.

8.8 Beam Loading and Long-Range Wake Fields

In the quest for obtaining ever-increasing total beam currents, both newly constructed and future accelerators have in common bunch trains consisting of multiple, closely spaced, high-current bunches. In both linear and circular accelerators this may lead to a relative phase shift between the bunches of a bunch train (an example was shown in Fig. 8.13).

When a beam passes through an accelerating cavity, it induces a voltage $V_{b,m}$ in each mode m of the cavity. The induced voltage is always retarding; that is, the beam-induced voltage always acts to decelerate the beam. Expressed another way, the beam always takes energy away from the cavity. This is refered to as beam loading.

The fundamental theorem of beam loading [47] is relevant on short time scales (i.e., for a single pass through an accelerating cavity). The theorem states that the voltage that a test particle would experience when trailing a (point) source particle at time $t_0 > 0$ is exactly twice its beam induced voltage at $t_0 = 0$. More generally, the induced voltage, or wake potential, V_δ is given by

$$
\begin{aligned}
V_\delta(t_0) &= 0 \quad \text{for} \quad t_0 < 0 \\
&= -kq \quad \text{for} \quad t_0 = 0 \\
&= -2kqe^{-\frac{t_0}{\tau_f}} \cos\omega_{\mathrm{rf}} t_0 \quad \text{for} \quad t_0 > 0 \,,
\end{aligned}
\tag{8.29}
$$

where $\tau_f = 2Q/\omega_{\mathrm{rf}}$ is the fill time of the cavity and Q is the cavity quality factor. Here, the variable

$$
k = \frac{\omega}{2}\left(\frac{R}{Q}\right) \tag{8.30}
$$

is called the loss parameter which tends to be determined by the structure geometry close to the beam. In practice, k is often calculated for each cavity mode using numerical programs, e.g., MAFIA. The wake potential describes the electromagnetic field that the point-like beam generates as it interacts with its surroundings and how this field acts back on the beam, thus perturbing its motion.

We consider next the effect of multi-bunch or multi-turn beam loading, which is caused by the interaction of the beam with the fundamental accelerating mode of the rf cavity on successive bunch passages. In addition, there is also a short-range effect, which consists of the superposition of many higher-order cavity modes. These higher-order modes are usually damped, so that the short-range wake field quickly decays. The fundamental mode on the other hand is used for the acceleration, and, hence, it necessarily exhibits a weak damping, or a high Q value.

In a circular accelerator containing a single accelerating cavity and a single particle bunch, the steady-state ($t \gg \tau_f$) beam-induced voltage V_b is given by summing over the contribution from all previous turns. For short bunches, using (8.29) with T_{rev} denoting the revolution frequency,

$$\begin{aligned} V_b &= -kq - 2kq \sum_{n=1}^{\infty} e^{-\frac{t_0}{\tau_f}} \cos \omega t_o \delta(t_0 - nT_{\text{rev}}) \\ &= -2kq \left[\sum_{n=0}^{\infty} e^{-\frac{t_0}{\tau_f}} \cos \omega t_o + \frac{1}{2} \right] \delta(t_0 - nT_{\text{rev}}) \\ &= -2kq \left[\sum_{n=0}^{\infty} e^{-\frac{nT_{\text{rev}}}{\tau_f}} \cos n\omega T_{\text{rev}} + \frac{1}{2} \right] . \end{aligned} \tag{8.31}$$

Driving the cavity near its resonance frequency, i.e., taking $\omega = \omega_{\text{rf}}$, and noting that $\cos(n\omega T_{\text{rev}}) = 1$ for all n, we can simplify this expression and find

$$\begin{aligned} V_b &= -2kq \left[\sum_{n=0}^{\infty} e^{-\frac{n\omega_{\text{rf}}}{2Q} T_{\text{rev}}} + \frac{1}{2} \right] \\ &= -2kq \left[\frac{1}{1 - e^{-\frac{n\omega_{\text{rf}}}{2Q} T_{\text{rev}}}} + \frac{1}{2} \right] . \end{aligned} \tag{8.32}$$

Neglecting the small self-loading term (factor 1/2), and applying

$$\frac{1}{1 - e^{-x}} \approx \frac{1}{x} \quad \text{for } x = \frac{\omega_{\text{rf}} T_{\text{rev}}}{2Q} \ll 1 , \tag{8.33}$$

we then obtain

$$\begin{aligned} V_b &= -2kq \left(\frac{2Q}{\omega_{\text{rf}} T_{\text{rev}}} \right) , \quad k = \frac{\omega_0}{2} \left(\frac{R}{Q} \right) \\ &= -2q f_{\text{rev}} R \\ &= -I_b R . \end{aligned} \tag{8.34}$$

This simple result shows that with the cavity tuned to resonance, the beam-induced voltage is simply given by the beam current at the resonance frequency ($I_b = 2I_{dc}$) times the cavity impedance. (Note that this is the loaded impedance, i.e., the combined impedances of the cavities, the connecting wave guides and rf power source, and the possible influence of rf feedback loops. The minus sign, again, indicates that the beam takes energy away from the accelerating cavity.

Application of (8.29 and 8.34) to the transport of high current particle beams is a subject of great interest in modern accelerators. In the extreme short-range limit ($t < \sigma_z\omega/c$), the variable t_0 may represent the time interval between particles within a single bunch in which case, by causality, the charge q represents the charge of all preceding *particles within the bunch.* As the beam current is increased, eventually, as many experimental and theoretical studies have shown, the ensuing motion can become unstable. BNS damping in a linac is used, for example, to avoid the beam breakup instability associated with short-range wake fields.

At the SLC high beam currents were achieved using BNS damping and single, widely spaced bunches each of high charge (about 4×10^{10} particles per bunch). At such high bunch charges, the effects of long-range wake fields ($t \sim \tau_{bb}$, where τ_{bb} is the spacing between bunches) were also observed. Shown in Fig. 8.21 is a measurement [48] showing the orbit oscillation induced after changing τ_{bb} by 1 (of about 170) rf buckets. Using similar measurements with variable bunch spacings, the frequency of the driving mode could be roughly determined and compared with numerical models. Interestingly, the simplicity of the SLC (FODO) lattice and the different sign of accelerated particle species was such that the deflections due to the higher-order wake fields added coherently along the SLC linac. To minimize this resonant build-up, the symmetry of the focussing was changed using a so-called 'split-tune' lattice [48].

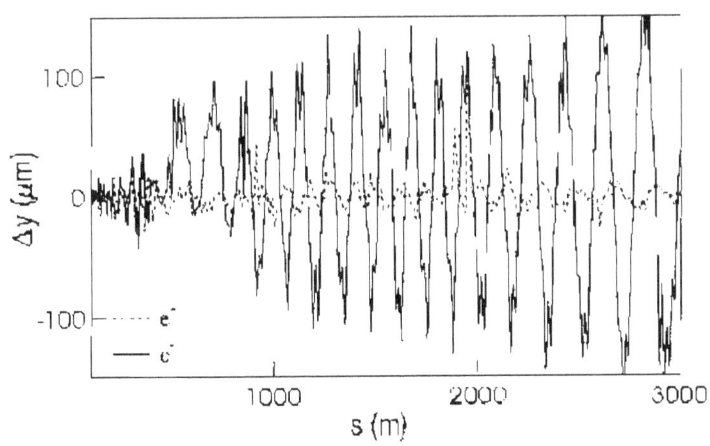

Fig. 8.21. Measured vertical difference orbit of the trailing electron bunch (*solid line*) after displacement of the leading positron bunch by 1 rf bucket (*dashed line*) in the SLC linac

Here the horizontal and vertical phase advances were changed so as not to be equal. In this way, resonant excitation of the leading positron bunch had less influence on the trajectory of the trailing electron bunch (the horizontal phase advance of the electrons equals the vertical phase advance of the positrons). The pulse-by-pulse jitter of the electron bunch was reduced [48] by about 15% horizontally (from $0.4\sigma_x$ to about $0.35\sigma_x$) and 30% vertically (from $0.75\sigma_y$ to about $0.50\sigma_y$).

The preferred way nowadays to achieve even higher beam currents, while avoiding intrabunch beam instabilities arising from increased single-bunch beam currents, uses multiple bunches (often called a bunch train) each with lower single-bunch beam currents. In this case, t_0 in (8.29) refers to the spacing between appropriate bunches. Now the beam-induced voltage experienced by a particular bunch is given by the sum of the voltages induced by *all preceding bunches* each obtained by evaluating (8.29) at the appropriate time $t_0 = N\tau_{bb}$, where N denotes the number of bunch spacings to the preceeding bunch of interest. In future linear collider designs with multi-bunch beams, research and development in the design of the accelerating structures aims towards minimizing the strength of the offending modes. An example,

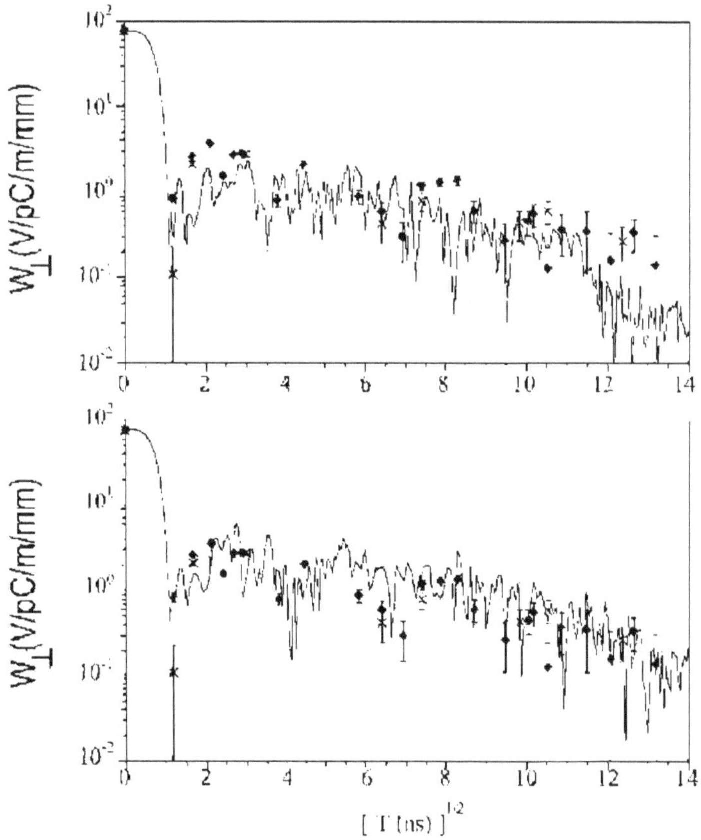

Fig. 8.22. Horizontal (*crosses*) and vertical (*diamonds*) wake field measurements and predictions (*solid lines*) in the X-band structure test at ASSET at SLAC (courtesy C. Adolphsen, 2002)

from the Accelerator Structure Setup (ASSET) facility at SLAC, is shown in Fig. 8.22 [49]. Here again, the spacing between the drive and trailing bunches was varied and the particle orbits were recorded and analyzed to give deflection seen by the trailing bunch and so determine the strength of the transverse wake field.

In existing circular accelerators operating already with high current, multi-bunch beams, much effort has been devoted to carefully developing and testing new cavity and feedback designs. At KEKB novel ARES [50] energy-storage cavities, super conducting cavities, and multi-bunch feedback are used to minimize the effect of wake fields. At PEP2 higher-order mode dampers and both multi-bunch and rf feedback are used. While bunch-to-

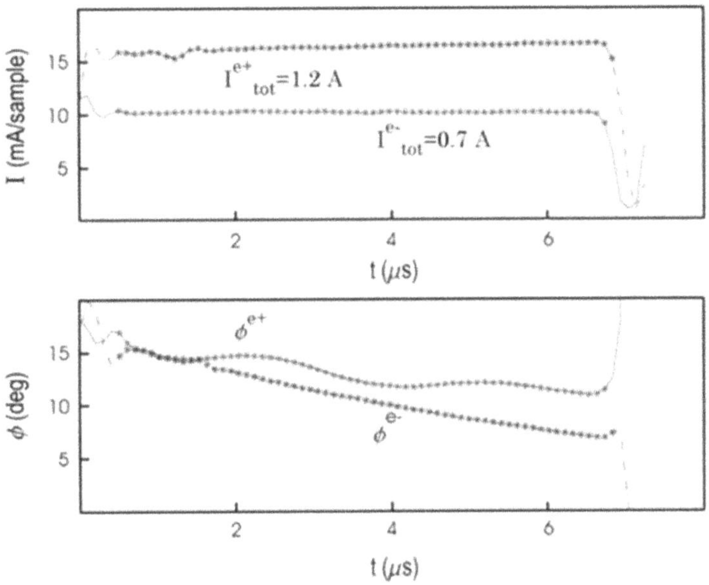

Fig. 8.23. Measured beam currents and beam phases for the PEP-II electron and positron beams (Courtesy P. Corredoura, 2000)

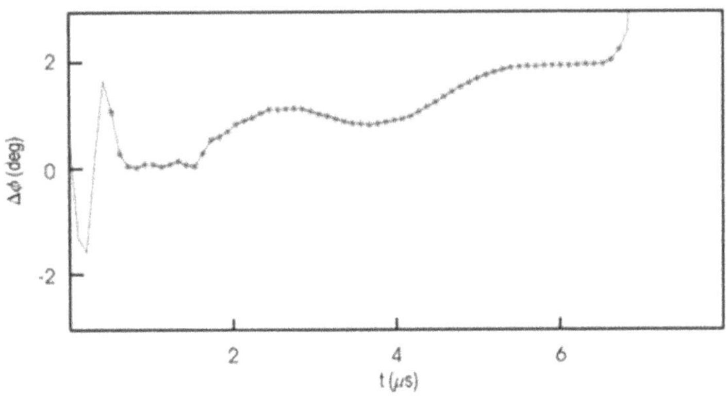

Fig. 8.24. Measured phase difference between the PEP-II electron (HER) and positron (LER) beams (Courtesy P. Corredoura, 2000)

bunch stability at both accelerators is excellent, even at total currents in the range of Amperes, beam loading is still an issue.

As an example, measured phase changes across the bunch fill patterns are shown in Fig. 8.23 from the PEP2 B-factory. The bunch distribution was uniform with a 5% gap in the fill pattern. This nonuniformity in the fill results in a change in cavity voltage along the bunch train and, from (7.5), a change in synchronous phase. This phase variation leads to a reduction in gain of the feedback loops. In addition, the phase difference between the two trains must be minimized to ensure the desired (longitudinal) collision point and hence the highest possible luminosity. The difference in the phase transients of the two beams is shown in Fig. 8.24.

8.9 Multi-Bunch Energy Compensation

Two methods, known as Δf and Δt compensation have been proposed to combat multibunch phase transients in linear accelerators. Shown in Fig. 8.25 is the principle of Δt compensation [51]. Here the voltage V_k represents the voltage response of a finite bandwidth accelerating structure to a step function input pulse. The lower curve represents the beam-induced voltage V_b of the entire bunch train. By injecting the beam prior to the time the linac structure is at peak voltage, the vector sum V_k+V_b is observed to be flat over the duration of the bunch train. The projected energy spread is therefore minimized and the phase relationship between the bunches is constant.

The principle of the Δf compensation is illustrated [52] in Fig. 8.26. In this design from the ATF in Japan, some fraction of the many accelerating structures are slightly detuned by $\pm\delta f$. The different bunches therefore obtain a different energy gain which depends on their location within the train.

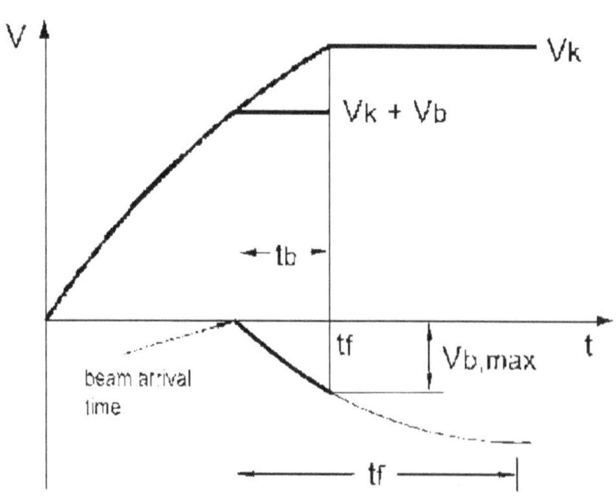

Fig. 8.25. Conceptual diagram illustrating multi-bunch, Δt beam loading and energy compensation

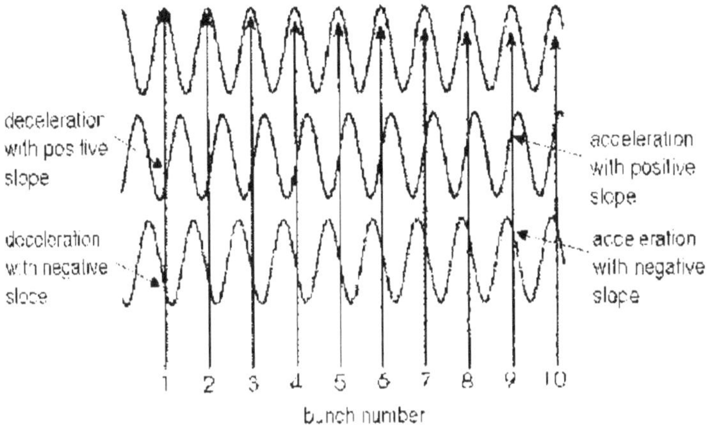

Fig. 8.26. Conceptual diagram illustrating multi-bunch, Δf beam loading and energy compensation [53] (Courtesy J. Urakawa, 1999)

With some structures detuned by $+\delta f$ and some by $-\delta f$ the position-energy correlations introduced by the slope of the rf cancels. In this way, not only is the projected energy spread along the bunch train minimized, but the energy spread of each bunch is also preserved.

Beam loading compensaton using the Δt method may be advantageous since the correction may be applied locally at each accelerating section. On the other hand it is anticipated [51] that about 10% more power is required relative to the Δf compensation scheme.

8.10 Damping Partition Number Change via RF Frequency Shift

The generation of small emittance beams is a key issue for synchrotron light sources, collider rings, and for future linear colliders. Dedicated accelerators have been designed to produce such beams, but techniques to further reduce the design emittances would yield immediate improvements. At injection into such accelerators, the transverse beam emittances are large and often fill a large fraction of the dynamic aperture. For lepton beams, at later times after the beam has radiation damped, the horizontal damping time and equilibrium emittance may be reduced by shifting the rf frequency, such that the particle orbit moves inwards. By passing off-center through the quadrupoles in regions of nonzero dispersion, the horizontal partition number J_x is changed. This reduces both the horizontal damping time and equilibrium emittance. In addition, to the extent that the vertical emittance is determined by betatron coupling, the reduction in horizontal emittance may be accompanied by a corresponding reduction in vertical emittance.

The horizontal damping time and beam emittance are both inversely proportional to the horizontal partition number $J_x = 1 - \mathcal{D}$, where [54]

$$\mathcal{D} = \frac{\int D_x G(G^2 + 2k)\,\mathrm{d}s}{\int G^2\,\mathrm{d}s} . \tag{8.35}$$

Here D_x is the horizontal dispersion, $G = 1/\rho$ and k describe the inverse bending radius in m^{-1} and the quadrupole focusing gradient m^{-2}, respectively, and the integrals are evaluated around the ring circumference. For the non-combined function magnets in the SLC damping ring, $Gk \approx 0$.

For a beam orbit offset Δx in the quadrupoles, the change in $\mathcal{D}$ is given approximately by

$$\Delta\mathcal{D} \approx \frac{2k^2 D_{x,q} L_q N_q}{2\pi/\rho} \Delta x , \tag{8.36}$$

where $k = \frac{ec}{E}(\frac{\partial B}{\partial x})$ with $e = 1.6 \times 10^{-19}$ C, $c = 3 \times 10^{10}$ m/s, $E = 1.19$ GeV, ρ is the local bending radius, $D_{x,q}$ is the dispersion at the quadrupoles, and L_q and N_q are respectively the quadrupole length and number of quadrupoles.

The orbit may be offset in the quadrupoles by either changing the accelerating frequency or by physically displacing the magnet support girders. Emittance optimization using the accelerating frequency has been used in e^+e^- storage rings previously [55] and is used routinely at LEP [56]. The effect of changing the geometric ring circumference was already dicussed in Sect. 4.3.1. The circumference adjustment is applicable provided that the transverse acceptance is not limited and that the injected beam energy spread is small compared to the energy acceptance. At the SLC, the electron damping ring was 'stretched' [57] in 1992 by 9 mm for a 15% increase in J_x. In doing so, the energy aperture at injection was reduced yet without any loss in transmitted beam current. For the case of the positron damping ring, the incoming beam filled the entire aperture so stretching the accelerator was not an option.

Shown in Fig. 8.27 is a calculation of the horizontal emittance $\gamma\epsilon_x$ as a function of time for 4 different frequency offsets for the case of the SLC damping rings. It is assumed that the beam is injected at the nominal rf frequency of 714 MHz with an initial emittance of 20×10^{-5} m-r. The accelerating frequency is increased after 1 ms (dashed line) for which the longitudinal emittance has damped by about a factor of 2. The simulations (using SAD [58]) with a trapezoidal approximation for the bending magnet fringe fields show a half unit reduction (i.e., 15–20%) in normalized emittance with a 100 kHz frequency change, while the damping time reduces from 3.4 ms to 3.0 ms.

At storage rings and colliders, there is no tight tolerance on maintaining the desired rf frequency. In a damping ring, the time required to reset the frequency and relock the beam phase to the desired extraction phase is critical since the frequency must be ramped back to nominal just before extraction in order not to introduce any energy and/or phase errors in downstream subsystems (in our example, the SLC bunch compressor and the SLC linac). Minimizing this time [59] is critical since reverting to the nominal rf frequency is associated with corresponding antidamping of the beam.

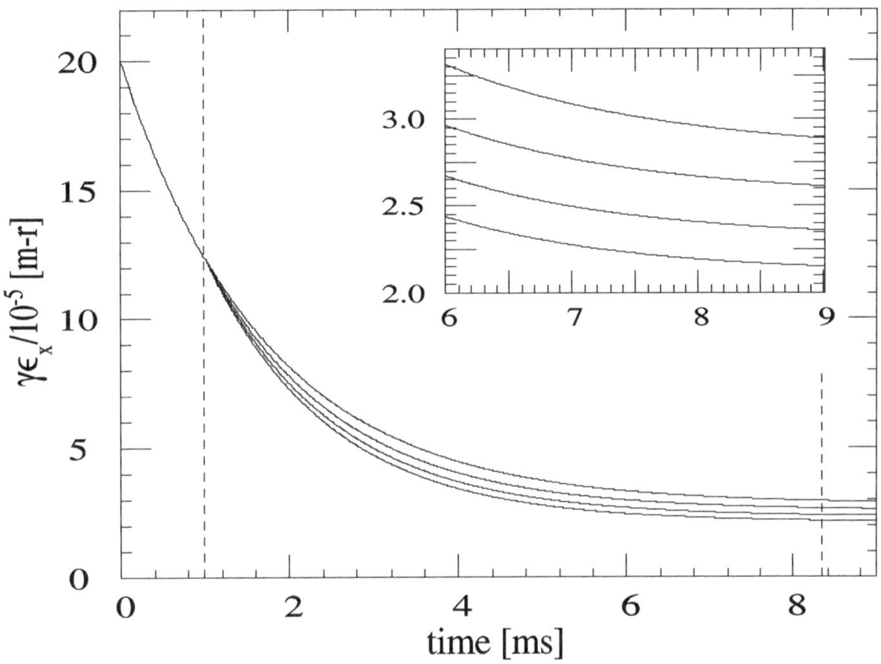

Fig. 8.27. Normalized horizontal beam emittance as a function of store time for different frequency shifts calculated for the SLC damping rings. The full store time is shown with an expanded view near beam extraction at 8.33 ms shown in the *insert*. The *curves*, when viewed *from top to bottom*, correspond to frequency shifts of 0, 50, 100, and 150 kHz, respectively

As a practical point, with a change in accelerating frequency, the accelerating cavities are detuned by an amount characterized by the tuning angle ϕ_z which is given by

$$\phi_z = \tan^{-1}\left[2Q\left(\frac{f_0 - f_{\rm rf}}{f_0}\right)\right] , \tag{8.37}$$

where Q is the loaded cavity quality factor, f_0 is the resonant frequency of the cavity, and $f_{\rm rf}$ is the frequency of the applied rf. With the cavity tuners *fixed*, the new tuning angle $\phi_z{}'$ corresponding to the new applied rf frequency $f_{\rm rf}{}' = f_{\rm rf} + \delta f_{\rm rf}$ is given by

$$\tan\phi_z{}' = 2Q\left[1 - (1 - \frac{1}{2Q}\tan\phi_z)\frac{f_{\rm rf}{}'}{f_{\rm rf}}\right] . \tag{8.38}$$

An example is shown in Fig. 8.28(a). Typically, the tuning angle is set for minimum reflected power:

$$\phi_z|_{\phi_l=0} = -\frac{I_b R}{\hat{V}}\sin\phi_b , \tag{8.39}$$

where I_b is twice the dc beam current, R is the total loaded impedance, $\hat{V}$ is the total cavity voltage in units of V, and ϕ_b is the synchronous beam phase

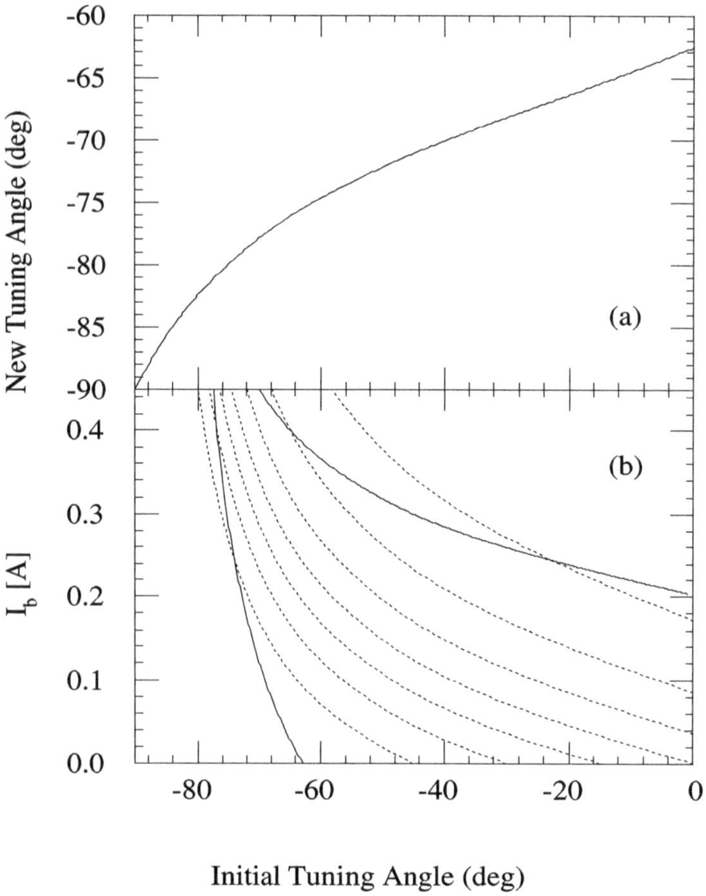

Fig. 8.28. Cavity tuning considerations. The initial and final tuning angles at $f_{\rm rf}$ and $f_{\rm rf}' = f_{\rm rf} + 100$ kHz, respectively, are shown in (**a**). The stability boundary for cavity voltage regulation at 600 kV is given by the *two solid curves* in (**b**). The *dashed curves* show the loading angle ϕ_l which is measured and regulated by the tuner feedback loops. At zero beam current the loading angle is equal to the tuning angle

measured with respect to the crest of the accelerating rf. The loading angle ϕ_l is related to the tuning angle ϕ_z by

$$\tan\phi_z = \left(1 + \frac{I_b R}{V_c}\cos\phi_b\right)\tan\phi_l - \frac{I_b R}{V_c}\sin\phi_b\,. \tag{8.40}$$

In the case of the SLC, at zero beam current the cavity detuning exceeded the power capabilities of the power source as indicated by the lower solid curve in Fig. 8.28(b).

Assuming sufficient frequency aperture, which may be restricted, e.g., by transverse betatron resonances that can be encountered during the frequency ramp with nonzero chromaticity or by a physical aperture in a region with nonzero dispersion, the maximum frequency change may be limited by either available rf power as discussed above or by the damping poles [60] at which the damping rate becomes zero.

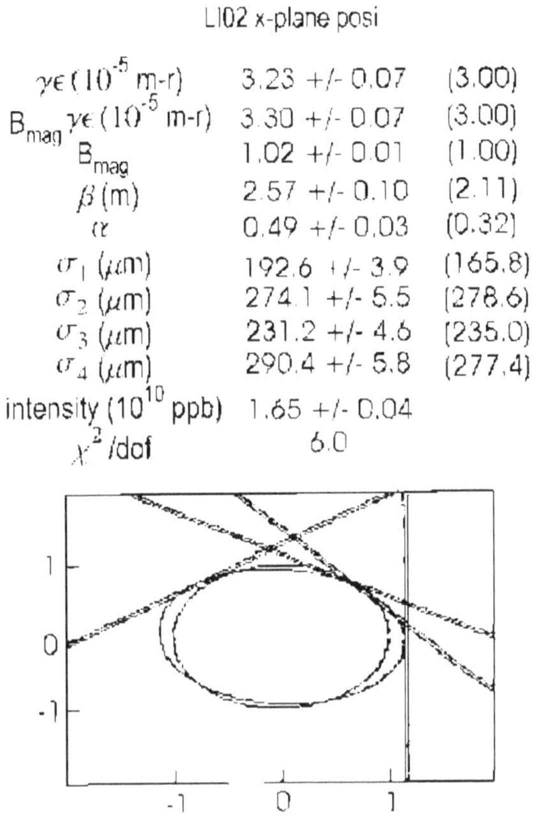

Fig. 8.29. Nominal emittance at injection to the SLC linac without an rf frequency shift in the upstream damping ring. The measured normalized emittance was 3.30 ± 0.07 m

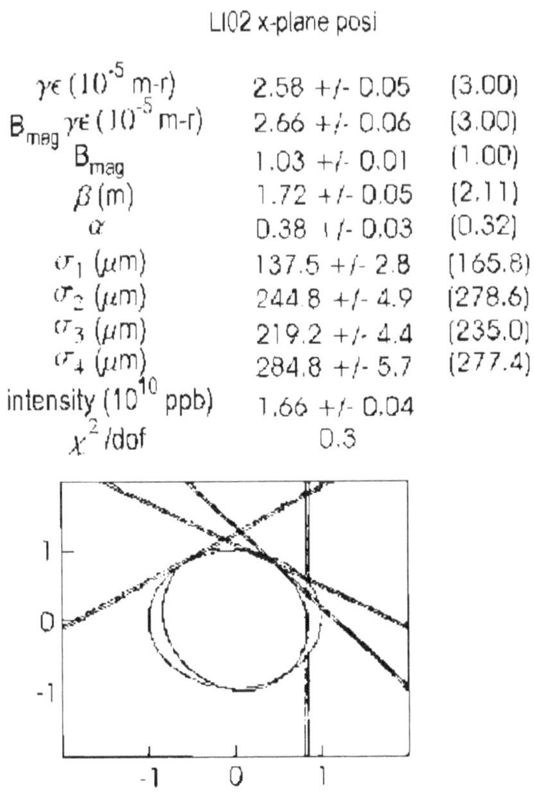

Fig. 8.30. Normalized emittance measurement at injection to the SLC linac with a 62.5 kHz frequency shift in the upstream damping ring. The normalized emittance was reduced from 3.30 ± 0.07 m to 2.66 ± 0.06 m

Measurements [61] showing the effect of a frequency ramp in the SLC positron ring are given in Figs. 8.29 and 8.30. In this experiment, the downstream compressor was turned off in order to more cleanly detect the effect of the frequency shift on the beam emittance. From Fig. 8.30, with a 62.5 kHz shift, the reduction in normalized emittance was from 3.30 ± 0.07 m to 2.66 ± 0.06 m. For the electron damping ring, the frequency shift reduced the horizontal normalized emittance from 3.22 ± 0.08 m to 2.93 ± 0.07 m. We expected that the emittances at the entrance of the final focus would be reduced by a similar amount to $\gamma\epsilon_x = 5.2 \times 10^{-5}$ m and $\gamma\epsilon_y = 1.1 \times 10^{-5}$ m, and, using the 1997 SLC interaction point parameters with rms angular divergences of $\theta_x = 450$ μrad and $\theta_y = 250$ μrad, the corresponding rms beam sizes at the collision point were estimated to be $\sigma_x = 1.3$ μm and $\sigma_y = 0.5$ μm. With a 1 mm bunch length and 4×10^{10} particles per bunch, the luminosity was computed by the code Guinea-Pig [62] to be 4.3×10^{32} m^{-2} per collision. This corresponded to an estimated increase of over 40% in luminosity by application of the frequency shift in the damping rings. Unfortunately, no rigorous study was undertaken to experimentally quantify the effect of the rf frequency shift on the SLC luminosity.

Decreasing the beam emittance by changing the damping partition numbers is also part of the HERA luminosity upgrade [63]. To maintain matched beam sizes, with a reduced proton beam size resulting from the modified optics at the interaction points of HERA, the electron beam size must be reduced. The approach that was taken is twofold: stronger focussing in the arcs, which reduces the horizontal equilibrium emittance, and a +200 Hz rf frequency shift, which changes the damping partition numbers.

Exercises

8.1 Phase Tolerances in a Bunch Compressor

Consider a 1.2 GeV electron beam passing through a 2856 MHz compressor cavity and a transport line with $R_{56} = -0.6$ m. Sketch the error in the final phase as a function of injection phase error for different accelerating voltages. At which compressor voltage is the final beam phase, behind the compressor region, minimially sensitive to initial phase errors?

8.2 Bunch Precompression

a) Describe (or sketch) the motion of the beam centroid in longitudinal phase space for the following process:

$$\begin{aligned}\hat{V} &= V_0 \quad \text{for} \quad t < t_0 \\ &= 0.75V_0 \quad \text{for} \quad t_0 < t < t_1 = \frac{\tau_{s,l}}{8} \\ &= V_0 \quad \text{for} \quad t_1 < t < t_2 = \frac{\tau_{s,h}}{4}\end{aligned}$$

$$= 0.75V_0 \quad \text{for} \quad t_2 < t < t_3 = \frac{\tau_{s,l}}{8}$$

$$= V_0 \quad \text{for} \quad t_3 < t < t\,, \tag{8.41}$$

where $\tau_{s,l}$ is the synchrotron period at the lower voltage and $\tau_{s,h}$ is the synchrotron period at the higher voltage.

b) Taking into consideration a multiparticle beam, describe the particle distribution in phase space during the process given by comparing initial and final states. Show that the the bunch length is compressed (at the expense of increased energy spread).

8.3 Harmonic Cavities

a) Verify the equations given in 8.20.

b) Sketch the relative phase between the two rf systems of Fig. 8.10 as the ratio of radiative losses to primary rf voltage varies from zero (proton beam limit) to slightly less than one.

8.4 Minimum Voltage Required for Beam Storage

The power P_γ radiated due to synchrotron radiation per turn by an electron or positron may be expressed [54] as

$$P_\gamma = \frac{cC_\gamma}{2\pi}\frac{E^4}{\rho^2}\,, \tag{8.42}$$

where c is the speed of light, $C_\gamma = 8.85 \times 10^{-5}$ m-GeV^{-3}, E is the beam energy in GeV, and ρ is the local radius of curvature of the bending magnets.

a) For an accelerator (without insertion devices) with $\rho = 2$ m and $E = 1$ GeV, what is the total radiated power for 10^{11} particles?

b) With a 100 ns particle revolution period, at what voltage could the beam no longer be captured? What is the synchronous phase at this voltage? Assume that there are no other energy loss sources.

c) For low current beams, the electron bunch length scales with the total accelerating voltage V as $V^{-\frac{1}{2}}$. What is the disadvantage of lowering the cavity voltage for increased bunch length compared with the use of harmonic cavities?

8.5 Phase Shift along a Bunch Train

The cavity fill time τ_f describes the time evolution of the cavity voltage in response to a step function. For example, if a cavity initially at amplitude V_0 has suddenly its power source turned off, then the cavity voltage decays as

$$\hat{V}(t) = V_0 e^{-\frac{t}{\tau_f}}\,, \quad \text{where} \quad \tau_f = \frac{2Q}{\omega_{\text{rf}}}\,. \tag{8.43}$$

For the case of a storage ring with fast feedback, estimate the change in synchronous phase across a 500 ns long bunch train of 100 mA average beam current along the train, and cavities with a loaded Q of $Q = 3000$, a total impedance of 5 MΩ, an rf frequency of 476 MHz, and an external rf voltage of 10 MV.

9 Injection and Extraction

In transferring the beam from one accelerator to another, preservation of the beam properties is essential. Injection should be accomplished with minimum beam loss and often minimal emittance dilution. Single-turn injection, in which a single bunch of particles is injected into a single empty rf bucket, is usually straightforward. In many cases, however, to attain higher bunch currents, one may also wish to accumulate beam in a storage ring by reinjecting different beam pulses into the same rf bucket. This is called multi-turn injection. In addition to conventional schemes, there are several new or more exotic injection techniques, devised to control and improve the properties of the stored beam.

Extraction refers to the removal of beam from an accelerator. It is roughly the reverse process of injection. One difference is that usually at extraction the beam energy is higher. Thus space charge effects are less important, but the hardware requirements for the septa and kicker magnets are more challenging. A high extraction efficiency is necessary to avoid activation of accelerator components and also to make optimum use of the accelerated beam, e.g., to achieve the maximum luminosity. Which extraction procedure is chosen depends on the specific application. Fast one-turn extraction is used for transferring bunches between different circular machines in an accelerator chain. For fixed-target experiments, slow extraction by the controlled excitation of nonlinear betatron resonances is a common technique, which provides a slow uniform depletion of particles in the ring, i.e., 'spill'. Again, several novel extraction schemes are being studied, for example, extraction using a bent crystal.

A good overview of conventional beam injection and extraction can be found in [1] and [2].

9.1 Transverse Single-Turn Injection

For single-turn injection, the beam is brought onto the central orbit using a septum magnet and a fast kicker element, as illustrated in Fig. 9.1. In the following, we assume that the injection is performed horizontally. The expressions derived can be extended easily to the vertical case, or to a combined horizontal and vertical injection.

M. G. Minty et al., *Measurement and Control of Charged Particle Beams*,
https://doi.org/10.1007/978-3-662-08581-3_9

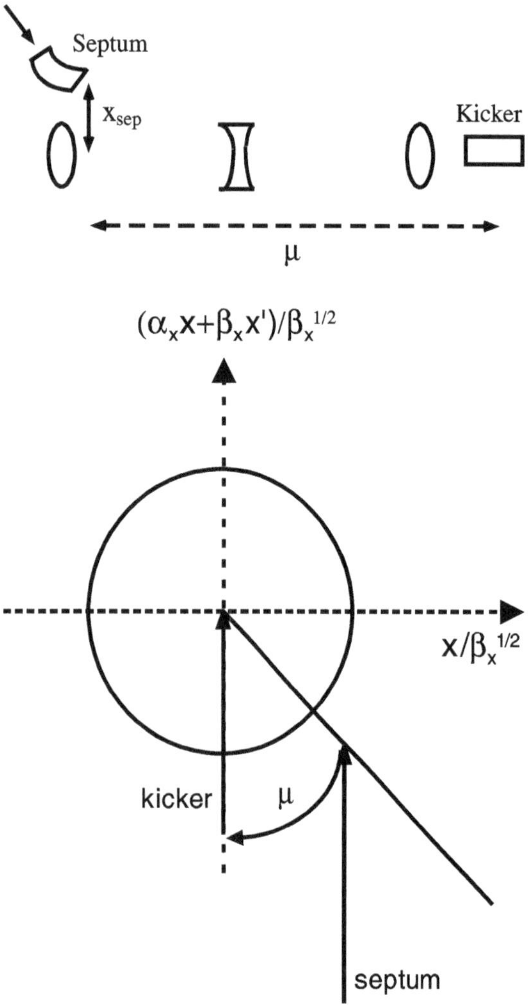

Fig. 9.1. Schematic of single-turn injection with septum and kicker: (*top*) magnet arrangement, (*bottom*) phase-space diagram. The septum strength is adjusted to place the beam onto the line describing an angle μ (the phase advance from septum to kicker) with respect to the vertical [1]

At the exit of the septum, the injected beam must be at a horizontal distance $x_{\rm sep}$ from the center of the machine aperture:

$$x_{\rm sep} \geq n_{xi}\sigma_{xi} + n_{xs}\sigma_{xs} + D_x \left(\frac{\Delta p}{p}\right)_i + x_{\rm co,rms} + x_{i,\rm rms} + d_{\rm sep} \ , \qquad (9.1)$$

where σ_{xi}, σ_{xs} are the rms beam sizes of the incoming beam and of the stored beam, respectively, n_{xs} is the beam stay-clear in the ring required for sufficient beam lifetime or negligible injection losses in units of the rms beam size of the stored beam (e.g., reasonable values may be $n_x \geq 8$ for electron rings, and $n_x \geq 4$ for proton rings), n_{xi} is the beam stay-clear for the injected beam also in units of its rms beam size, $(\Delta p/p)_i$ the relative momentum deviation of the injected beam with respect to the ring energy, D_x the dispersion function, $x_{\rm co,rms}$ the rms closed-orbit offset at the location of the septum, $x_{i,\rm rms}$ the rms orbit variation of the injected beam, and $d_{\rm sep}$ the thickness of the septum. For simplicity in (9.1) we have assumed that the injected and the stored beam have the same beam stay-clear, in units of their rms beam size.

The injected beam must be positioned at the center of the aperture when it reaches the kicker. Let R denote the 2×2 transport matrix between the septum and the kicker. Requiring that the beam be on-axis at the kicker, the condition $x_{\mathrm{kic}} = R_{11}x_{\mathrm{sep}} + R_{12}x'_{\mathrm{sep}} = 0$, where x_{kic} denotes the position of the beam with respect to the center of the kicker magnet, determines the correlation of angle x'_{sep} and offset x_{sep} of the injected beam at the exit of the septum:

$$x'_{\mathrm{sep}} = -\frac{R_{11}}{R_{12}} x_{\mathrm{sep}} = -\frac{\alpha_{\mathrm{sep}} + \cot\mu}{\beta_{\mathrm{sep}}} x_{\mathrm{sep}} , \tag{9.2}$$

where μ denotes the phase advance from septum to kicker, and α_{sep} and β_{sep} are the beta and alpha function at the septum. The angle x'_{sep} can be adjusted by changing the strength of the septum magnet, in order to meet condition (9.2).

Using (9.2) the slope of the injected beam at the kicker is

$$x'_{\mathrm{kic}} = R_{21}x_{\mathrm{sep}} + R_{22}x'_{\mathrm{sep}} = \left(R_{21} - \frac{R_{11}}{R_{12}}R_{22}\right) x_{\mathrm{sep}} = -\frac{1}{R_{12}} x_{\mathrm{sep}} . \tag{9.3}$$

To position the injected beam on the design orbit in the ring, the kicker must apply the opposite angular deflection, which is

$$\theta_{\mathrm{kic}} = -x'_{\mathrm{kic}} = \frac{x_{\mathrm{sep}}}{\sqrt{\beta_{\mathrm{sep}}\beta_{\mathrm{kic}}}\sin\mu} . \tag{9.4}$$

A large value of the beta function at the kicker, β_{kic}, reduces the kicker strength, and a large value of β_{sep} reduces the relative contribution to θ_{kic} which arises from the septum thickness d_{sep} (see (9.1) and recall that $\sigma_{xi}, \sigma_{xs} \propto \sqrt{\beta_{\mathrm{sep}}}$).

In case of a FODO lattice, the septum and kicker are best placed downstream of a focusing quadrupole, where the beta functions are close to maximum. In the particular case that the phase advance μ is $\pi/2$, the above formulae simplify to $x'_{\mathrm{sep}} = -\alpha_{\mathrm{sep}}x_{\mathrm{sep}}/\beta_{\mathrm{sep}}$, $\theta_{\mathrm{kic}} = x_{\mathrm{sep}}/\sqrt{\beta_{\mathrm{sep}}\beta_{\mathrm{kic}}}$.

The septum may consist of either dc septum magnets or dc electrostatic wires. In either case, the stray or leakage fields of the septum are a concern. These nonlinear fields can affect the quality of the stored beam. The stray fields may be reduced by a magnetic shielding, which extends beyond the septum ends.

The kicker magnets must be fast, since their rise and fall times often determine the minimum size of the gaps between bunches or between bunch trains in the ring, which in turn may depend on the repetition rate of the accelerator. Typical time constants are tens of nanoseconds, and voltage and current levels of 80 kV and 5000 A, with fields of 500 Gauss, are not uncommon. Frequently ferrites are used for field containment, and sometimes a ceramic vacuum chamber is inserted between ferrite and the beam, with a conducting layer deposited on the inside of the ceramic. The coated ceramic reduces the impedance seen by the beam, as well as the associated heating of the ferrite. It is remarkable that the conducting layer can be much smaller than the skin depth and still provide adequate shielding of the beam fields, since shielding occurs already when the thickness of the metal coating is larger than the square of the skin depth divided by the thickness of the ceramic [3]. The coupling impedance experienced by the beam should be measured prior to installation of the kicker chamber. As an example, measurements for a prototype LHC kicker chamber are documented in [4]. A non-flat shape of the kicker pulse results in unequal deflections for different bunches in the beam. If necessary, a double-kicker system can relax the tolerance on the pulse shape (see the discussion of the KEK/ATF extraction scheme in Sect. 9.7).

Other important injection issues include transient beam loading and phase-space matching. Reference [5] gives a thorough review of beam loading compensation in storage rings, including a discussion of direct rf feedback and of problems that can arise from klystron power limititions. Horizontal and vertical dispersion and beta functions, as well as the ratio of bunch length and energy spread must be matched to the ring (or linac) optics. The transverse optics is matched by varying the strength of quadrupole magnets in the injection line, and, in case of dispersion, possibly also the strength of steering magnets. The longitudinal matching can be achieved by optimizing the amplitude of the storage-ring rf voltage, or, in certain cases, via bunch rotation, bunch compression, or energy compression prior to injection (cf. Sect. 8).

9.2 Multi-Turn Injection

For many applications, the beam is first accumulated in a ring, to increase its intensity, before it is further accelerated or sent towards its final destination. Typically a current-limited cw beam from the injector is thereby converted into a pulsed beam at higher-intensity and energy. The accumulation requires multi-turn injection.

9.2.1 Transverse Multi-Turn Injection

Multi-turn injection usually employs a ramped orbit bump in the vicinity of the septum, i.e., a slow change in the position of the ring closed orbit from turn to turn, instead of a fast kicker. The injection scheme is different for electrons and for protons or heavy ions.

In case of electron rings, radiation damping is utilized. First, a single bunch is injected. Then the orbit bump is reduced over a few revolution periods. After a few damping times, when the beam size has shrunk to its small equilibrium value, the orbit bump is reintroduced, and another bunch is injected into the same bucket. Similar schemes, though usually in the synchrotron phase space (see below), are employed for accumulation of electron cooled proton beams and stochastically cooled antiproton beams.

For proton or heavy ions beams, the orbit bump is reduced slowly in time, and bunches are injected into different regions of the ring acceptance, so that the early bunches occupy the central region, and the later ones the outer parts of the acceptance. Some emittance dilution is inherent to this scheme. If beam is injected over N_t turns ('N_t-turn injection'), the final emittance can be estimated from the rough formula [1]:

$$\epsilon_f > 1.5 N_t \epsilon_i \ . \tag{9.5}$$

Much larger emittance dilutions arise at low beam energy or high intensity, when space charge effects are important.

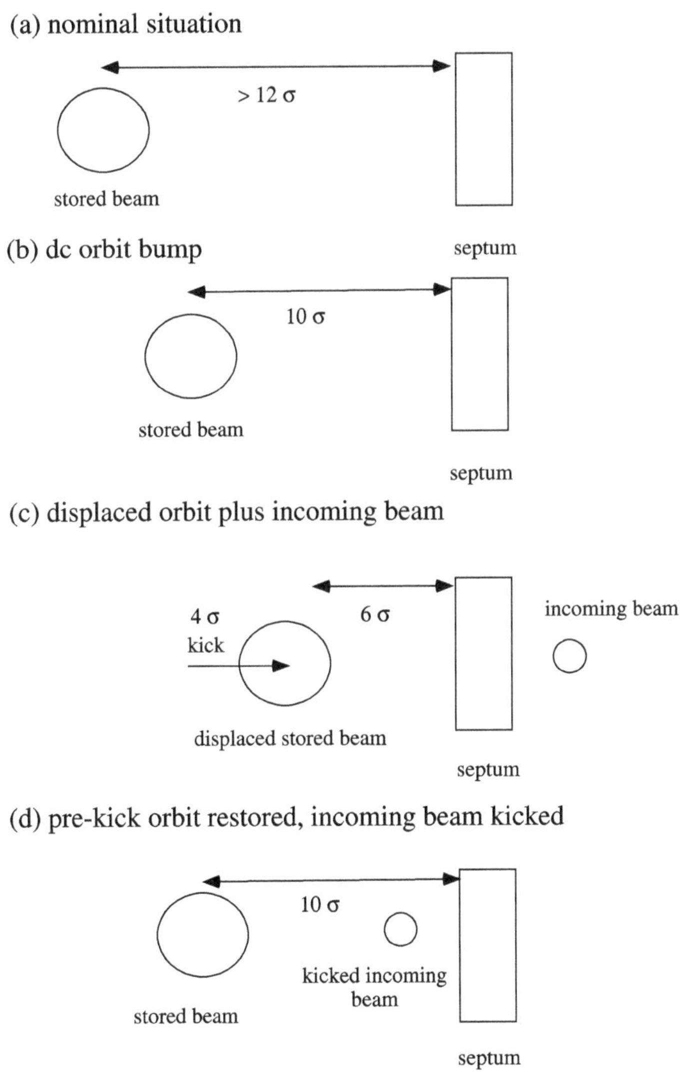

Fig. 9.2. Injection process for PEP-II [6] (Courtesy M. Donald, 2002)

It is also possible and indeed more elegant, to install two kickers in the ring, powered in parallel by the same pulser. They are arranged such that only the second kicker of the pair deflects the injected beam, while both kickers act on the stored beam. If the phase advance between the two kickers is π or 2π, and the sign of the kick appropriately chosen, the kicker deflections generate a closed bump for the circulating beam. The advantage is that in this case the rise and fall times of the kicker do not have to be smaller than the bunch spacing, but can be on the order of the revolution time. The requirements on the kickers can be further alleviated by a dc orbit bump, which brings the stored beam closer to the septum prior to the injection. Figure 9.2 illustrates such a scheme, which is used at PEP-II [6].

9.2.2 Longitudinal and Transverse Multi-Turn Injection

The accumulation efficiency can be increased by combining transverse and longitudinal injection. This option was studied for LEAR [7], where the injected bunches come from a linac, which allows for an easy variation of their

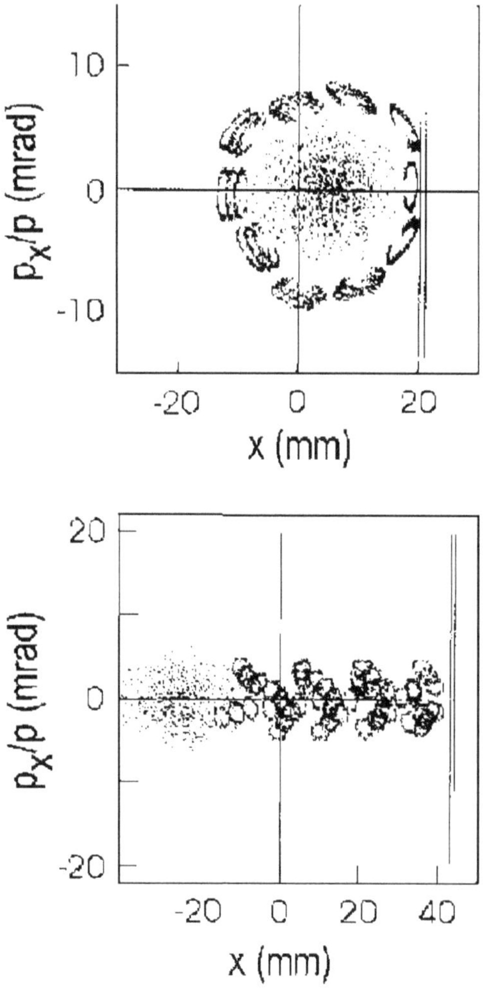

Fig. 9.3. Plots of the simulated horizontal phase space for multiturn injection into LEAR [7]: (*top*) 10 turns after start of purely transverse injection and (*bottom*) 20 turns after start of combined longitudinal transverse injection. Each bunch is represented by three ellipses with slightly different momentum deviations. The *two vertical lines* on the right represent the thickness of the septum (Courtesy Ch. Carli, 2002)

energy. As in the purely transverse multi-turn injection scheme, a local orbit bump is created and then decreased during the injection of successive bunches. At the same time the linac energy is ramped such that, at the injection septum, the closed orbit corresponding to the instantaneous linac energy remains constant. In other words, the change in the bump amplitude $x_{\mathrm{bump}}(t)$ and the simultaneous variation of the momentum $\delta(t)$ are related by

$$D_x \delta(t) = -x_{\mathrm{bump}}(t) + x_0 \ , \tag{9.6}$$

where x_0 is a constant and D_x the horizontal dispersion at the injection septum. In this scheme, the final transverse emittance is smaller than for the conventional multi-turn injection at the expense of an increased momentum spread. Figure 9.3 compares simulated phase space distributions [7] for transverse and combined injection into LEAR. Figure 9.4 shows the predicted improvement in the accumulation efficiency [7].

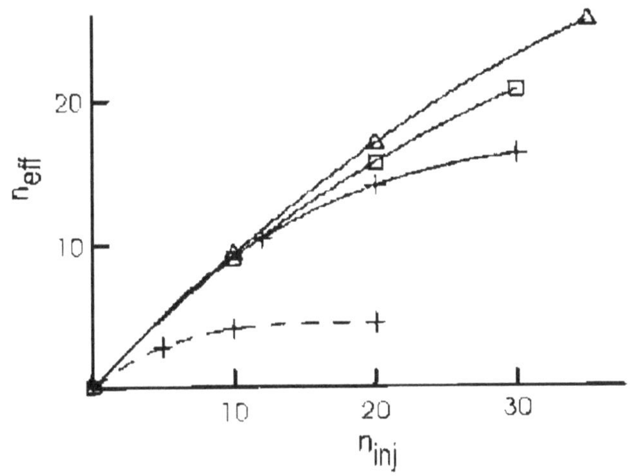

Fig. 9.4. Effective number of turns n_{eff} stored in LEAR as a function of the number of injected turns n_{inj} [7]. The *solid lines* represent the combined transverse longitudinal injection scheme, for three different LEAR optics (*crosses*: standard optics; *triangles*: improved optics with large beta functions at the electron cooler; *squares*: improved optics with intermediate beta functions at the electron cooler). For comparison, the *dashed line* is for a purely transverse injection and the standard optics (Courtesy Ch. Carli, 2002)

9.2.3 Longitudinal Multiturn Injection

If for an electron ring the time between subsequent injections is short compared with the radiation damping time, multiturn transverse injection becomes difficult. In such a case, longitudinal injection offers a solution. Here the circulating beam is brought close to the septum with an ac bump and the incoming beam is injected with a negative energy offset such that the product of this offset in energy and the dispersion is equal to the distance between the newly injected and the stored beam. The injected bunches execute slow synchrotron oscillations.

Consider as an example the injection scheme employed at LEP [8]. Half a synchrotron period after the first bunch is injected, the next injection occurs. At this time, the first bunch is at its maximum distance from the septum. The situation is illustrated in Fig. 9.5. Similarily one could conceive injecting every 1/4 oscillation period, thus accumulating 4 injected bunches in one rf bucket. An advantage of longitudinal injection is a factor two faster radiation damping of the injection oscillations, since the longitudinal damping partition number is twice the transverse ($J_z \approx 2J_x$; compare Sect. 4.3). A possible disadvantage is that the acceptable time separation of successive injections is constrained by the synchrotron frequency.

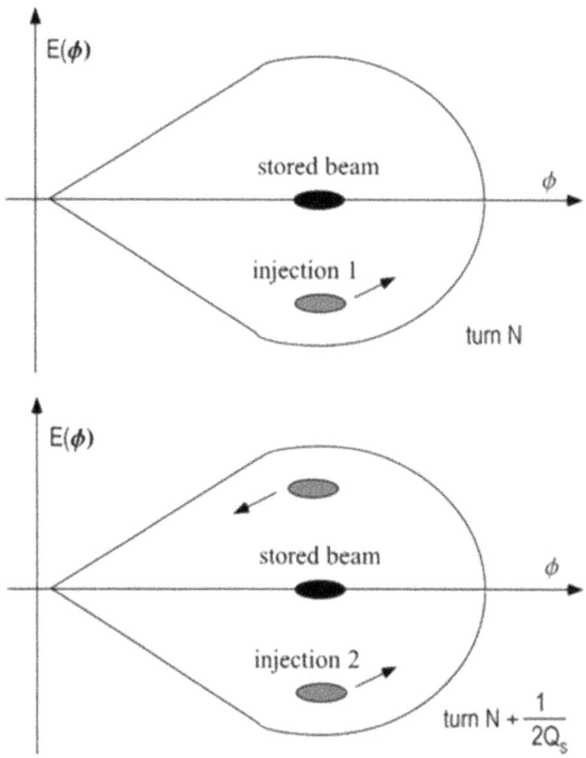

Fig. 9.5. Double injection into the same rf bucket; bunches are injected off-energy at a point with dispersion [8]. Time between the two injections is half a synchrotron oscillation period modulo a full period (Courtesy P. Baudrenghien and P. Collier, 2001)

9.2.4 Phase-Space Painting

For proton and ions beams, the multi-turn injection is often described as "phase-space painting" [9, 10]. This term refers to the injection of many small (linac) bunches into different spots of a 2- or 6-dimensional storage-ring phase space, so as to generate a desired beam distribution, e.g., an approximately uniform distribution with reduced space-charge effects. The combined longitudinal-transverse injection into LEAR discussed in Sect. 9.2.2 can be considered as an example of phase-space painting. In the simplest case the beam is injected at a fixed position in the longitudinal (or transverse) phase space, and the painting is done automatically by the synchrotron oscillations.

During the injection process, the position of the injected beam in phase space can also be moved adiabatically, i.e., at a speed which is slow compared with the synchrotron oscillations. The injected phase-space density $P(r)$ and its projection $p(x)$ are related via

$$p(x) = 2 \int_x^R \frac{rP(r)\,\mathrm{d}r}{\sqrt{r^2 - x^2}} , \tag{9.7}$$

where R denotes the maximum radius in phase space at which bunches are injected. From a desired function $p(x)$, the corresponding radial density $P(r)$ can be computed using (9.7). The radial increment in the injected beam position between two successive bunches is [9]

$$\Delta r \approx \frac{1}{2\pi r P(r) N_{\text{inj}}} , \tag{9.8}$$

with N_{inj} the total number of injected bunches.

More complicated schemes are frequently used. Similar to the above horizontal-longitudinal injection for LEAR, one can also combine horizontal and vertical painting. For example, at Rutherford Appleton Laboratory (RAL) a vertical steering magnet in the injection line is ramped, while the guide field in the ring is decreased [10]. Initially, there are small horizontal and large vertical oscillations, while at the end of the injection the situation is reversed. Instead of the vertical steering in the injection line, a programmable vertical orbit bump in the ring could be employed alternatively.

9.3 H^- Charge Exchange Injection

The principle of injection using H^- exchange originated in Novosibirsk [11]. It is now the preferred injection scheme for proton machines [1]. In this scheme H^- ions are accelerated by a linac and are stripped to protons, when they traverse a thin foil during injection into the ring [1, 12], as illustrated in Fig. 9.6.

The stripping of the H^- ions to protons occurs within the ring acceptance. Since during the stripping the particles change their charge, Liouville's theorem on the conservation of the beam density in phase space does not apply, and, thus, in principle, a high proton density could be attained by injecting successive bunches into the same region of phase space. In most practical applications, however, vertical steering in the injection line is combined with a

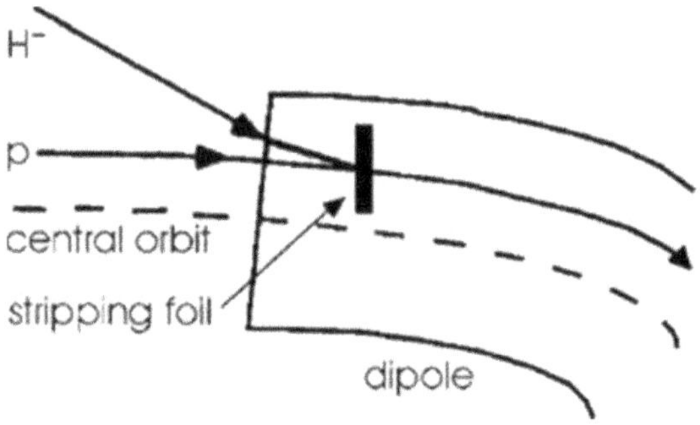

Fig. 9.6. Schematic of H^- stripping injection

ramped horizontal orbit bump in the ring in order to provide a very uniform filling of the phase space and to minimize space-charge effects.

The heating of the stripping foil and stripper scattering effects limit the foil thickness. Typical thicknesses range between 50 and 200 $\mu\mathrm{g\,cm}^{-2}$ (less than 1 μm), with stripping efficiencies of 98% for 50 MeV protons [1]. As foil materials, polyparaxylene, carbon and aluminum oxide have been used. The rms scattering angle for a single foil traversal is typically on the order 0.2 mrad. The total scattering angle increases as the square root of the average number of passages through the foil. The stripping foils are supported at three edges, so that vertical beam motion cannot does not reduce the number of foil traversals.

Stacking simultaneously in both betatron and synchrotron phase space reduces the number of foil traversals, however. This can be achieved by changing the magnitude of an orbit bump, while also ramping the frequency and phase of the rf system during the injection cycle. The following lattice parameters at the location of the foil are considered advantageous [1]: $D_x' = 0$, $\alpha_x = 0$, and $D(\Delta p/p) > \sqrt{A\beta}$, where $\Delta p/p$ is the momentum acceptance and A is the transverse acceptance (in emittance units). A location between two symmetric defocusing quadrupoles is suitable for meeting these conditions. The second quadrupole assists in deflecting the unstripped H^- ions.

Stripping foils are used for heavy ions as well. The final charge distribution of the ions depends on the foil thickness and on the particle energy [13]. In the extreme case, the ions can be fully stripped.

Foils are also used for extracting H^0 atoms or protons from H^- storage rings, where the neutral atoms are generated by capturing electrons from the foil. For these applications, also a gas jet or a laser beam [14] can facilitate the extraction in a similar way.

9.4 Resonant Injection

Another injection scheme worth mentioning is a proposal for resonant injection [15]. Here 'bumper' magnets with dipole, quadrupole and octupole fields are excited to produce a separatrix with two stable regions in phase space;

the stored beam is in one region, and a bunch is injected into the other region. Afterwards, the fields are adjusted to merge the two parts of the beam. Then the injection condition is reestablished.

9.5 Continuous Injection

Continuous injection has been proposed as a means for maximizing the luminosity of a circular collider [16]. The motivation is obvious; if the stored beam could be continually replenished so that the current per bunch stays constant, then the average luminosity would roughly equal the peak luminosity. Continuous injection also reduces fill-to-fill variations and avoids transient phenomena, e.g., transient beam loading, thereby establishing quasi-static conditions, which is of interest not only for colliders, but also for light sources. Finally, in a colliding-beam storage ring the beam lifetime τ decreases inversely with the luminosity L,

$$\frac{1}{\tau} = -\frac{1}{N}\frac{\mathrm{d}N}{\mathrm{d}t} \propto \frac{L}{N} . \tag{9.9}$$

Thus, continuous injection supporting a much reduced lifetime could provide a substantial gain in average luminosity.

As an example, taking all these effects together, continuous injection is estimated [16] to potentially increase the average luminosity of the PEP-II B factory by about a factor of 5, assuming that each bunch in both rings can be replenished every 2.1 s. In this example, a 67-ns long orbit bump would move the injected bunches transversely to about 4σ from the stored beam core. This is done so that the injected bunches have an unobstructed passage through the physics detector. The effective minimum beam lifetime which can be supported is given by [16]

$$\tau = \frac{N_b}{\Delta N_b}\,\Delta t , \tag{9.10}$$

where N_b is the nominal number of particles per bunch, ΔN_b is the number of particles added into a single bunch per injection, and Δt the time period between successive injections into the same bunch. Extreme parameters for PEP-II are [16] $N_b \approx 1.2 \times 10^{11}$, $\Delta N_b \approx 10^9$, and $\Delta t \approx 2.1$ s, yielding a minimum supportable beam lifetime of $\tau \approx 4.2$ minutes.

9.6 Injection Envelope Matching

At injection into a storage ring, if the incoming beam distribution is not properly matched to the ring optics, the beam envelope in phase space will rotate around the matched design envelope. This oscillation will result in turn-to-turn beam-size variations, which can be measured using a synchrotron light monitor and a fast-gated camera.

An injection-mismatch measurement from the SLC damping ring [17, 18] is shown in Fig. 9.7. The different pictures correspond to successive turns after injection, at the indicated turn number. Each picture is an average over 8 individual images. Clearly visible is a variation of the bunch shape from turn to turn.

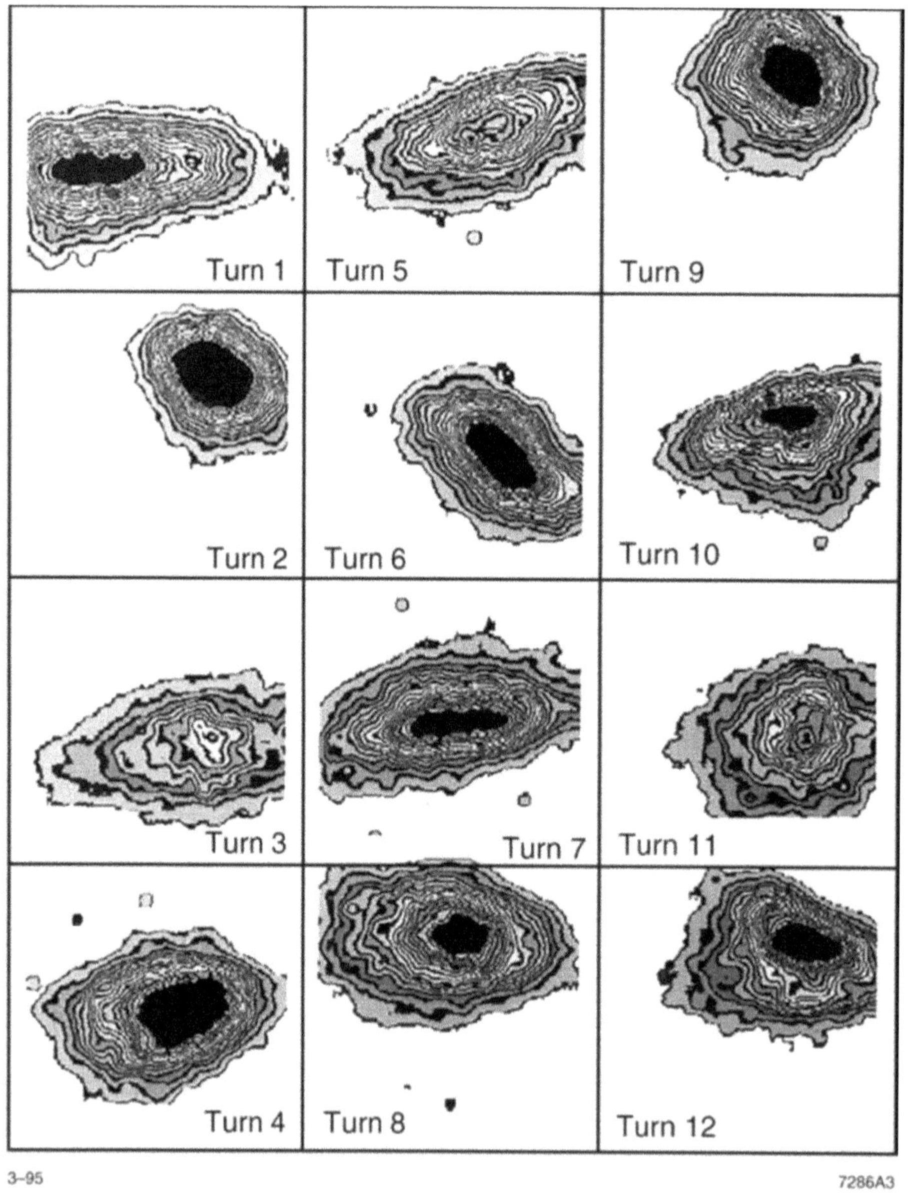

Fig. 9.7. Beam images of the first twelve turns after injection into the SLC damping ring, illustrating the effect of an injection mismatch [17, 18]. These are pictures from a synchrotron light monitor taken with a gated camera. Each image is an average over 8 beam pulses. The beam-size variation from turn to turn is an indication of injection mismatch

At the SLC damping ring, the matching of the injected beam distribution consists of minimizing the measured beam size after 1250 turns, by varying several quadrupoles at the end of the injection beam transport line. A number of 1250 turns was chosen, because at this time the initial beta and dispersion mismatch has completely filamented. Since, on the other hand, the time scale is much shorter than the radiation damping time, the emittance is given directly by $(\mathrm{B_{mag}} \cdot \epsilon)$, where ϵ is the emittance of the injected beam, and $\mathrm{B_{mag}}$ the mismatch factor defined in (4.59).

The beam size variation can be analyzed in the frequency domain by a Fourier transform to determine the amplitude of the mismatch factor and ultimately, for a well calibrated monitor, the beam emittance at injection. A beta mismatch will appear as a frequency line at twice the betatron tune, while a horizontal dispersion mismatch will be evident as a line at the betatron tune itself [18]. If only a beta mismatch is present, the ratio ρ of the dc Fourier component and the component at $2Q_x$ is equal to $(\mathrm{B_{mag}}/\sqrt{\mathrm{B_{mag}^2} - 1})$. From this, $\mathrm{B_{mag}} = 1/\sqrt{1-\rho^{-2}}$ can be determined [18, 19].

Figure 9.8 shows the beam size squared for the first 100 turns after injection, as well as the FFT (multiplied with γ/β_x where γ is the relativistic Lorentz factor and β_x the beta function). Clearly visible are peaks at $2Q_x$ in the horizontal signal and at $(1-2Q_y)$ in the vertical one. The final emittance after filamentation, $(\mathrm{B_{mag}} \cdot \epsilon)$, is given by the dc component of the FFT.

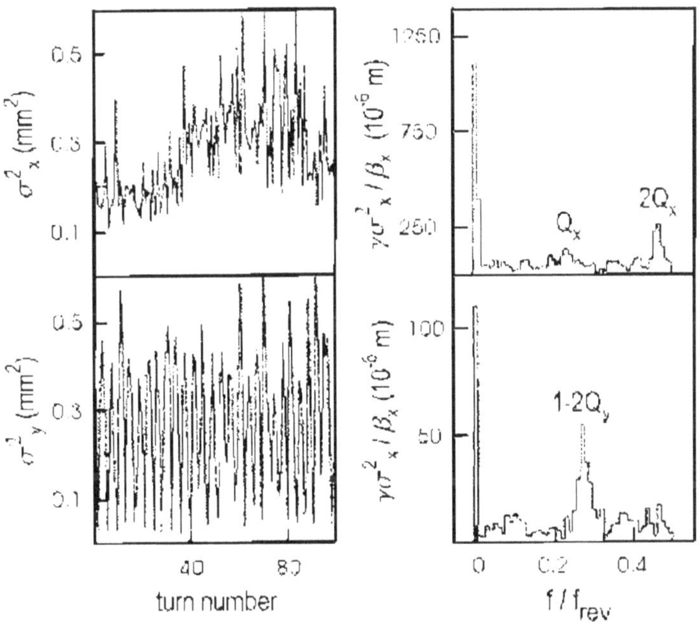

Fig. 9.8. Horizontal (*top*) and vertical (*bottom*) beam sizes for the first 100 turns after injection into the SLC damping ring (*left*) and their FFT (*right*) [18]. Clearly visible in the frequency spectra are lines at $2Q_x$ (*top*) and at $(1-2Q_y)$ (*bottom*), whose amplitude is a measure of the amount of beta mismatch

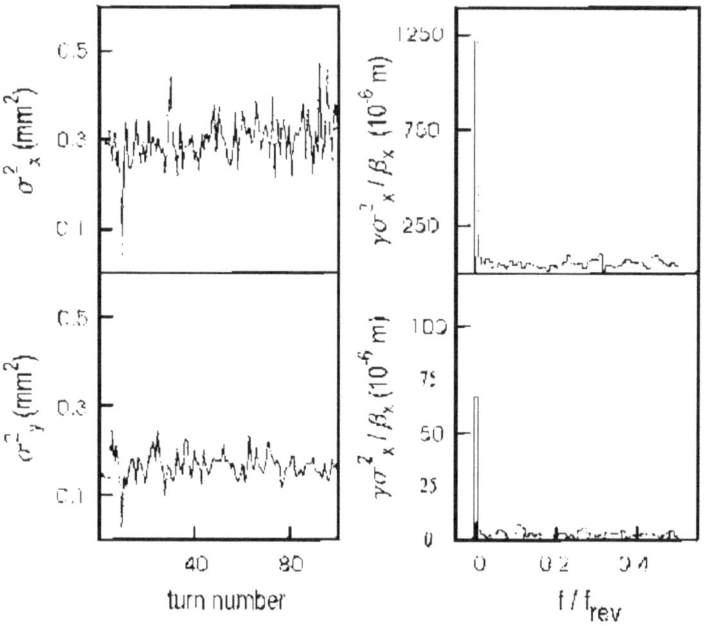

Fig. 9.9. Same as Fig. 9.8, after beta matching [18]. The peaks at twice the betatron tune have disappeared

The standard matching procedure reduces the FFT signals at $2Q_x$, $(1-2Q_y)$ and Q_x, as illustrated in Fig. 9.9.

9.7 Fast Extraction

Fast extraction is similar to single-turn injection. Orbit bumps are generated which move the stored beam close to a septum magnet. Then a fast kicker is fired, which deflects the next bunch, or group of bunches, into the extraction channel. If only one kicker is used, the kicker rise time must normally be smaller than the separation between two circulating bunches (or groups of bunches). The pulse length and fall time of the kicker are determined by the number of bunches to be extracted, and by the ring fill pattern. From (9.4), the minimum deflection angle required is

$$\theta_{\text{kic}} = \frac{x_{\text{sep}}}{\sqrt{\beta_{\text{sep}}\beta_{\text{kic}}}\sin\mu} , \tag{9.11}$$

where β_{sep} and β_{kic} are the beta functions at septum and kicker, μ is the phase advance between these two elements, and x_{sep} is the minimum displacement at the septum required for clean extraction. The initial orbit bump reduces the value of x_{sep}. In a FODO lattice with a phase advance per cell of, e.g., 90°, the kicker can be positioned just upstream of the focusing quadrupole, and the septum at the identical position one cell downstream. In this way, the beta functions are maximized, and so the deflection angle required for the kicker is minimized.

For extraction from the damping ring of a linear collider, it is extremely important that the deflection imparted by the kicker has a very small pulse-to-pulse fluctuation ('jitter') and is sufficiently flat over the length of a bunch train. In order to confine the orbit variation at the interaction point (IP) to $0.1\sigma^*$ (σ^* is the IP spot size), the orbit variation at the septum should be smaller than $0.1\sigma_{\text{sep}}$. The tolerance on the relative deflection error then is $\Delta\theta_{\text{kic}}/\theta_{\text{kic}} < 0.1\sigma_{\text{sep}}/x_{\text{sep}}$, where σ_{sep} is the rms beam size at the septum, and x_{sep} the transverse displacement of the kicked beam. Without fast orbit bumps, this can also be rewritten as [20]

$$\frac{\Delta\theta_{\text{kic}}}{\theta_{\text{kic}}} \leq \frac{1}{10} \frac{\sqrt{\epsilon_{\text{ext}}\beta_{\text{sep}}}}{d_{\text{sep}} + n_s\sqrt{\epsilon_{\text{inj}}\beta_{\text{sep}}}} , \tag{9.12}$$

where ϵ_{inj} and ϵ_{ext} are the injected and extracted beam emittances, β_{sep} the beta function at the septum, and n_s the distance between the closed orbit and the septum plate in units of the injected rms beam size, when the beam is largest (i.e., the injected beam size enters, because the aperture at the septum must be large enough to accommodate the injected beam). For electron rings, one must have $n_s \geq 7$. Using typical parameters for a linear-collider damping ring, the relative jitter tolerance for the kicker, $\Delta\theta_{\text{kic}}/\theta_{\text{kic}}$, is on the order of a few 10^{-4}, and it is mainly determined by the ratio of the extracted beam emittance to the injected emittance [20]. A possible solution is the use of a double kicker system, separated by a betatron phase advance of π, to cancel the jitter [20, 21]. This compensation scheme is illustrated in Fig. 9.10. One kicker would be placed before the septum and the other in the extraction line. If a pulser feeds both kickers in parallel, with appropriate cable delays, kicker pulse errors in the first magnet are canceled by those in the second

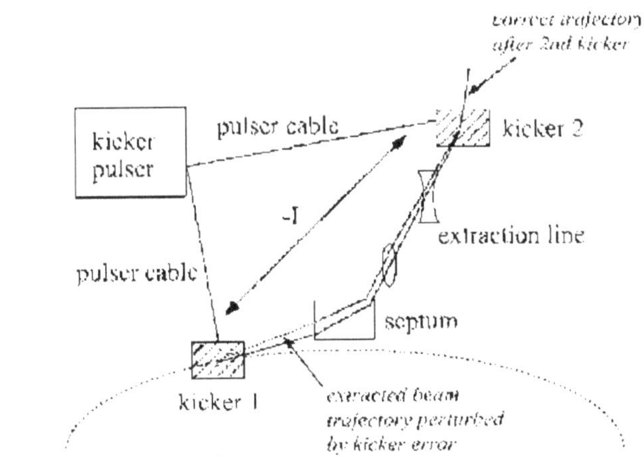

Fig. 9.10. Damping ring extraction with double kicker system, for reducing downstream beam orbit fluctuations. A change in the pulse shape alters the deflection from both kickers equally. The R matrix between the two kickers and cable delay times are chosen such that the effects of the two deflection errors cancel exactly, and the final beam trajectory is unchanged

magnet. A double-kicker system of this type has been built and installed at the KEK ATF damping ring [22].

Similar techniques can be applied to compensate for drifts of the septum field. For example, the NLC design contemplates the use of a compensating bending magnet in the extraction line, which is powered in series with the septum and placed such that field fluctuations will cancel [21].

9.8 Kickers

There are several different types of kicker magnets [6, 23] such as: (1) a current loop inside the vacuum, (2) a terminated transmission line inside the vacuum, (3) a ferrite magnet outside the vacuum, and (4) a multi-cell transmission line with ferrite flux returns [21]. As an example, Fig. 9.11 shows the ferrite kicker and the kicker pulser circuit adopted for PEP-II [6].

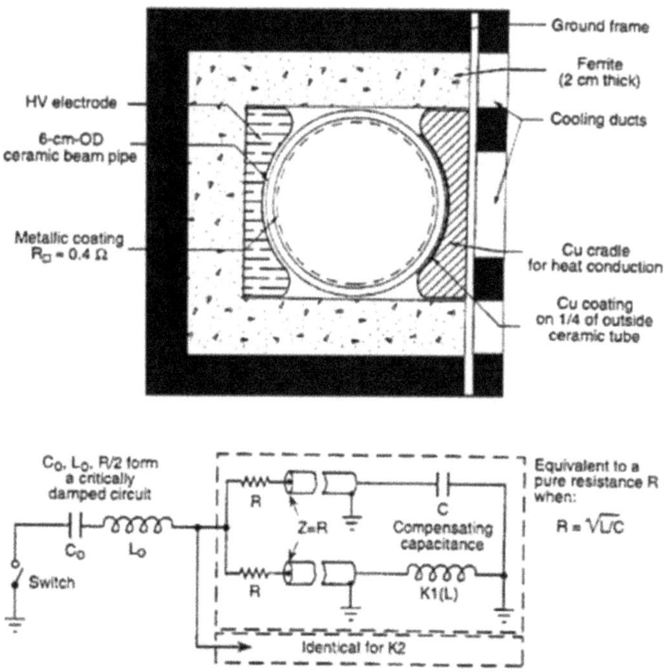

Fig. 9.11. Schematic of PEP-II kickers [6]: (*top*) kicker magnet cross section; (*bottom*) pulsing circuit with FET switch (Courtesy M. Zisman, 2002)

Typical kicker rise and fall times are 50–150 ns (SLC and NLC design: 60 ns, PEP-II: 120 ns). For a fast horizontal kicker with ferrite yoke, the characteristic time constant of the kicker scales as lw/g, where l is the length, w the width, and g the vertical gap of the kicker. This time constant can be reduced by dividing the kicker into several segments of shorter length. The kicker magnets are powered by kicker pulsers, usually based on thyratron cable discharges. The pulse shape can be modified by adding filters and capacitors in parallel with the charge line. Spark gaps and solid-state FETs, such as in Fig. 9.11, are thyratron alternatives with potentially shorter rise and fall times [21].

For many future applications with closely spaced bunch trains, shorter kicker time constants are desired. A very fast counter-travelling wave kicker was designed and built for the TESLA project [24]. This kicker scheme uses two parallel conducting plates or electrodes. These are excited by short pulses from a generator, generating an electromagnetic wave which travels in a direction opposite to the beam, and produces a horizontal kick. At the end of

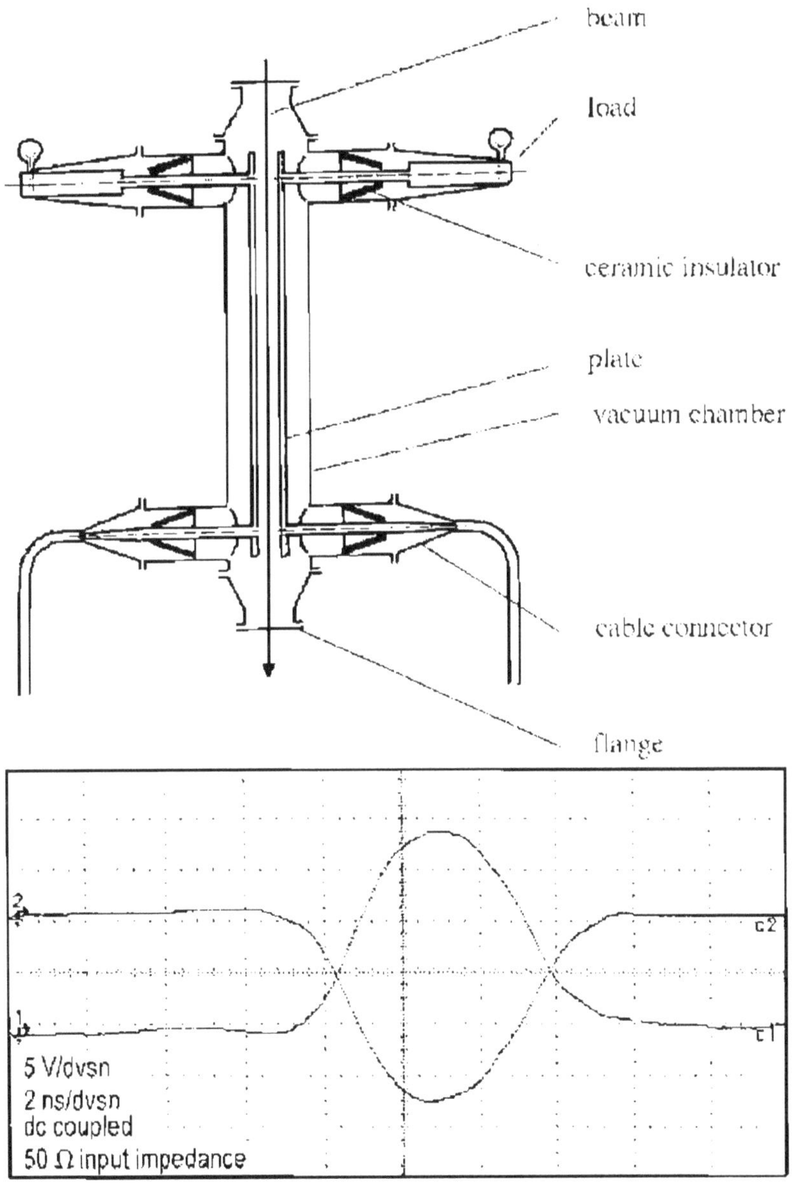

Fig. 9.12. Very fast kicker prototype [24]: (*top*) layout; (*bottom*) measured output rf pulses (Courtesy V. Shiltsev, 1998)

the kicker plates, the wave passes into two ceramic outputs, in which ideally it is fully absorbed without reflection. For a beam travelling opposite to the wave the effects of the magnetic and electric forces add, while they cancel each other for a beam moving in the same direction. The maximum integrated kicker strength in units of voltage is given by

$$S_0 \approx \frac{2U_m l}{a} , \tag{9.13}$$

where U_m is the maximum pulse voltage at each plate, a is the half aperture, and l the length. A kicker was tested with $U_m = 2$ kV, $a = 25$ mm, and total length l of 0.5 m. Figure 9.12 shows output pulses measured on this kicker prototype, demonstrating a zero-to-zero pulse length of about 6 ns. The maximum pulse height corresponds to the predicted kick strength of

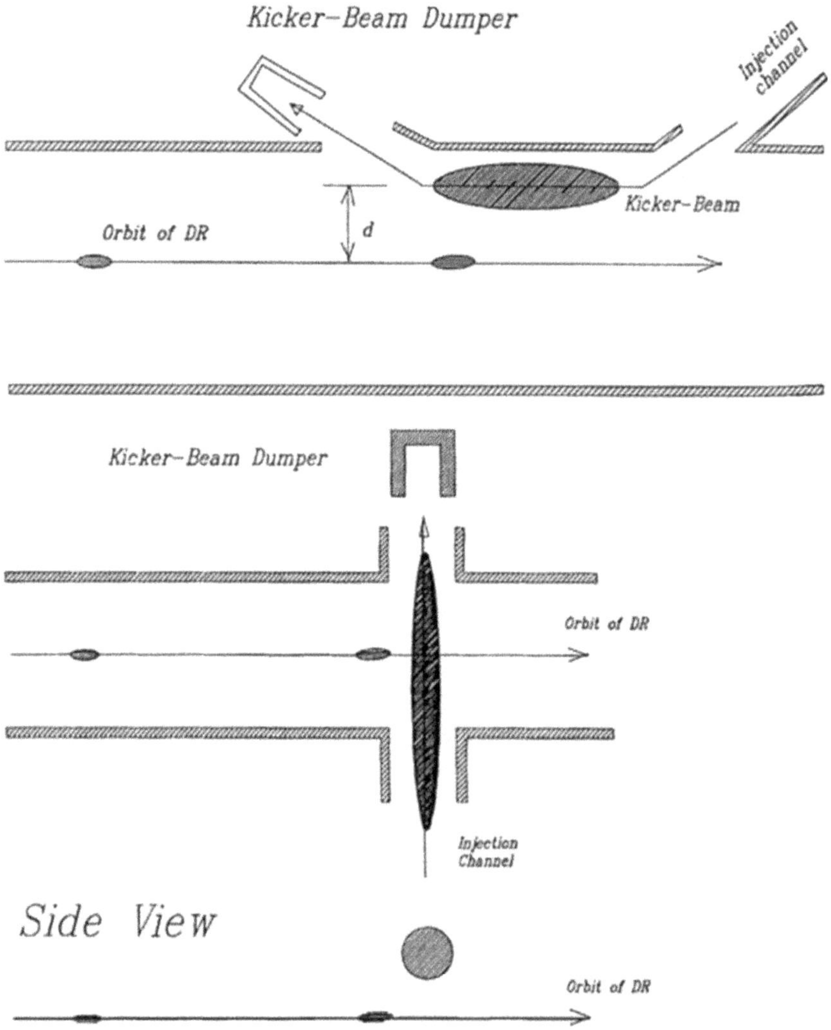

Fig. 9.13. Beam-beam kicker: 'head-on' (*top*) and 'cross' scheme (*bottom*) [25] (Courtesy V. Shiltsev, 1998)

80 kV, or 2.76 Gm. For electron or positron beams with $E = 3.3$ GeV, this would amount to a deflection angle of 24 μrad.

An ultrafast beam-beam kicker was proposed [25], in order to provide even faster kicker pulses. Here, a wide high-charge low-energy bunch traverses the beam pipe either parallel to the beam direction or perpendicular to it. The electro-magnetic or electric field of this bunch is used to deflect (and extract) a bunch circulating in the ring. The pulse length of the beam-beam kicker is determined by the length of the low-energy bunch and can be on the order of 2 ns. Figure 9.13 illustrates two possible geometries.

9.9 Septa

As we have seen in (9.1) and (9.4), a small septum thickness d_{sep} reduces the requirements on the kicker and increases the extraction efficiency. For this reason, electrostatic wire septa have been employed since many years, for example, during fast extraction at the Fermilab Tevatron [26]. The Tevatron electrostatic septum consists of two 354 cm long sections with 86 cm space in between, made from 75% tungsten and 25% rhenium wires of 0.002 inch diameter and 0.1 inch spacing with an angle of 25 μrad between sections. The voltage of 93 kV results in an electric field of 83 kV/cm [26]. Very similar electrostatic deflectors have been proposed for the muon collider [27]. At high energies, the integrated strength of a wire septum often cannot provide a deflection angle large enough for clean extraction, and, in such cases, an additional thin septum magnet is positioned immediately downstream.

In general, two types of septum magnets are widely used [28]: Lambertson iron septum dipoles and current-carrying septum dipoles. A Lambertson magnet is illustrated in Fig. 9.14. The triangular cut-out in the window frame leaves space for the circulating beam. As shown, a kicker deflects the beam horizontally into the septum, by which it is then bent vertically.

Figure 9.15 depicts a current sheet septum. A current carrying septum with thickness d and current density J generates a field $B = \mu_0 J d$. For d of the order of a millimeter, the septum is used in a pulsed mode to provide

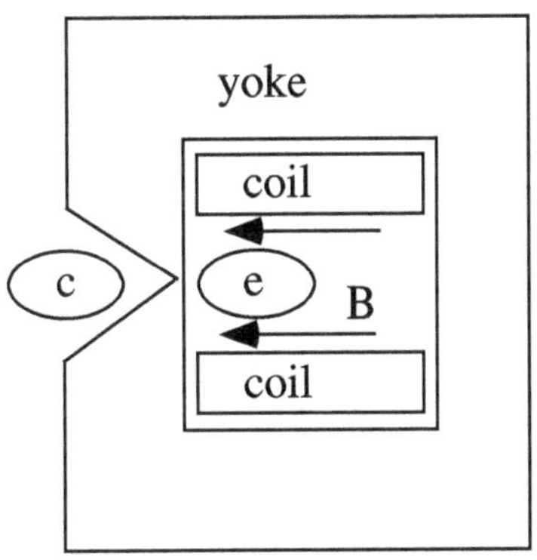

Fig. 9.14. Schematic of Lambertson septum iron magnet [2]; the symbol c represents the circulating beam, the symbol e the extracted beam

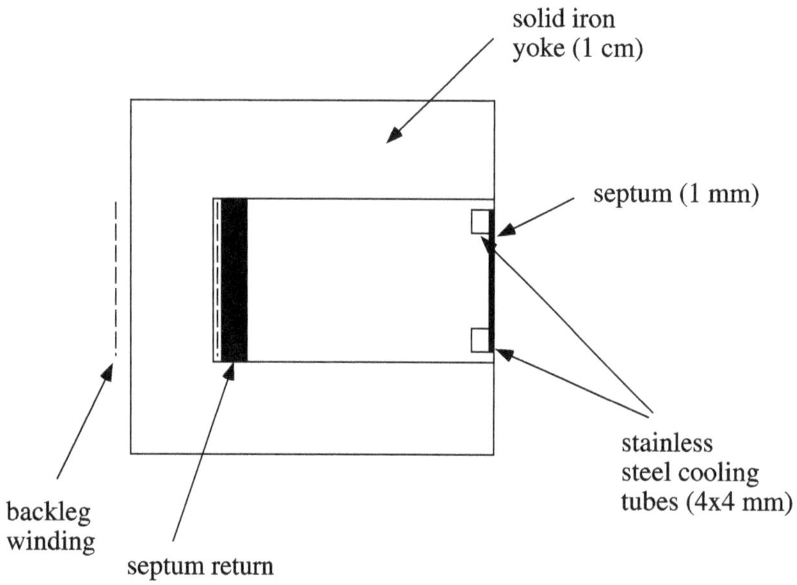

Fig. 9.15. Cross section of current sheet septum [6] (Courtesy M. Zisman, 2002)

enough field strength. For larger thicknesses, dc operation is common. Septum leakage fields which affect the circulating beam are a concern. In addition to normal dipole and higher order fields, the septum stray field may contain a skew quadrupole component.

9.10 Slow Extraction

The beam can be slowly extracted by exciting a third order nonlinear resonance, using sextupoles. Also a second order linear resonance can be used, in combination with octupoles. The extraction efficiency depends on the ratio of the betatron amplitude growth per turn and the septum thickness. It can be improved with a high-beta insertion at the septum.

Figure 9.16 depicts the phase space near the 3rd order resonance, excited by sextupole magnets. Particles inside the inner triangle are stable. Outside the triangle the oscillation amplitude of a particle grows exponentially. Therefore, particles in this region are rapidly lost, along a particular direction in phase space (in this example, towards the right). The size of the triangle depends on the strength of the sextupoles and on the betatron tune.

Near the third-integer resonance, $(3Q-q) \approx 0$, with integer q, the particle motion can be described by a Hamiltonian of the form

$$H(I,\psi,\vartheta) = (Q - q/3)I + \frac{1}{24}(2I)^{3/2}|\tilde{K}_s| \sin(3\psi + \theta_0) , \qquad (9.14)$$

where ϑ is the azimuthal position around the ring, which acts as the time-like variable, and I and ψ are the action-angle variables, which are related to the

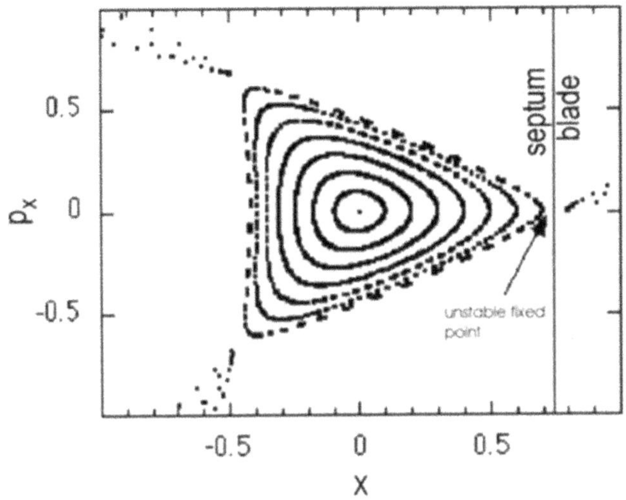

Fig. 9.16. Phase space schematic for slow extraction near the 3rd integer resonance (axes in arbitrary units). The sextupole excitation pattern around the ring is chosen such that the maximum excursion in the horizontal coordinate x occurs at the location of the septum. The position of the septum wire is indicated as a vertical line close to the unstable fixed point

transverse particle coordinates at the septum via $x_{\text{sep}} = \sqrt{2\beta_{\text{sep}} I}\cos\psi$ and $x'_{\text{sep}} = -\sqrt{2I/\beta_{\text{sep}}}\sin\psi - \alpha\sqrt{2I/\beta}\cos\psi$. The term $|\tilde{K}_s|$ is the absolute value and θ_0 the complex phase of the resonant Fourier harmonic of the sextupole distribution around the ring appropriately weighted by the beta function:

$$|\tilde{K}_s|e^{-iq\theta_0} = \frac{1}{2\pi}\int_0^{2\pi} k_s(\theta)\beta^{3/2}(\theta)e^{-iq\theta}\ \mathrm{d}\theta\ . \tag{9.15}$$

Here θ is the azimuthal angle around the ring, and the sextupole strength (in units of m^{-3}) is given by $k_s(\theta) = \partial^2 B_z(\theta)/\partial x^2/(B\rho)$, with $B\rho$ the magnetic rigidity.

Suppose the tune is slightly below the 3rd integer resonance, $(3Q-q) < 0$. Then a corner point of the separatrix coincides with the horizontal position coordinate x_{sep} at the septum, if $\theta_0 = \pi/2$. Above the resonance, $(3Q-q) > 0$, the optimum choice would be $\theta_0 = -\pi/2$. The value of θ_0 can be adjusted by changes to the sextupole configuration, or by changes to the ring optics. The particles arrive at the septum with a large amplitude on every 3rd turn. The amplitude growth over three turns, for a particle near the unstable fixed point (at the asymptotic angle $\psi = \pi/6$), is approximately

$$\Delta x_{\text{sep}} \approx \frac{3\pi x_{\text{sep}}^2|\tilde{K}_s|}{4\beta_{\text{sep}}^{1/2}}\ . \tag{9.16}$$

This shows that large sextupole strengths and a large beta function at the septum (since $x_{\text{sep}} \sim \sqrt{\beta_{\text{sep}}}$) are advantageous.

A slow spill can be controlled by adjusting either the strength of the sextupoles or the betatron tune. Extraction may also involve beam steering. Also making use of chromaticity, particles of different momenta can progressively be brought onto the resonance. Extraction starts when the beam particles at one end of the momentum distribution fill the triangular stable area in phase space. The stable area then shrinks to zero for these particles, and subsequently particles of different momenta are extracted.

To achieve a slow extraction efficiency greater than 98%, the thickness of the septum must typically be of the order of 100 μm.

9.11 Extraction via Resonance Islands

A novel method for multi-turn extraction from a circular particle accelerator was explored at the CERN PS [29, 30]. Here, in addition to fast or slow extraction an intermediate extraction mode is needed, which is called multi-turn extraction. The PS serves as injector to the SPS. The latter has an 11 times larger circumference. In order to fill the SPS ring with only two 'shots' from the PS, each PS beam is extracted over 5 turns.

The conventional technique used for this extraction is called the 'continuous transfer' (CT). The principle is illustrated in Fig. 9.17. The tune is moved closed to the quarter integer resonance. Then the beam is deflected so that a fraction of it is shaved off at the electrostatic septum blade. Three other slices are transferred on subsequent turns. The central part is extracted last during the fifth turn, by applying a larger deflection. Since during extraction the beam is cut into 5 transverse pieces, the slices transferred ideally have a five times lower transverse emittance than the original beam.

However, there are various problems with this approach: (1) beam losses at the septum are unavoidable; (2) the extracted slices do not match the natural circular shape of the phase-space trajectories which implies emittance growth

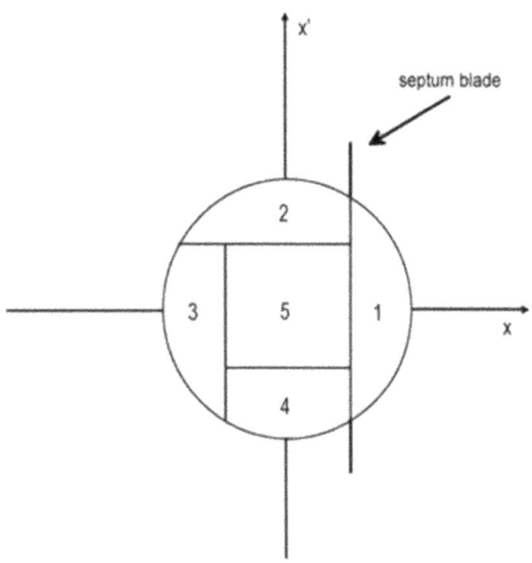

Fig. 9.17. Principle of the conventional 'continuous-transfer' extraction from the CERN PS ring. The beam is shaved by means of an electrostatic septum [29, 30]. The horizontal tune of the PS is set to 6.25 (Courtesy M. Giovannozzi, 2001)

in the downstream SPS; (3) unequal emittances of the extracted slices. For these reasons the CT extraction is not particularly suitable for the CERN neutrino to Gran Sasso (CNGS) proton beam [31].

The alternative novel scheme [29] makes use of stable nonlinear resonance islands. By exciting sextupole and octupole magnets, islands are created in phase space. The position and width of the islands are controlled by moving the betatron tune across the quarter-integer resonance. Initially, the islands are introduced adiabatically near the origin. The beam is thereby split into 5 components, all round in shape, and well matched to the circular phase space structure. Then the tune is shifted away from the resonance, so that the islands separate and approach larger amplitudes. Now the beam can be deflected as in the conventional CT scheme described above.

In this case, however, beam losses can be avoided by deflecting an empty region of phase space (between the resonance islands) onto the septum blade. Each slice is well matched, and, hence, the emittance growth is negligible. Finally, by properly adjusting the island parameters, the slices can be equally populated and be produced such that their emittances are equal. Figure 9.18 shows the proposed tune evolution for this extraction scheme, and Fig. 9.19 the simulated beam distribution at various times of the trapping process. At the end of the process, the islands are well separated. In the simulation, no particles are lost, neither during the island creation ('capture process') nor when shifting the island positions.

Open questions concern the quantitative relation between slice emittance and island parameters, the optimization of the tune change, and the robustness against perturbing effects such as tune modulation, e.g., caused by power-supply ripple.

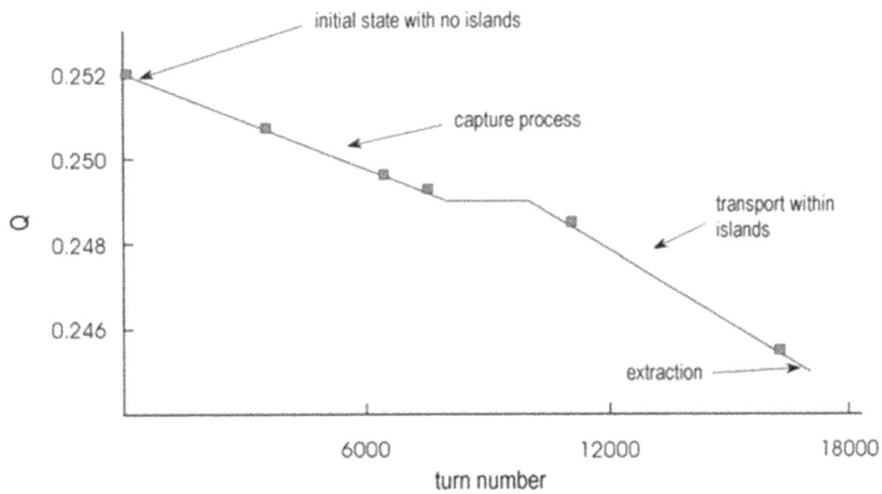

Fig. 9.18. Variation of the small-amplitude tune Q as a function of turn number n during resonant multiturn extraction [29]. The *solid squares* refer to tune values for which phase-space portraits are shown in Fig. 9.19 (Courtesy M. Giovannozzi, 2001)

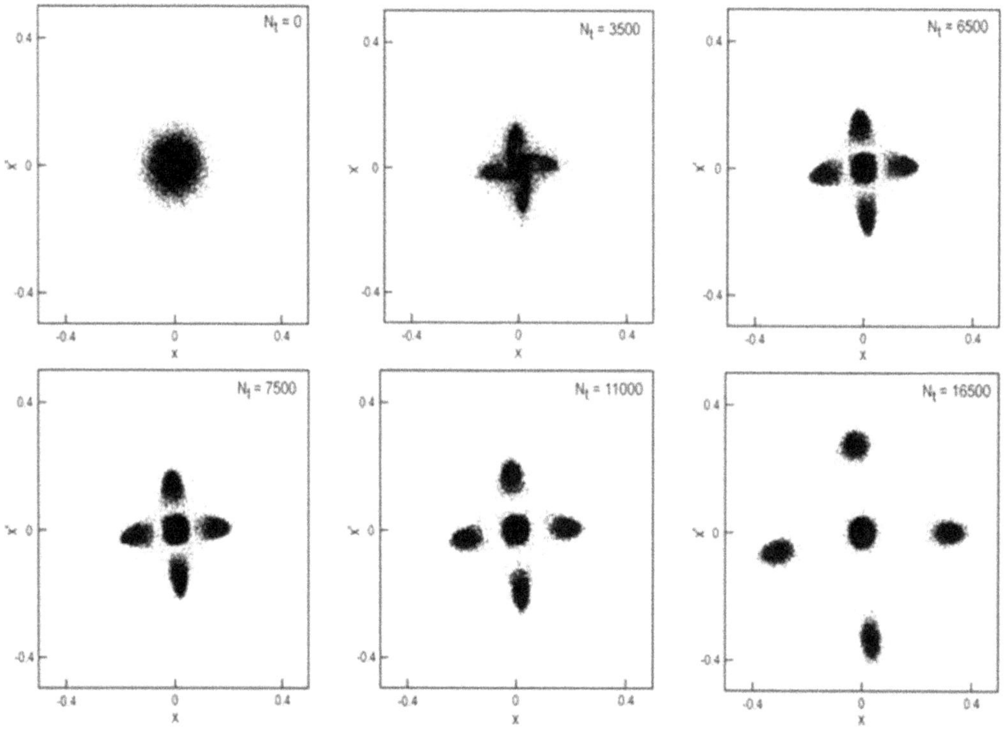

Fig. 9.19. Simulated evolution of the beam distribution during the trapping process of resonant multiturn extraction (axes in arbitrary units) [29]. The different plots correspond to the tune values which are represented by solid squares in Fig. 9.18. Each plot represents 2.25×10^4 points (Courtesy M. Giovannozzi, 2001)

9.12 Beam Separation

A problem similar to injection and extraction is the beam separation near the collision point of a collider, or the beam (re-)combination before and after the arcs of a recirculating linac.

To illustrate the concept and a possible approach, we describe a design example for horizontal beam separation at a Very Large Lepton Collider (VLLC) [32]. The purpose of the beam separation is to feed the two beams into the two separate magnet channels of the collider arcs. The separation is launched in the straight section close to the collision point. The optical lattice in the straight is assumed to be the same FODO lattice as in the arcs, with a cell length L_p and quadrupole focal length f.

An electrostatic separator is placed next to a horizontally focusing quadrupole. Its field is chosen such that the beam are offset by $\pm N_\sigma$ rms beam sizes at the next quadrupole downstream, which is horizontally defocusing. The separation $N_\sigma \sigma_D$ (where σ_D denotes the rms beam size at the quadrupole), the deflection angle ϕ_e and the integrated strength of the septum are related via [32]

$$\phi_e = \frac{2 N_\sigma \sigma_D}{L_p} , \tag{9.17}$$

where

$$E_x l_e = \phi_e E \ , \tag{9.18}$$

in which E_x denotes the (horizontal) electric field and E the beam energy. A LEP separator consists of 4-m long electrodes, and provides a nominal field of 2.5 MV over a gap of 0.11 m [33]. About two of these separators would be needed for the VLLC application (beam energy 184 GeV) [32]. The defocusing quadrupole enhances the slope between the two beams. Downstream are dc magnetic septum magnets with opposite vertical fields on either side of a current sheet. They add enough slope to the beam that they may be brought into separate channels at the next focusing quadrupole. We denote the half separation of the two channels at that (focusing) quadrupole by d_m, the deflection angle of the magnetic septum by ϕ_m, its integrated magnetic field by Bl_m, and the electrostatic separation at the intermediate defocusing quadrupole with focal length f by x_s. The required deflection by the magnetic septum is then given by

$$\phi_m = \frac{2(d_m - x_s)}{L_p} - (\phi_e + x_s/f) \ . \tag{9.19}$$

The deflection ϕ_m is generated as

$$\phi_m = Bl_m c/E_b \ , \tag{9.20}$$

where B is the septum field and l_m the septum length. For a length of 10 m, a septum field of about 0.25 T is required, which could be produced by a septum of thickness 7 mm, assuming a current density of 60 A/mm^2 in the septum sheet, which is the operation value for the d.c. septum at the SPS. The entire beam separation is illustrated schematically in Fig. 9.20. Dispersion generated by the separation is compensated in the arc dispersion suppressors.

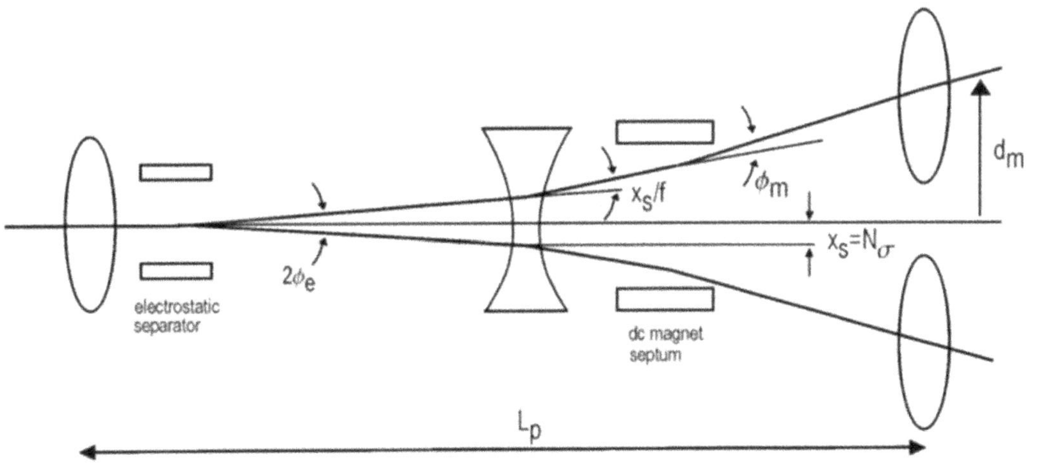

Fig. 9.20. Schematic of beam separation at the VLLC as described in [32]

9.13 Crystal Extraction

Crystal extraction is conceptually quite different from the extraction methods. It was first studied at Dubna and Protvino [34], and later tested extensively at the CERN SPS [35, 36] and at the Fermilab Tevatron [37]. Here, particles in the transverse beam halo, entering a crystal placed close to the beam, are trapped between the crystalline planes [38]. If the crystal is slightly bent, the particles can be deflected outwards, and subsequently be transported to a fixed-target experiment. Figure 9.21 shows a schematic view of crystal extraction.

Crystal extraction is foreseen as an option for the LHC. It would be parasitic to the normal collider operation, and re-utilize the halo particles which do no longer contribute to the collider luminosity.

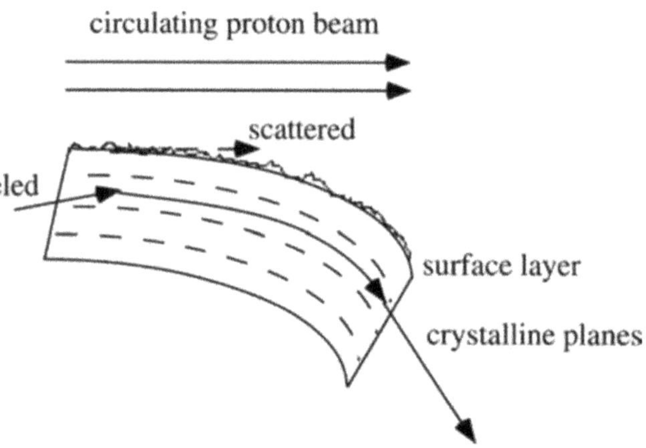

Fig. 9.21. Extraction from the transverse halo of a circulating proton beam by means of a bent crystal. Particles incident with a large impact parameter are channeled and deflected outwards. Particles hitting the inefficient crystal surface layer experience multiple scattering, and may be channeled on a later turn; this is called multi-pass extraction

Channeling occurs if the incident angle of the particles is smaller than the Lindhard critical angle [38]. The critical angle depends on the orientation of the crystal and on the material. The crystal orientation can be defined with respect to an axial direction $[ijk]$ or with respect to a plane (ijk)[1]. For the (110) planar direction in silicon the critical angle is [38]

$$\Psi_{\text{crit}} \approx 5\ \mu\text{rad}\ \frac{\sqrt{Z}}{\sqrt{p[\text{TeV}/c]}}\,, \tag{9.21}$$

where p is the momentum of the incident particle and Z is its charge in units of the electron charge.

[1] In a crystal with cubic symmetry, a vector $[ijk]$ with components along the three orthogonal symmetry axes has a perpendicular atomic plane, which is denoted (ijk).

Thermal vibrations, the discreteness of the crystal lattice, and the presence of the electrons in the target all increase the transverse energy of a channeled particle, and can ultimately lead to dechanneling. This is approximated by an exponential depletion of the number n of channeled particles with the traversed distance z:

$$n = n_0 \exp(-z/L_0) \ . \tag{9.22}$$

The empirical parameter L_0 is called the dechanneling length, and it increases linearly with momentum. For silicon, we have [38] $L_0 \approx 0.9$ m p[TeV/c]. Since scattering on nuclei is an important dechanneling process, the channeling can be improved by cooling of the target.

Another concern are imperfections on the crystal surface. These give rise to an inefficient surface layer, typically a few micrometers thick, in which no channeling takes place. To be extracted in a single pass, a particle must enter the crystal with an impact parameter larger than the thickness of the surface layer. On the other hand, particle passing through the surface area experience multiple scattering, and can re-enter the crystal on subsequent revolutions, this time at a larger impact parameter and under the right conditions to be channeled and extracted.

Channeling is possible only for bending radii larger than a minimum 'critical' radius, whose value depends on the crystal, its orientation, and the beam energy. For a proton beam incident parallel to the (110) plane[2] of a silicon crystal, this critical radius is [38]

$$R_c \approx 0.4 \text{ m } p \ [\text{TeV}/c] \ . \tag{9.23}$$

The efficiency of cystal extraction is defined as the number of particles extracted divided by the number of particles lost. Proton extraction efficiencies up to 18% have been obtained [36]. Using a crystal coated with a 30 μm amorphous SiO layer, pure multi-pass extraction with an efficiency of 4–7% was demonstrated [36]. The importance of multi-pass extraction implies that not only the initial impact parameter, but also machine parameters such as the beta function at the crystal and the betatron tune play an important role for the overall efficiency.

Finally, in addition to protons also heavy ions can be extracted by a bent crystal. For fully stripped Pb ions ($Z = 82$) at 22 TeV, an extraction efficiency of 10% was achieved at the SPS [36]. This value was slightly lower than for protons of equivalent energy per nucleon.

---

[2] In a crystal with cubic symmetry, every vector $[ijk]$ defines a perpendicular atomic plane (ijk).

Exercises

9.1 Septum Fields for Injection and Extraction

Suppose that the minimum beam separation at the septum is $x_{\rm sep} > n_s\sigma_x$. Derive an expression for the integrated kick strength $B_{\rm kic}L_{\rm kic}$, with $B_{\rm kic}$ the kicker magnetic field and $L_{\rm kic}$ its length, as a function of normalized emittance $\epsilon_{x,N}$ and energy for a proton beam.

Assume $n_s = 10$, $\beta = 100$ m, $\mu = \pi/2$, a kicker length of $L_{\rm kic} = 5$ m, and a normalized emittance $\epsilon_{x,N} = 4$ μm. Compute the magnetic field $B_{\rm kic}$ required at a beam energy of 10 GeV and at 10 TeV.

9.2 Emittance Dilutions due to Injection Errors

Consider injection into a storage ring with a 1 mm orbit error at $\beta = 100$ m, in both betatron phases.

a) Estimate the corresponding (growth in) normalized emittance after complete filamentation for proton, muon and electron beams at 10 GeV and at 1 TeV. Compare this with the design normalized emittances of the LHC (3.75 μm), the multi-TeV Muon Collider (50 μm), and the NLC (3 nm vertically). Note that $m_ec^2 = 511$ keV, $m_pc^2 = 938$ MeV, and $m_\mu c^2 = 105.7$ MeV.

b) In general the orbit error results in emittance growth comparable to the design emittance when it is of the same order as the rms beam size. Calculate the rms beam sizes for a 7-TeV proton beam (LHC), a 500-GeV electron beam (NLC) and a 2-TeV muon beam (MC) at $\beta = 100$ m.

9.3 Filamentation

Consider a point bunch which is injected somewhere in phase space at a radius r. Compute the projected beam density $p(x)$, normalized to unity, after filamentation.

9.4 Particle Impact for Slow Extraction

Derive (9.16).

9.5 Crystal Channeling

What is the maximum bending angle over a length of 3 cm, for the LHC beam energy of 7 TeV?

10 Polarization Issues

The study of spin dynamics in synchrotrons has evolved over the years as has the desire for achieving polarized particle beams of the highest possible beam energies. A selection of reviews of the dynamics of polarized beams may be found in [1]–[9]. In this chapter, we focus on experimental data and describe spin transport in circular accelerators and transport lines. Except where explicitly mentioned, radiative effects in electron accelerators or very high energy proton accelerators are not treated here. We begin with a review of the Thomas-BMT equation for spin motion. This will be given in terms of the SU(2) spinor representation. Spinor algebra will be introduced and applied in the description of techniques used for preserving the beam polarization during acceleration through depolarizing resonances at moderate beam energies.

10.1 Equation of Spin Motion

The concept of particle spin was first introduced by Uhlenbeck and Goudsmit in 1926 to explain certain features of atomic spectra. They presupposed that the (in this case) electron of mass m and charge e, possessed both a magnetic moment $\boldsymbol{\mu}$ and the spin angular momentum $\boldsymbol{s}$, related to one another by

$$\boldsymbol{\mu} = \frac{ge}{2m}\boldsymbol{s} , \tag{10.1}$$

where g is the gyromagnetic ratio whose value was empirically taken to be 2 for electrons to explain certain experimental observations. In 1927 Thomas [10] showed that once a relativistic kinematic effect was taken into account, the value of $g = 2$ was consistent with the atomic spectra measurements.

The equation of motion for the spin angular momentum in an external magnetic field [5] is given, in the particle rest frame, by

$$\begin{aligned} \frac{\mathrm{d}\boldsymbol{s}}{\mathrm{d}t} &= \boldsymbol{\mu} \times \boldsymbol{B} \\ &= \boldsymbol{\Omega} \times \boldsymbol{s} , \end{aligned} \tag{10.2}$$

M. G. Minty et al., *Measurement and Control of Charged Particle Beams*, https://doi.org/10.1007/978-3-662-08581-3_10

where the angular velocity of the spin precession is

$$\boldsymbol{\Omega} = -\frac{ge}{2m}\boldsymbol{B}. \tag{10.3}$$

In the above equations, the spin angular momentum $\boldsymbol{s}$ of a single particle takes on discrete values of magnitude $|\boldsymbol{s}| = \hbar/2$ for spin-$\frac{1}{2}$ particles (i.e. electrons and protons). It is convenient to normalize $\boldsymbol{s}$ and work with the spin vector $\boldsymbol{S}$, with $|\boldsymbol{S}| = 1$, defined as the normalized spin expectation value in the rest frame.

10.2 Thomas-BMT Equation

In the laboratory frame, the spin precession for a relativistic particle in external electromagnetic fields is given by the Thomas-BMT equation [10]–[13]:

$$\frac{\mathrm{d}\boldsymbol{S}}{\mathrm{d}t} = -\frac{e}{\gamma m}\left[(1+a\gamma)\ \boldsymbol{B}_{\perp} + (1+a)\ \boldsymbol{B}_{\parallel} + \left(a\gamma + \frac{\gamma}{\gamma+1}\right)\frac{\boldsymbol{E}\times\boldsymbol{v}}{c^2}\right] \times \boldsymbol{S}\ , \tag{10.4}$$

where $\boldsymbol{B}_{\perp}$ and $\boldsymbol{B}_{\parallel}$ represent the magnetic fields perpendicular and parallel to the particle velocity respectively, $\boldsymbol{\beta} = \boldsymbol{v}/c$ with $\boldsymbol{v}$ the particle velocity, and $\gamma = 1/\sqrt{1-\beta^2}$ the Lorentz factor or ratio of the particle energy to mass.

The factor a in (10.4) is the gyromagnetic anomaly of the electron. It is

$$a = \frac{g-2}{2} = 0.00115966\ , \tag{10.5}$$

and deviates from zero due to radiative corrections. For protons, which are composite, we replace a with the symbol G, where

$$G = \frac{g-2}{2} = 1.792846\ . \tag{10.6}$$

For many practical applications there are no significant electric fields[1], and the Thomas-BMT equation is simply

$$\frac{\mathrm{d}\boldsymbol{S}}{\mathrm{d}t} = -\frac{e}{\gamma m}\left[(1+a\gamma)\boldsymbol{B}_{\perp} + (1+a)\boldsymbol{B}_{\parallel}\right]\ \times \boldsymbol{S}\ . \tag{10.7}$$

From this equation applied to protons, the spin precession due to transverse magnetic fields depends on the particle energy through the factor $1/\gamma + G$ while the amount of precession due to a longitudinal magnetic field scales as $(1+G)/\gamma$. We will see later that this has implications for spin rotator design.

[1] More precisely, the term $\boldsymbol{\beta} \times \boldsymbol{E}$ is nearly zero since the electric fields in an accelerator are usually parallel to the particle velocity.

10.3 Beam Polarization

The polarization **P** of a bunch is defined as the ensemble average over the spin vectors $\boldsymbol{S}$ of the individual particles:

$$\mathbf{P} = \langle \boldsymbol{S} \rangle = \left| \frac{1}{N} \sum_{p=1}^{N} \boldsymbol{S}_p \right| , \tag{10.8}$$

where N denotes the number of particles in the bunch.

As an illustration of the formulae presented so far, we calculate the beam depolarization due to spin precession and energy spread in a transport line neglecting radiation effects. We suppose a transverse magnetic field bends an electron orbit by the angle θ. Then, according to (10.4) one finds that the electron spin direction precesses by $\phi = (a\gamma)\theta$ relative to the orbit in the laboratory frame. Considering a relativistic electron beam of many particles with a finite energy spread, the spin vectors of different electrons in the beam precess by different angles, since ϕ depends on the particle energy. If the initial beam polarization is longitudinal with magnitude P_0 and the beam is bent horizontally by an angle θ, not only is the final polarization vector $\mathbf{P}_f$ rotated by an angle $a\gamma_0\theta$, but also its magnitude is reduced as

$$P_f = P_0 e^{-(a\gamma_0\sigma_\delta\theta)^2/2} , \tag{10.9}$$

where γ_0 is the Lorentz factor corresponding to the mean energy of the beam, and σ_δ the rms relative momentum spread assuming a Gaussian momentum distribution. Equation (10.9) is strictly valid, if the orbit is bent in one plane. The same formula was also used to model the spin transport and the depolarization in the SLC North Collider Arc [14, 15], which constituted a nonplanar transport line with horizontal and vertical bends used to follow the terrain of the SLAC site. In this application, an 'effective' bending angle θ, entering in (10.9), was determined by measuring the final direction of the polarization vector as a function of the beam energy [15].

10.4 Spinor Algebra Using SU(2)

We can use methods of quantum mechanics [5] to describe spin transport. It is mathematically advantageous to do so since transporting 2 component spinors (Ψ) is simpler than transporting the 3-dimensional spin vector $\boldsymbol{S}$. The relationship between the two is given by

$$S_i = \Psi^\dagger \sigma_i \Psi , \tag{10.10}$$

with the Pauli matrices defined[2] as

[2] Caution. Different authors adopt different conventions; here we adopt the convention (specified in the introduction; c.f. Fig. 1.1) that x is radial, s is longitudinal, and y is vertical. A cyclic permutation may be used to transform between conventions.

$$\sigma_x = \begin{pmatrix} 0 & 1 \\ 1 & 0 \end{pmatrix}, \quad \sigma_s = \begin{pmatrix} 0 & -\mathrm{i} \\ \mathrm{i} & 0 \end{pmatrix}, \quad \sigma_y = \begin{pmatrix} 1 & 0 \\ 0 & -1 \end{pmatrix}. \tag{10.11}$$

Together with the 2×2 identity matrix I_2, these 4 matrices generate an irreducible representation of the SU(2) group.

10.5 Equation of Spin Motion

For generality we can reexpress [1] the equation of spin motion (10.2) in terms of a time-like variable θ defined as

$$\theta = \int_0^s \frac{\mathrm{d}s'}{\rho(s')}, \tag{10.12}$$

which is equal to the accumulated bending angle or so-called orbital angle. Then, as we will show below, the equation of spin motion (10.2) is equivalent to

$$\frac{\mathrm{d}\Psi}{\mathrm{d}\theta} = \frac{i}{2} H\Psi, \tag{10.13}$$

where $H = \tilde{\boldsymbol{\Omega}} \cdot \boldsymbol{\sigma}$ denotes the effective spin Hamiltonian, which is represented here as a matrix [5] with

$$\tilde{\boldsymbol{\Omega}} = \boldsymbol{\Omega} / \frac{\mathrm{d}\theta}{\mathrm{d}t}. \tag{10.14}$$

In the absence of depolarizing resonances (see Sect. 10.7), for a particle circulating in the horizontal plane under the influence of vertical magnetic fields only, H may be expressed as

$$H = \begin{pmatrix} -\kappa & 0 \\ 0 & \kappa \end{pmatrix}, \tag{10.15}$$

with $\kappa = G\gamma$ for protons and $\kappa = a\gamma$ for electrons.

Equation (10.13) may be expressed as

$$\frac{\mathrm{d}\Psi}{\mathrm{d}\theta} = -\mathrm{i}\frac{\lambda}{2}(\boldsymbol{\sigma} \cdot \hat{\boldsymbol{b}})\Psi, \tag{10.16}$$

where $\lambda = |\tilde{\boldsymbol{\Omega}}|$ denotes the amplitude of the precession frequency and depends in general on the particle coordinates (to concentrate on the underlying principles, we defer the explicit expressions until needed in Sect. 10.12) and $\hat{\boldsymbol{b}} = \hat{\boldsymbol{\Omega}}$ is a unit vector aligned with the precession axis. The vector $\hat{\boldsymbol{b}}$ and λ depend on the longitudinal coordinate and on the position of the particle in the six-dimensional phase space [7]. Therefore $\hat{\boldsymbol{b}}$ may have different orientations at fixed orbital angle θ on subsequent turns.

The solution of (10.16), which defines the spinor transformation matrix M, is

$$\Psi(\theta) = M\Psi(0). \tag{10.17}$$

If $\lambda\hat{b}$ is constant along a section of the orbit, then

$$M = e^{-i(\frac{\lambda}{2})(\boldsymbol{\sigma}\cdot\hat{\boldsymbol{b}})\theta} . \tag{10.18}$$

Using the algebra of the σ matrices, after expanding the exponential, the solution for the spinor is

$$\Psi(\theta) = \left[\cos\left(\frac{\lambda\theta}{2}\right) - i(\boldsymbol{\sigma}\cdot\hat{\boldsymbol{b}})\sin\left(\frac{\lambda\theta}{2}\right)\right]\Psi(0) . \tag{10.19}$$

To complete this section we use the Pauli algebra to confirm that the equation of motion for the spinor Ψ (10.13) indeed leads to the equation of motion for the spin vector $\mathbf{S}$. Since the components of the spin are given by the expectation value of the Pauli spin matrices ($\mathbf{S} = \Psi^{\dagger}\boldsymbol{\sigma}\Psi$), the equation of spin motion for the polarization vector is obtained by taking the derivative of the latter expression

$$\frac{\mathrm{d}\mathbf{S}}{\mathrm{d}\theta} = \frac{\mathrm{d}\Psi^{\dagger}}{\mathrm{d}\theta}\boldsymbol{\sigma}\Psi + \Psi^{\dagger}\boldsymbol{\sigma}\frac{d\Psi}{d\theta} . \tag{10.20}$$

Using (10.16) and its Hermitian conjugate, $\frac{\mathrm{d}\Psi^{\dagger}}{\mathrm{d}\theta} = \mathrm{i}\frac{\lambda}{2}\Psi^{\dagger}(\boldsymbol{\sigma}\cdot\hat{\boldsymbol{b}})$, one finds

$$\frac{\mathrm{d}\mathbf{S}}{\mathrm{d}\theta} = i\frac{\lambda}{2}\Psi^{\dagger}[(\boldsymbol{\sigma}\cdot\hat{\boldsymbol{b}})\boldsymbol{\sigma} - \boldsymbol{\sigma}(\boldsymbol{\sigma}\cdot\hat{\boldsymbol{b}}))]\Psi . \tag{10.21}$$

Making use of the Pauli algebra $\sigma_i\sigma_j = 1$ for $i = j$ and $\sigma_i\sigma_j = i\epsilon_{ijk}\sigma_k$ for $i \neq j$, one can show that

$$[\boldsymbol{\sigma}\cdot\hat{\boldsymbol{b}}, \boldsymbol{\sigma}] = 2\mathrm{i}(\hat{\boldsymbol{b}}\times\boldsymbol{\sigma}) . \tag{10.22}$$

Therefore, we may write

$$\frac{\mathrm{d}\mathbf{S}}{\mathrm{d}\theta} = -\lambda\Psi^{\dagger}(\hat{\boldsymbol{b}}\times\boldsymbol{\sigma})\Psi = -\lambda\hat{\boldsymbol{b}}\times(\Psi^{\dagger}\boldsymbol{\sigma}\Psi) = -\lambda(\hat{\boldsymbol{b}}\times\mathbf{S}) , \tag{10.23}$$

which may be compared with (10.4).

10.6 Periodic Solution to the Equation of Spin Motion

The first step for studying spin motion in a circular accelerator is to find the periodic solution to the equation of spin motion on the closed orbit. Here we write a spin transfer matrix M as a product of n precession matrices, each of which characterizes a section of constant magnetic field causing spin precession; i.e., $M = M_1M_2\cdots M_n$.

The transfer matrix corresponding to one turn around the accelerator is refered to as the one turn spin transfer map denoted by M_0. For the closed orbit M_0 is periodic: $M_0(\theta + 2\pi) = M_0(\theta)$. Because the norm of the vector in the precession equation is an invariant, M_0 is unitary so that it may be expressed as

$$M_0 = e^{-i\pi\nu_0(\boldsymbol{\sigma}\cdot\hat{\boldsymbol{n}}_0)} = I_2 \cos\pi\nu_0 - i(\boldsymbol{\sigma}\cdot\hat{\boldsymbol{n}}_0)\sin\pi\nu_0 \;, \tag{10.24}$$

or

$$M_0 = \begin{pmatrix} \cos\pi\nu_0 - \mathrm{i}\sin\pi\nu_0\cos\alpha_y & -\sin\pi\nu_0\cos\alpha_s - \mathrm{i}\sin\pi\nu_0\cos\alpha_x \\ \sin\pi\nu_0\cos\alpha_s - \mathrm{i}\sin\pi\nu_0\cos\alpha_x & \cos\pi\nu_0 + \mathrm{i}\sin\pi\nu_0\cos\alpha_y \end{pmatrix} , \tag{10.25}$$

The unit vector $\hat{\boldsymbol{n}}_0$ is the precession axis for the one turn map M_0. It is periodic and fulfills the Thomas-BMT equation [5]; a spin set parallel to $\hat{n}_0$ at orbital location θ will, after one turn $(\theta + 2\pi)$, also be parallel to $\hat{n}_0$. Thus $\hat{n}_0$ is refered to as the 'stable spin direction' on the closed orbit and may be described by direction cosines:

$$\hat{\boldsymbol{n}}_0 = (\cos\alpha_x, \cos\alpha_s, \cos\alpha_y) \tag{10.26}$$

with normalization $\cos^2\alpha_x + \cos^2\alpha_s + \cos^2\alpha_y = 1$.

The parameter ν_0, called the spin tune, represents the number of times the spin of a particle on the closed orbit precesses about the stable spin direction in one turn around the ring. The fractional part of the spin tune may be obtained from the trace of the (periodic) spin precession matrix:

$$\begin{aligned} \mathrm{Tr}\ M_0 &= 2\cos\pi\nu_0 \quad \text{or} \\ \nu_0 &= \frac{1}{\pi}\cos^{-1}\left(\frac{\mathrm{Tr}\ M_0}{2}\right) . \end{aligned} \tag{10.27}$$

For the spin motion of particles not on the closed orbit, M is in practice not periodic since accelerators are not typically operated on resonance (for which the particle returns to the same point in phase space after an integer number of turns). The unit vector $\hat{\boldsymbol{b}}$ in the equation of spin motion (10.16) may therefore have different orientations at fixed orbital coordinate θ on subsequent turns $(\theta, \theta + 2\pi, \theta + 4\pi, \ldots)$.

10.7 Depolarizing Resonances

The spin of a particle executing synchro-betatron motion around the closed orbit is perturbed by magnetic fields sampled at the betatron and synchrotron frequencies which are characteristic of the particle motion. Since the particles within a bunch have generally different amplitudes and phases, the perturbation to the spin is different for different particles resulting in a spread of the particle spins and thus a lower polarization.

Depolarizing resonances occur whenever the spin tune beats with any of the natural oscillation frequencies of the orbital motion; that is when the spin tune, ν_0, equals a resonance tune, ν_{res}, by satisfying

$$\nu_0 = \nu_{\mathrm{res}} \equiv m + qQ_x + rQ_y + sQ_s \ , \tag{10.28}$$

where Q_x and Q_y are the horizontal and vertical betatron tunes, Q_s is the synchrotron tune, while $m, q, r,$and s are integers[3]. Here m is the product of an integer times the periodicity of the lattice. The quantity $|q| + |r| + |s|$ is called the order of the resonance.

The general resonance condition specifies the criteria for many different types of resonances. Imperfection depolarizing resonances, for which

$$\nu_0 = \nu_{\mathrm{res}} = m = \text{integer} \ , \tag{10.29}$$

arise, for example, from horizontal magnetic fields experienced by the orbiting particle due to magnet imperfections, dipole magnet rotations about the beam direction, and to vertical quadrupole magnet misalignments. Lowest order intrinsic resonances, which result from the horizontal fields of quadrupoles, occur if

$$\nu_0 = \nu_{\mathrm{res}} = m + rQ_y \ . \tag{10.30}$$

In practice, the above two types of resonances have proven the most significant in the energy regimes of existing accelerators with polarized beams.

Other higher order spin depolarizing resonances may occur for any combination of integers which satisfy (10.28). Studies have shown that higher-order intrinsic resonances of the form

$$\nu_0 = \nu_{\mathrm{res}} = m + rQ_y + sQ_s \ . \tag{10.31}$$

become increasingly important at higher beam energies. Due to the interaction with the particle longitudinal momentum, such resonances are also refered to as synchrotron sideband depolarizing resonances.

Resonant spin motion has been observed in many accelerators. Interestingly, it was observed in the SLC collider arcs [14], through which bunches pass only once. The 1 mile arcs were used to transport 45.6 GeV polarized electrons from the linac to the interaction point where they collided head-on with positrons. The arc consists of 23 achromatic sections with a 108° phase advance per cell. The vertical beam trajectory and the components of a spin vector are shown in Fig. 10.1 assuming an initial vertical offset of 0.5 mm and random quadrupole misalignments. For the nominal SLC operating energy, the phase advances of (betatron) orbit and spin in the SLC arcs were almost identical. In Fig. 10.1 this equality is seen to contribute to a net tilt of the spin vector as evidenced by the increase in the vertical component of a spin along the arc. In practice, vertical bumps were used to optimally align the polarization to be longitudinal at the interaction point.

[3] For particles in the bunch with large orbital amplitudes, ν_0 in (10.28) should be replaced by spin tune of the individual particle ν [7]–[9].

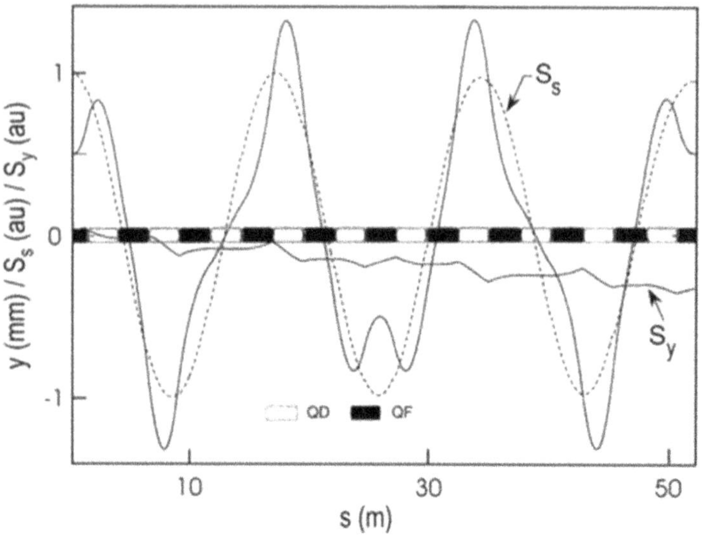

Fig. 10.1. Beam orbit (*solid line*) and spin transport in the SLC collider arc (Courtesy P. Emma, 1999)

10.8 Polarization Preservation in Storage Rings

For proton and deuteron accelerators, for which the polarization of the beam is produced at the particle source, the first requirement of maintaining a beam's polarization is preserving this polarization at injection into downstream accelerators. Ignoring synchro-betatron motion, this is easily achieved by orienting (using upstream spin rotators) the beam polarization so that it is aligned with the stable spin direction of the downstream transport line or storage ring.

A mismatch at injection results in a cosine-like reduction of the time-averaged beam polarization. Letting $(\cos\alpha_x, \cos\alpha_s, \cos\alpha_y)$ denote the orientation of the injected polarization $\mathbf{P}_{\text{inj}}$ and $(\cos\beta_x, \cos\beta_s, \cos\beta_y)$ the orientation of the stable spin direction $\hat{\boldsymbol{n}}_{\mathbf{0}}$ in the laboratory reference frame, then the projection of the injected polarization vector $\mathbf{P}_{\text{inj}}$ onto $\hat{\boldsymbol{n}}_{\mathbf{0}}$ is

$$\begin{aligned}\|\mathbf{P}\| &= \mathbf{P}_{\text{inj}} \cdot \hat{\boldsymbol{n}}_{\mathbf{0}} \\ &= P_{\text{inj}}(\cos\beta_x \cos\alpha_x + \cos\beta_s \cos\alpha_s + \cos\beta_y \cos\alpha_y)\,. \end{aligned} \tag{10.32}$$

The components of the time-average polarization one would measure at the injection point are then obtained by projecting the polarization onto the three coordinate axes:

$$P_y = \|\mathbf{P}\| \cos\beta_y, \quad P_x = \|\mathbf{P}\| \cos\beta_x, \quad P_s = \|\mathbf{P}\| \cos\beta_s\,. \tag{10.33}$$

A conceptual illustration is given in Fig. 10.2. At other locations in the ring, the measurable polarization components are obtained by performing a second projection using $\|\mathbf{P}\|\hat{\boldsymbol{n}}_{\mathbf{0}}(\boldsymbol{s})$ where $\hat{\boldsymbol{n}}_{\mathbf{0}}(\boldsymbol{s})$ is the stable spin direction at the point of interest.

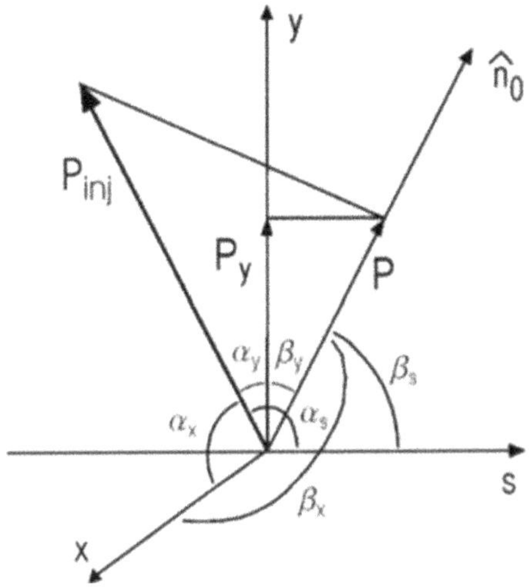

Fig. 10.2. Graphical representation showing the projection of the injected polarization P_{inj} onto the stable spin direction $\hat{n}_0$ at the injection point. The time-averaged vertical polarization one would measure at that point in denoted by P_y

Once the beam is successfully injected without loss of polarization, it must be ramped to high energy thereby encountering numerous depolarizing resonances along the energy ramp. With considerable effort, polarized proton beams were accelerated through many intrinsic and imperfection depolarizing resonances to GeV energies at the ZGS [16], Saturne [17], the AGS [18], and KEK [19]. The methods employed were based on overcoming each depolarizing resonance individually. In this section we review techniques used to overcome these resonances. In the next section we describe 'siberian snakes' which are used for preserving the beam polarization during the energy ramp. Proof of principle experiments with snakes were initially carried out at the Indiana University Cyclotron Facility (IUCF) [20]–[22]. Siberian snakes are now used in routine operation during acceleration of polarized protons at RHIC [23] and constitute the preferred method of polarization preservation for future high energy hadron accelerators.

10.8.1 Harmonic Correction

Harmonic correction of imperfection depolarizing resonances was used at the AGS [18] to ramp vertically polarized proton beams to about 22 GeV. There 96 correction dipoles were employed whose currents were programmed during the acceleration process such that the Fourier harmonics of the radially outward and longitudinal fields in the measured particle trajectories for the most nearby resonances were minimized. The Fourier harmonics are given by $a_n \cos n\theta + b_n \sin n\theta$, where n denotes the resonance harmonic of interest. As will be shown later, the resonance strength depends on the vertical beam displacement (in quadrupoles, for example, the nominally vertical polarization experiences a radial precession field with an off-axis beam). To eliminate depolarization, the coefficients a_n and b_n were experimentally adjusted to minimize the horizontal magnetic fields causing each imperfection resonance. Shown in Fig. 10.3 are traces on an oscilloscope from the AGS [18] showing

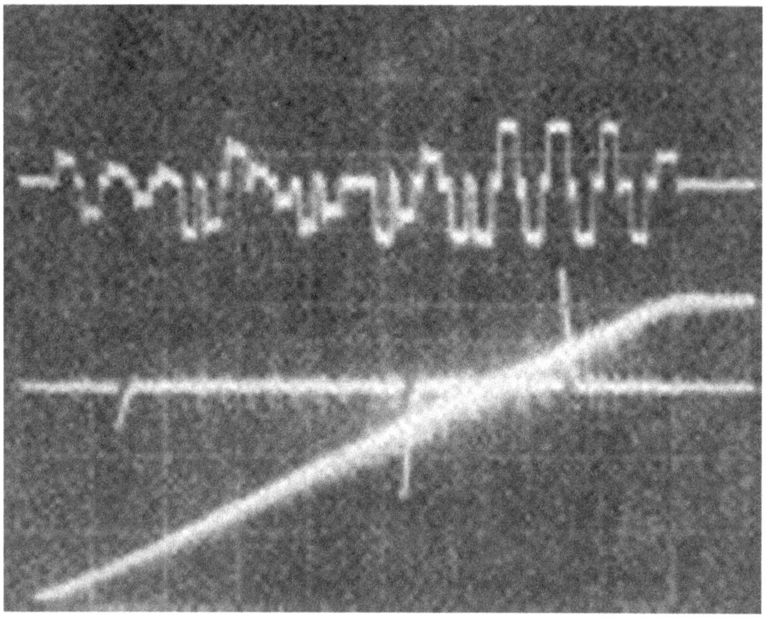

Fig. 10.3. Oscilloscope traces showing the currents in pulsed magnets during the energy ramp to 16.5 GeV/c at the AGS. Shown are the pulsed dipole currents (*top*), the pulsed quadrupoles currents (*middle*), and the main dipole current (*bottom*) (Courtesy A. Krisch, 1999)

the corrector dipole currents (top trace) and the main dipole current (bottom trace), which is proportional to the beam energy.

Harmonic correction methods [24]–[26] were also applied in the case of high energy electron beams at both HERA [27]–[29] and, deterministically, at LEP [30]. In electron accelerators, an initially unpolarized beam may after some time become polarized due to the emission of synchrotron radiation. This is known as the Sokolov–Ternov effect [31] which also predicts a maximum possible beam polarization of 92.4% for electrons. In practice however this level of polarization is not reached due to spin-orbit coupling caused by the trajectory oscillations which result from photon emission [3]. Minimizing the strength of the harmonics nearest the beam energy thus minimizes the

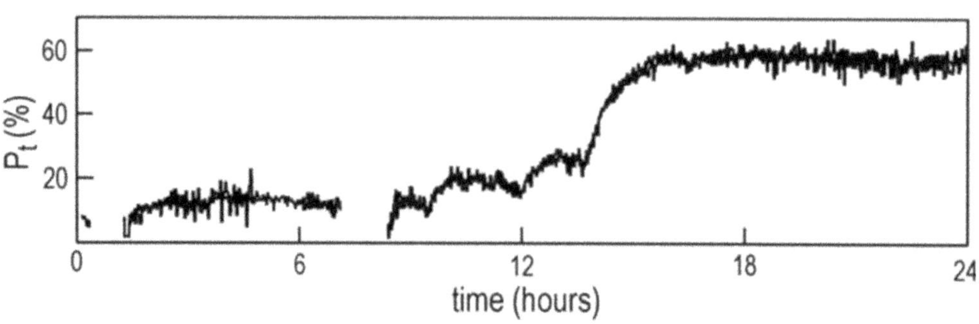

Fig. 10.4. Measured transverse beam polarization at HERA showing improvements gained using harmonic correction (Courtesy of the HERMES experiment, 2002)

influence of the depolarizing effects due to synchrotron radiation. At HERA and LEP, instead of empirically varying the whole closed orbit, closed orbit bumps were used to minimize the strength of the nearest imperfection resonances. A recent result from HERA is shown [29] in Fig. 10.4 obtained after implementing a new optic as intended for the HERA upgrade.

10.8.2 Adiabatic Spin Flip

The method of adiabatic spin flip, which was also used at the AGS, is based on the results of Froissart and Stora [32]. The Froissart–Stora formula, which describes the spin transport through a single, isolated imperfection or lowest-order intrinsic resonance, is

$$P_y(\infty) = \left(2e^{-\frac{\pi|\epsilon|^2}{2\alpha}} - 1\right) P_y(-\infty)\,, \tag{10.34}$$

where ϵ is the resonance strength (see Sect. 10.12 and [1]), $\alpha = \mathrm{d}\nu_0/\mathrm{d}\theta$, and $P_y(-\infty)$ or $P_y(\infty)$ refer to the initial and final polarizations, respectively.

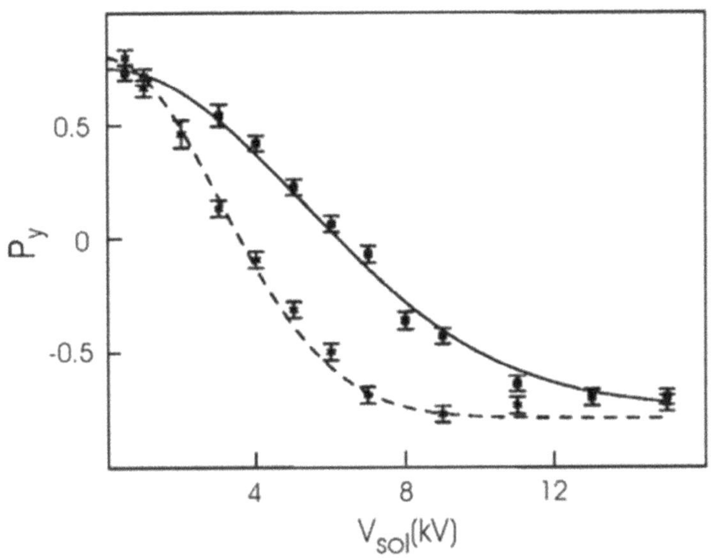

Fig. 10.5. Spin flip of a vertically polarized beam at the IUCF with kinetic energy 105.4 MeV and a frequency sweep range of 2 kHz. The estimated resonance strength at 15 kV applied voltage is 1.9×10^{-4} (Courtesy A. Krisch, 1999)

The Froissart–Stora formula mandates that the spin of the orbiting particle will flip if the passage is slow and the resonance is strong (the argument in the exponent of (10.34) is large). This behavior has been verified by experiment [17, 33, 34]. An example taken from the IUCF cooler ring [34] is shown in Fig. 10.5. Here a solenoid was used to produce a sinusoidally varying longitudinal magnetic field of frequency f_{rf} and peak amplitude V_{sol}. This served to create a depolarizing resonance such that

$$\nu_0 \approx c_1 \pm \frac{f_{\rm rf}}{f_{\rm rev}} = c_2 \pm \left(\frac{f_{\rm rev} - f_{\rm rf}}{f_{\rm rev}} \right) , \tag{10.35}$$

where c_1 and c_2 are integers. In these measurements a frequency sweep of span 2 kHz centered about the revolution frequency was used. After each data point, the beam was dumped and a new beam was injected. The curves in Fig. 10.5 show the prediction by the Froissart–Stora equation computed for the measured resonance strength for two different ramp rates.

10.8.3 Tune Jump

Intrinsic depolarizing resonances were overcome at the AGS using the method of tune jump [18]. From the Froissart–Stora relation, if the resonance is crossed sufficiently quickly (the exponent approaches zero), then the polarization will be preserved. Therefore, as the spin tune $\nu_0 = G\gamma$ increases during acceleration, the resonance can be traversed without loss of polarization by rapidly shifting the vertical betatron tune, ν_y. A classic example [18] from the AGS was shown in Fig. 10.3 in which the current of the pulsed quadrupoles (middle trace) is depicted. To achieve this, strong pulsed quadrupoles and special ceramic beam pipes (to allow passage of the field) were required.

10.9 Siberian Snakes

The above mentioned correction schemes were anticipated to be of limited applicability when accelerating polarized beams to very high energies. The harmonic correction employed at the AGS was complicated and time consuming; the empirically found corrections also depended on the closed orbit of the accelerator, which was observed to drift with time and change between running periods. At very high energies, where the resonances will be overlapping due to an increase in the resonance strength with increasing energy (see (10.49) below), the method of adiabatic spin flip fails [35]. The method of tune jump is stopband limited since, for a very strong intrinsic resonance, the vertical betatron tune shift required to overcome the resonance may exceed the separation between the machine betatron resonances. Finally, the number of resonances to be crossed increases with energy. At the SSC, where there would have been more than 10^4 imperfection and first-order intrinsic resonances, overcoming each resonance individually clearly would have been impractical.

An ingenious arrangement of magnets was proposed [36, 37] by Derbenev and Kondratenko in 1976 (see also [13]). Use of this technique[4] would simultaneously overcome all imperfection and lowest-order intrinsic resonances by making the spin tune energy independent. A so-called type-1 snake rotates

[4] This scheme was dubbed 'siberian snake' by E. Courant.

the spin of each proton by 180 degrees about the longitudinal axis on each turn around the ring without changing the closed orbit outside of the snake. This forces the spin tune to be 1/2 and the stable spin direction $\hat{\boldsymbol{n}}_0$ to lie in the horizontal plane. The resonance condition of (10.29) and (10.30) can therefore never be satisfied regardless of the beam energy: the condition for imperfection resonances with integer spin tune (10.29) is never satisfied and, for betatron tunes not equal to 1/2 (corresponding to half integer orbit resonances), intrinisic resonances (10.30) are also avoided. A type-2 siberian snake precesses the spin about the radial direction. A type-3 siberian snake precesses the spin about the vertical direction.

The most cost-effective construction of a siberian snake depends upon the energy range of interest. A siberian snake consisting of a solenoidal field (and skew quadrupoles for coupling correction) requires a field strength (given here for protons) of

$$\int B_{\parallel}\ \mathrm{d}l = \frac{mc\beta\gamma}{(1+G)e}\psi\,, \tag{10.36}$$

in which ψ ($=\pi$ for a full siberian snake) is the angle through which the spin is precessed. In this case, the required field integral depends linearly on γ. Due to technical constraints the field strength and magnet length cannot be increased indefinitely. Therefore siberian snakes made with solenoids are better suited for low energy operation.

Alternatively a ‘conventional’ siberian snake consisting of eight transverse field dipoles each of which precesses the spin by $\frac{\pi}{2}$ requires a field strength of

$$\int B_{\perp}\ \mathrm{d}l = \frac{mc\beta}{Ge}\psi \quad \text{(transverse snake)}\,, \tag{10.37}$$

which is independent of γ. This type of siberian snake therefore has the advantage that a single set of dipole operating currents suffices for all beam energies once the relativistic factor β is close to 1. However, in a dipole magnet, the orbit deflection angle is $\psi/G\gamma$ and so depends on the energy. For low beam energies ($\gamma < 10$), a siberian snake consisting of dipoles would thus require the construction of large and costly dipoles. Siberian snakes consisting of dipole magnets are therefore more suitable for operation at high beam energies.

With a single type-1 siberian snake, the one-turn spin transfer matrix evaluated at the orbital angle θ is

$$M = e^{-\mathrm{i}\pi\nu_0(\hat{\boldsymbol{n}}_0\cdot\boldsymbol{\sigma})} = \left[e^{-\mathrm{i}\frac{G\gamma}{2}(\pi-\theta)\sigma_{\mathbf{y}}}e^{-\mathrm{i}\frac{\eta}{2}\sigma_{\mathbf{s}}}\right]\left[e^{-\mathrm{i}\frac{G\gamma}{2}(\pi+\theta)\sigma_{\mathbf{y}}}\right]\,, \tag{10.38}$$

where η gives the spin precession about the longitudinal in radians. Taking the trace of (10.38), the particle spin tune is

$$\cos\pi\nu_0 = \cos\left(\pi G\gamma\right)\cos\left(\frac{\eta}{2}\right)\,. \tag{10.39}$$

If the siberian snake is off ($\eta = 0$), then $\nu_0 = G\gamma$ as expected. With the snake fully turned on ($\eta = \pi$), then $\cos \pi\nu_0 = 0$ and $\nu_0 = 1/2$ (modulo 2π).

For a siberian snake design optimum for high beam energies, the use of transverse magnetic fields for spin precession has the unfortunate consequence of also deflecting the particle orbit. The design of a siberian snake therefore uses closed horizontal and vertical bumps so that the orbit outside of the snake region is unchanged (optical transparency). However this is achieved only at the expense of increased snake length which may become costly. Many different snake designs have been proposed. Some of the earlier designs by Steffen are given in [38]–[40] and the design currently used at RHIC is described in [41, 42] have been proposed. Shown in Fig. 10.6 is one such design consisting of alternating horizontal and vertical dipoles. This design is conveniently expressed as

$$\mathrm{V}\left(\frac{\pi}{2}\right)\mathrm{V}\left(-\frac{\pi}{2}\right)\mathrm{H}\left(\frac{\pi}{2}\right)\mathrm{V}\left(-\frac{\pi}{2}\right)\mathrm{H}(-\pi)\mathrm{V}\left(\frac{\pi}{2}\right)\mathrm{H}\left(\frac{\pi}{2}\right), \tag{10.40}$$

where H and V represent horizontal and vertical dipoles rotating the spin through the angle of the argument.

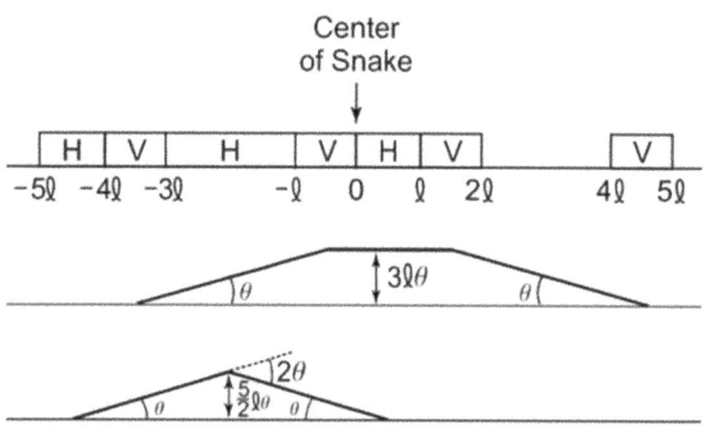

Fig. 10.6. A design of a type-1 siberian snake showing the geometry (*top*) and the design orbit in the vertical (*middle*) and horizontal (*bottom*) planes. The beam moves from the left to the right (Courtesy A. Chao, 1999)

In addition to preserving the polarization, it is often desirable to rotate the spin from vertical to longitudinal [4] at one of possibly many interaction points in a storage ring as is done at HERA [43], for example. An optically transparant spin rotator which does that was proposed by Montague [5, 44]. The spin rotation matrix R is given by

$$R = \mathrm{V}(\alpha)\,\mathrm{V}(-\alpha)\,\mathrm{V}\left(-\frac{\pi}{4}\right)\mathrm{H}(-\pi)\,\mathrm{V}\left(\frac{\pi}{4}\right)X\mathrm{V}\left(-\frac{\pi}{4}\right)\mathrm{H}(\pi)\,\mathrm{V}\left(\frac{\pi}{4}\right), \tag{10.41}$$

where α is an arbitrary precession angle and 'X' indicates the location of the interaction point. For transverse magnetic fields, the orbital bending angles are obtained by the precession angle divided by $G\gamma$. In contrast to spin transformations for which the transfer matrices of each magnet do not commute, the small orbital deflections do essentially commute and are seen to sum to zero.

Radiation in strong vertical bending magnets, used in spin rotators for example, in an electron storage ring can cause excitation of vertical betatron motion which may then lead to depolarization due to the radial component of quadrupole fields [5]. In addition, when rotators are used to precess the polarization into the longitudinal direction, the horizontal motion excited by radiation in the arcs can cause depolarization in the vertical fields of the quadrupoles in the interaction regions. These depolarization mechanisms may be avoided by invoking spin matching conditions to obtain spin transparency [4, 26, 45].

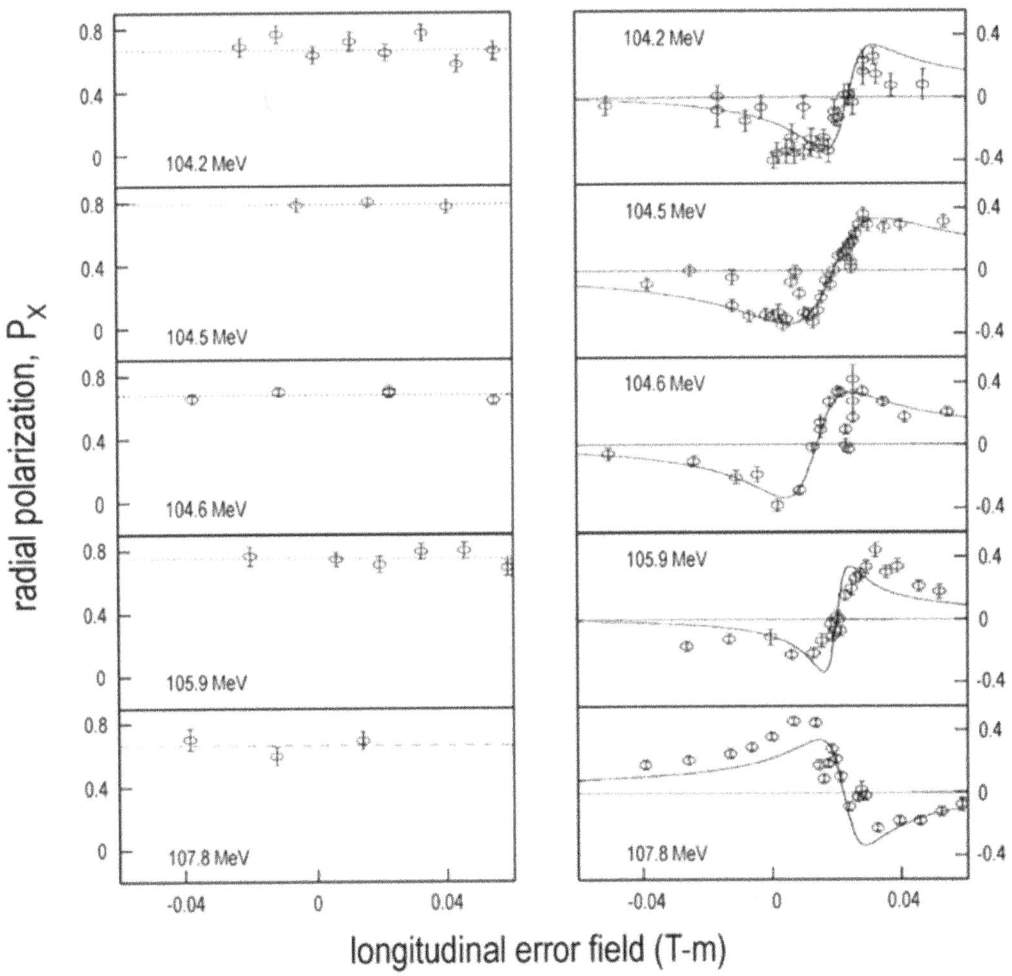

Fig. 10.7. Measurements of time-averaged radial polarization at the IUCF near a $G\gamma = 2$ imperfection resonance (at about 106.4 GeV) at 5 different beam kinetic energies with a 100% siberian snake on (*left*) and off (*right*)

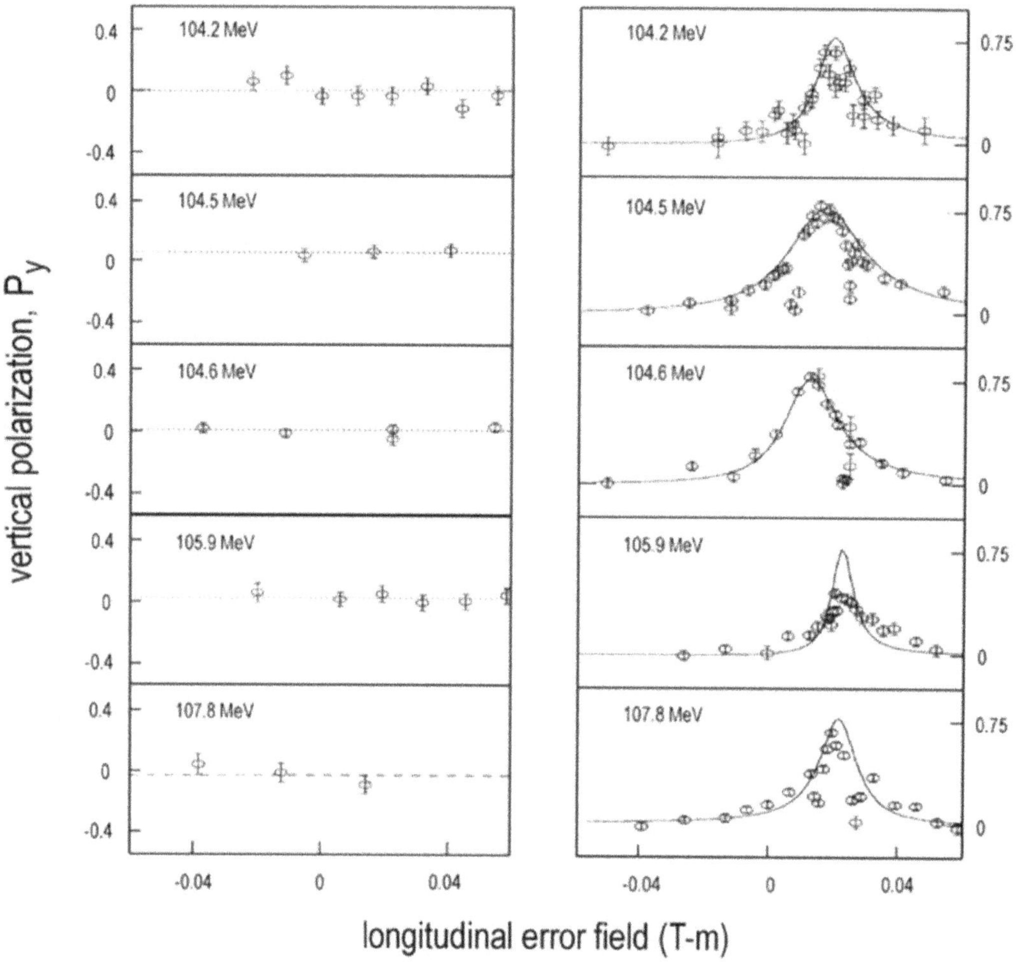

Fig. 10.8. Measurements of time-averaged vertical polarization at the IUCF near a $G\gamma = 2$ imperfection resonance (at about 106.4 GeV) at 5 different beam kinetic energies with a 100% siberian snake on (*left*) and off (*right*)

Interestingly, it was not until 1989 that the siberian snake concept was tested experimentally [20, 21]. Shown in Figs. 10.7 and 10.8 are measurements which demonstrated the use of siberian snakes for control of the beam polarization. Plotted are the time-averaged vertical and radial polarization measured at different beam kinetic energies in the vicinity[5] of the $G\gamma = 2$ imperfection resonance. The horizontal axis shows the strength of an error field introduced diametrically opposite to the siberian snake. In these measurements, the error field was set, the beam was injected, the polarization was measured, and then the beam was dumped. The polarization of the injected beam was always oriented to match the stable spin direction at the injection point; i.e. in the horizontal plane with the snake turned on or vertically with

[5] From the figures, the resonance was found to lie between 105.9 MeV and 107.8 MeV rather than at kinetic energy $(2 - G)/mG = 108.4$ MeV one might expect using $G\gamma = 2$. This was later explained by the presence of an unintentional type-3 snake in the accelerator.

no snake. From Fig. 10.7, with the snake turned on, the radial polarization was not affected by the nearby resonance. From Fig. 10.8, with the snake turned off, the measured time-averaged vertical polarization was observed to be less near the resonance. The curves in Figs. 10.7 and 10.8 were obtained using the periodic solution to the equatio of spin motion taking into account the presence of the type-3 snake.

It is worth mentioning that the presence of a siberian snake in an accelerator may introduce a new kind of resonance dubbed a snake resonance [46]. The snake resonance condition is given by $u\nu_{\mathrm{res}} = \nu_0 + n$ in which ν_0 $(=\frac{1}{2})$ is the spin tune determined by the snake and u and n are integers. These are relevant when the fractional betatron tune Q is 1/integer, which is not a condition under which an accelerator is typically operated. Nonetheless, the presence of snake resonances has been both predicted and verified experimentally [47].

10.10 Partial Siberian Snakes

In 1989 Roser [48] proposed an elegant and intuitive variant of the siberian snake called a partial siberian snake. These are magnetic devices which enable the polarization to be maintained when accelerating through imperfection resonances. They rotate the spin by a small fraction of π around a horizontal axis and have the advantage of reduced cost and less required space. At low beam energies, a partial snake consisting of dipoles does not induce too large an orbit excursion, while a solenoidal partial snake requires significantly less magnetic field than a full snake. As mentioned above the disturbance to the polarization at imperfection resonances can be minimized by minimizing certain harmonics in the fields on the closed orbit. Partial snakes imply the opposite approach, namely the snake induces a large predetermined imperfection onto the spin motion which dominates all other imperfections. Then during normal acceleration with the large resonance strength, the Froissart–Stora formula guarantees a full spin flip without loss of polarization. The polarization will flip again after passage through the next imperfection resonance.

The expression for the dependence of the spin tune on energy and the precession per turn was given in (10.39) and is shown in Fig. 10.9 for different percentages of applied longitudinal field ($\eta = \pi$ denotes a full snake which is designated in the figure by 100%). The diagonal line shows the spin tune with no snake. As can be seen, even a relatively weak ($\sim$ few %) snake can cause a significant deviation of the spin tune from $G\gamma$ during acceleration and a large spin tune gap thus avoiding the imperfection resonance at $\nu_0 = n =$ integer.

An example showing the first demonstration of the ability of partial siberian snakes to avoid imperfection depolarizing resonances, in this case at fixed beam energy, is shown in Fig. 10.10. With a 10% snake full polarization was maintained despite the applied error field. At the AGS a 5%

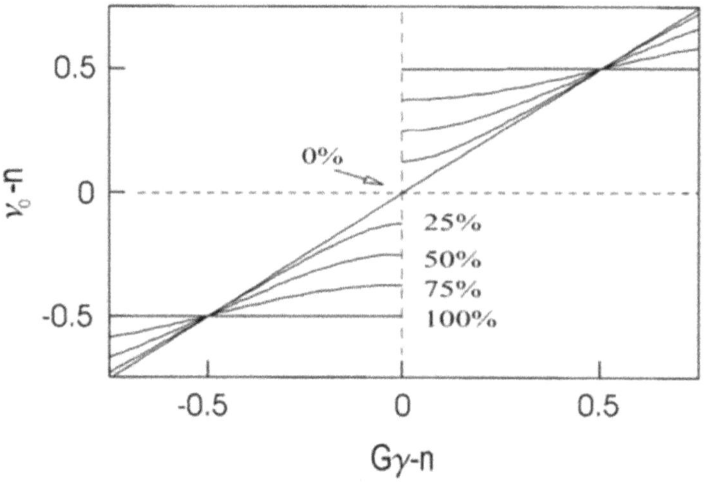

Fig. 10.9. The dependence of the spin tune on $G\gamma$ for various snake strengths (indicated by percentage of full 180 degree spin precession)

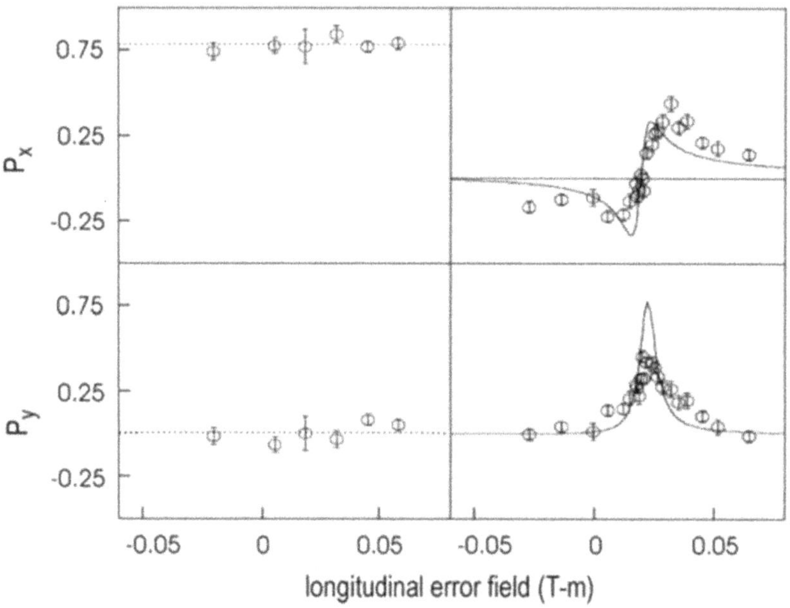

Fig. 10.10. Demonstration from the IUCF of polarization preservation in the vicinity of an imperfection resonance using a 10% partial siberian snake (*left*) and loss of polarization without the partial snake (*right*). In these measurements the kinetic energy of the beam was fixed at 105.9 MeV

partial siberian snake has been installed and is used routinely to ramp polarized protrons through numerous imperfection resonances to the required transfer energy ($\sim$ 25 GeV) for injection into RHIC [49, 50]. Note that for fixed beam energy, once the betatron tunes are known, the strength of the partial siberian snake could be set such that a nearby intrinsic resonance is also avoided.

10.11 RF Dipole

At the AGS, the many intrinisic resonances not avoidable the energy ramp are now overcome using an rf dipole magnet [51] (see also Sect. 3.10). From the Froissart–Stora equation (10.34), full spin flip may be expected if the resonance strength is large. The resonance strength may be artificially increased (see (10.48) below) by exciting a coherent vertical betatron oscillation thus inducing a complete spin flip [52]. If the location of externally induced resonance is placed close to an intrinsic resonance, then depending on the relative tune separation, phase, and strength, the induced resonance can be made to dominate the spin motion and full spin flip can be achieved [51]. At the AGS, the rf dipole is adiabatically turned on and off to avoid the emittance growth that would be observed with a pure impulse excitation [51].

10.12 Single Resonance Model

Until now we have avoided the use of complicated formulas and have presented basic concepts useful for practical applications of spin transport and preservation. In this section we define the resonance strength parameter ϵ used previously and show explicitly, for the case of an isolated resonance in the absence of siberian snakes, its effect on the beam polarization.

We rely on the work of Courant and Ruth [1], who expressed the magnetic fields in the Thomas-BMT equation in terms of the particle coordinates. They found that the equation of spin motion reduces to

$$\frac{\mathrm{d}S_x}{\mathrm{d}\theta} = -\kappa S_s - rS_y; \quad \frac{\mathrm{d}S_s}{\mathrm{d}\theta} = +\kappa S_x - tS_y; \quad \frac{\mathrm{d}S_y}{\mathrm{d}\theta} = +rS_x + tS_s\ , \tag{10.42}$$

where κ, r, and t are functions of the transverse coordinates of the particle orbit. In the cartesian coordinate system (with $\hat{x}$ radially outward, $\hat{s}$ along the beam direction, and $\hat{y}$ vertical) and ρ the local radius of curvature of the reference orbit, these variables are given by [1]

$$\begin{aligned} \kappa &= G\gamma - (1+G\gamma)\rho x'' \approx G\gamma \\ r &= (1+G\gamma)y' - \rho(1+G)\left(\frac{y}{\rho}\right)' \\ t &= (1+G\gamma)\rho y''\ , \end{aligned} \tag{10.43}$$

where the derivatives are with respect to the longitudinal coordinate s.

Equation (10.42) can then be transformed [1] into an equivalent spinor representation, for which

$$\frac{\mathrm{d}\Psi}{\mathrm{d}\theta} = \frac{\mathrm{i}}{2}H\Psi\ . \tag{10.44}$$

Here H is the spinor precession matrix given by

$$H = \begin{pmatrix} -\kappa & -t - \mathrm{i}r \\ -t + \mathrm{i}r & \kappa \end{pmatrix} , \tag{10.45}$$

and Ψ is a two component complex spinor. In first approximation the function H is uniquely determined by the properties of the synchrotron. In the preceding analyses we have assumed that the function H is piecewise constant. As before the spin components are obtained by taking the expectation value of the Pauli matrix vector, $\boldsymbol{\sigma}$, i.e.,

$$S_i = \Psi^\dagger \sigma_i \Psi . \tag{10.46}$$

Due to the periodic nature of a synchrotron, the coupling term of (10.45) may be expanded in terms of the Fourier components, i.e.,

$$t + \mathrm{i}r = \sum_k \epsilon_k e^{-\mathrm{i}\nu^{\pm}{}_{\mathrm{res},k}\theta} , \tag{10.47}$$

in which θ is the particle's orbital angle, $\nu^{\pm}{}_{\mathrm{res},k} = k$ for imperfection resonances, $\nu^{\pm}{}_{\mathrm{res},k} = k \pm Q_y$ for the first order intrinsic resonances, and ϵ_k is the resonance strength which is given by the Fourier amplitude

$$\epsilon_k = \frac{1}{2\pi} \int (t + \mathrm{i}r) e^{\mathrm{i}\nu^{\pm}{}_{\mathrm{res},k}\theta} \, \mathrm{d}\theta . \tag{10.48}$$

For the case of an imperfection resonance, the resonance strength is given approximately by summing over the radial error fields encountered by a particle on the closed orbit in one turn around the ring:

$$\epsilon_k \approx \frac{1 + G\gamma}{2\pi} \sum_l \Delta s_l \, \frac{\partial B_x / \partial y}{B\rho} y e^{\mathrm{i}\nu_{\mathrm{res},k}\theta} , \tag{10.49}$$

where y is the transverse amplitude with respect to the magnet center, and Δs_l is the length of the lth integration step around the ring.

As an illustration of the previous results, we now show that a transverse imperfection resonance can also shift the spin tune ν_0. In the single resonance approximation [46, 53], the spin equation in the laboratory frame is given by

$$\frac{\mathrm{d}\Psi}{\mathrm{d}\theta} = -\frac{\mathrm{i}}{2} \begin{pmatrix} G\gamma & -\zeta \\ -\zeta^* & -G\gamma \end{pmatrix} \Psi, \quad \text{with} \quad \zeta = \epsilon \cdot e^{-\mathrm{i}\nu_{\mathrm{res}}\theta} , \tag{10.50}$$

in which ϵ is the resonance strength, ν_{res} is the resonance tune, and θ is the particle orbital angle around the accelerator. The spin motion near the imperfection resonance can be visualized by transforming the spin equation to the resonance precession frame (i.e., the reference frame in which the polarization vector does not precess if the spin tune is exactly equal to the resonant tune). Thus, considering

$$\Psi_k = e^{\mathrm{i}\frac{\nu_{\mathrm{res}}\theta}{2}\sigma_y} \Psi , \tag{10.51}$$

we obtain an expression for the effect of the imperfections,

$$\frac{\mathrm{d}\Psi_k}{\mathrm{d}\theta} = -\frac{\mathrm{i}}{2}\begin{pmatrix} -\delta & \epsilon \\ \epsilon^* & \delta \end{pmatrix}\Psi_1 , \tag{10.52}$$

with $\delta \equiv (\nu_{\mathrm{res}} - G\gamma)$. This can also be written as

$$\frac{\mathrm{d}\Psi_k}{\mathrm{d}\theta} = \frac{\mathrm{i}}{2}(\delta\sigma_y + \epsilon_R\sigma_x - \epsilon_I\sigma_s)\Psi_k , \tag{10.53}$$

where σ_i are the Pauli matrices and $\epsilon = \epsilon_R + i\epsilon_I$ the complex resonance strength. Equation (10.53) can be integrated easily to yield

$$\Psi_k(\theta_f) = e^{\frac{\mathrm{i}}{2}(\delta\sigma_y+\epsilon_R\sigma_x-\epsilon_I\sigma_s)(\theta_f-\theta_i)}\Psi_k(\theta_i) . \tag{10.54}$$

Then transforming back to the laboratory frame, we obtain

$$\begin{aligned}\Psi(\theta_f) &= e^{-\mathrm{i}\frac{\nu_{\mathrm{res}}\theta_f}{2}\sigma_y} e^{\frac{\mathrm{i}}{2}(\delta\sigma_y+\epsilon_R\sigma_x-\epsilon_I\sigma_s)(\theta_f-\theta_i)} e^{\mathrm{i}\frac{\nu_{\mathrm{res}}\theta_i}{2}\sigma_y}\Psi(\theta_i) \\ &= T(\theta_f,\theta_i)\Psi(\theta_i) .\end{aligned} \tag{10.55}$$

By expanding the exponential, the spin transfer matrix $T(\theta_f, \theta_i)$ for a single resonance may be calculated [46, 53]:

$$T(\theta_f,\theta_i) = \begin{pmatrix} ae^{\mathrm{i}\left(c-\frac{\nu_{\mathrm{res}}(\theta_f-\theta_i)}{2}\right)} & ibe^{-\mathrm{i}\left(d+\frac{\nu_{\mathrm{res}}(\theta_f+\theta_i)}{2}\right)} \\ ibe^{\mathrm{i}\left(d+\frac{\nu_{\mathrm{res}}(\theta_f+\theta_i)}{2}\right)} & ae^{-\mathrm{i}\left(c-\frac{\nu_{\mathrm{res}}(\theta_f-\theta_i)}{2}\right)} \end{pmatrix} , \tag{10.56}$$

with

$$\begin{aligned} &b = \frac{|\epsilon|}{\lambda}\sin\frac{\lambda(\theta_f-\theta_i)}{2}, \quad a = \sqrt{1-b^2}, \\ &c = \arctan\left[\frac{\delta}{\lambda}\tan\frac{\lambda(\theta_f-\theta_i)}{2}\right], \quad d = \arg(\epsilon^*), \\ &\delta = \nu_{\mathrm{res}} - G\gamma, \quad \lambda = \sqrt{\delta^2 + |\epsilon|^2} . \end{aligned} \tag{10.57}$$

The spin tune on the closed orbit can be obtained from the trace of the one turn transfer map, $T(\theta + 2\pi, \theta)$ of (10.56), i.e.,

$$\cos\pi\nu_0 = a\cos(c - \nu_0\pi) . \tag{10.58}$$

Figure 10.11 shows the spin tune shift, $\delta\nu = G\gamma - \nu_0$, as a function of $G\gamma - 2$ for the special cases where $|\epsilon| = \sqrt{\epsilon_R^2 + \epsilon_I^2} = 0.0008$ and $|\epsilon| = 0.0015$. In both cases, for $G\gamma$ far away from the resonance tune, ν_{res}, one has $\delta \gg |\epsilon|$ and $a \approx 1$, so that $\nu_0 \approx G\gamma$. As $G\gamma$ approaches the resonance tune, the spin tune is shifted from $G\gamma$ by $\Delta\nu_0 = -|\epsilon|$ below the resonance and by $\Delta\nu_0 = |\epsilon|$ above the resonance, i.e., the spin tune is always shifted away from the resonance tune. Therefore at a given energy, the observed width of the vertical

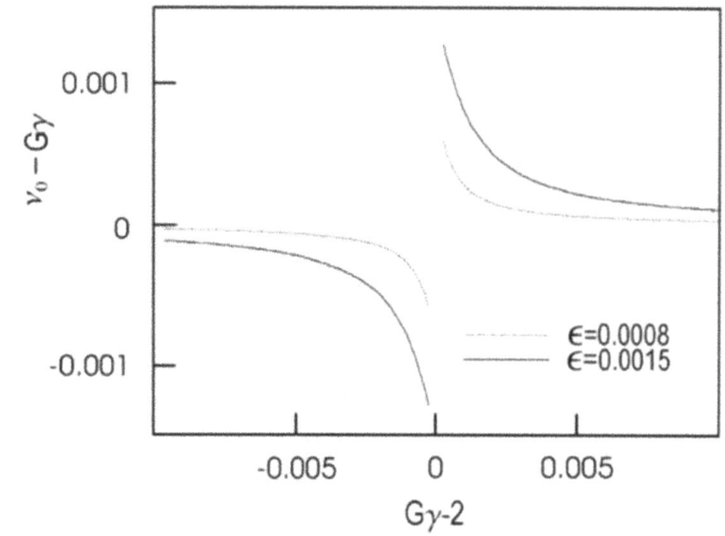

Fig. 10.11. Spin tune shift ($\nu_0 - G\gamma$) versus ($G\gamma - 2$) near an imperfection resonance, according to (10.58). The *dashed curve* corresponds to a resonance strength of magnitude 0.0008. The *solid curve* results for a resonance strength of 0.0015

polarization (in Fig. 10.8, for example) would always be increased when the effect of the imperfection resonance is included. The observed slope of the radial polarization at the fully compensated field value would also be lessened in magnitude. Figure 10.11 indicates that the effect of the imperfection resonance becomes important only in very close proximity to the resonance.

The solution of (10.52), decomposed into two eigenmodes, is

$$\Psi_\pm = e^{\pm i \frac{\lambda\theta}{2}} \begin{pmatrix} \frac{\epsilon}{|\epsilon|}\sqrt{\frac{\lambda\pm\delta}{2\lambda}} \\ \mp\sqrt{\frac{\lambda\mp\delta}{2\lambda}} \end{pmatrix} , \tag{10.59}$$

where $\lambda = \sqrt{\delta^2 + |\epsilon|^2}$. The particle spin is given by a linear combination of the eigensolutions,

$$\Psi_1(\theta) = C_+\Psi_+ + C_-\Psi_- , \tag{10.60}$$

normalized such that $|C_+|^2 + |C_-|^2 = 1$. The component along the y axis is

$$\begin{aligned} S_y &= \Psi^\dagger \sigma_y \Psi \\ &= {\Psi_1}^\dagger \sigma_y \Psi_1 \\ &= \frac{\delta}{\lambda}\left(|C_+|^2 - |C_-|^2\right) + \frac{2|\epsilon|}{\lambda}\mathrm{Re}[C_+ {C_-}^* e^{i\lambda\theta}] . \end{aligned} \tag{10.61}$$

For an initially vertically polarized particle, the time-averaged vertical polarization is found to be

$$\langle S_y \rangle = \frac{\delta}{\lambda}(|C_+|^2 - |C_-|^2) = \frac{\delta^2}{\lambda^2} = \frac{\delta^2}{\delta^2 + |\epsilon|^2} , \tag{10.62}$$

which is less than the initial polarization by an amount that depends on the resonance strength ϵ.

Exercises

10.1 Electrostatic Lenses and Muon Storage Rings

The muon anomolous magnetic moment, a_μ, now recognized to be about 0.001166, can be measured very accurately using electrostatic lenses with a transverse electric field. In the rotating reference frame, the spin precession is given by

$$\boldsymbol{\Omega} = -\frac{e}{m\gamma}\left(a_\mu B_\perp \gamma + (1+a_\mu)B_\parallel + \left(a_\mu\gamma - \frac{\gamma}{\gamma^2-1}\right)\frac{\boldsymbol{E}\times\boldsymbol{v}}{c^2}\right) . \tag{10.63}$$

Show that even the transverse electric field $\boldsymbol{E}$ does not contribute to spin precession when the Lorentz factor is[6]

$$\gamma = \sqrt{1+\frac{1}{a_\mu}} . \tag{10.64}$$

10.2 Spinors

This exercise provides practice with spinor-matrix algebra.

a) Using (10.10) find the spinor wave function for the spin basis $\boldsymbol{S} = [\boldsymbol{S}_x, \boldsymbol{S}_s, \boldsymbol{S}_y]$ with $\boldsymbol{S}_x = [1\ 0\ 0]$, $\boldsymbol{S}_s = [0\ 1\ 0]$, and $\boldsymbol{S}_s = [0\ 0\ 1]$.

b) Show that the Pauli matrices are unitary ($\sigma_j {\sigma^\dagger}_j = I$) and Hermitian (${\sigma^\dagger}_j = \sigma_j$) with $\boldsymbol{\sigma}^\dagger \cdot \boldsymbol{\sigma} = 3I$.

c) Verify the compact form of the commutation relations:

$$\sigma_j\sigma_k = \delta_{jk}I + i\Sigma_m \epsilon_{jkm}\sigma_m , \tag{10.65}$$

where

$$\begin{aligned} \delta_{jk} &= 1 \ \text{ if } \ j=k \\ &= 0 \ \text{ if } \ j \neq k \end{aligned} \tag{10.66}$$

and ϵ_{jkm} is the Levi-Civita tensor defined by

$$\begin{aligned} \epsilon_{jkm} &= 0 \text{ if any two indices are equal} \\ &= +1 \text{ for even permutation of indices} \\ &= -1 \text{ for odd permutation of indices .} \end{aligned} \tag{10.67}$$

10.3 Spin Precession in Solenoidal Fields

Consider a vertically polarized beam traversing ($\boldsymbol{\beta} = \beta\hat{s}$) a longitudinal solenoid of field $\boldsymbol{B} = B_z\hat{s}$ of length l in the absence of any electric fields.

---

[6] Adapted from lecture notes of A. Chao (1999)

a) Show that the spin precession ϕ after traversal of the solenoid is given by

$$\phi = -\frac{e}{\gamma mc^2}\frac{l}{\beta}(1+G)B_z \ . \tag{10.68}$$

b) Suppose this solenoid is in a circular accelerator. By equating the centrifugal and Lorentz forces on the particle show that the magnetic rigidity is

$$B\rho = \frac{\beta E}{ec} \ , \tag{10.69}$$

where B is the vertical magnetic dipole field, and reexpress (10.68) in terms of the rigidity.

c) For spin polarization in a storage ring we have seen that spin precession by π per turn helps cancel various spin resonances. For the case of a proton beam with 100 MeV kinetic energy specify the required integrated field strength to achieve this.

10.4 Periodic Spin Motion

Using the expansion of the exponential

$$e^{i\alpha\sigma_j} = \cos\alpha + i\sigma_j \sin\alpha \quad \text{where } j = x, s, y \ , \tag{10.70}$$

verify (10.25).

10.5 SLC '3-state experiment'

Assuming no resonant depolarization (that is pure spin precession) in the SLC arcs, show that the magnitude of the polarization at the interaction point (IP) can be obtained from three successive measurements of the longitudinal polarization at the IP by proper orientation of the incoming polarization with each measurement.

10.6 Type-3 Snakes

Let L represent a precession about the longitudinal axis.

a) Show that the configuration

$$\left[\mathrm{V}\left(-\frac{\phi}{2}\right)\mathrm{L}\left(+\frac{\chi}{4}\right)\mathrm{V}\left(+\frac{\phi}{2}\right)\right]\mathrm{L}\left(-\frac{\chi}{2}\right)\left[\mathrm{V}\left(-\frac{\phi}{2}\right)\mathrm{L}\left(+\frac{\chi}{4}\right)\mathrm{V}\left(+\frac{\phi}{2}\right)\right] \tag{10.71}$$

is does not introduce a net deflection, but produces a net spin precession about the vertical axis.

b) Draw the spin orientation and the particle orbit for the given magnet configuration.

11 Cooling

Many applications of particle accelerators require beam cooling, which refers to a reduction of the beam phase space volume or an increase in the beam density via dissipative forces. In electron and positron storage rings cooling naturally occurs due to synchrotron radiation, and special synchrotron-radiation damping rings for the production of low-emittance beams are an integral part of electron-positron linear colliders. For other types of particles different cooling techniques are available. Electron cooling and stochastic cooling of hadron beams are used to accumulate beams of rare particles (such as antiprotons), to combat emittance growth (e.g., due to scattering on an internal target), or to produce beams of high quality for certain experiments. Laser cooling is employed to cool ion beams down to extremely small temperatures. Here the laser is used to induce transitions between the ion electronic states and the cooling exploits the Dopper frequency shift. Electron beams of unprecedentedly small emittance may be obtained by a different type of laser cooling, where the laser beam acts like a wiggler magnet. Finally, designs of a future muon collider rely on the principle of ionization cooling. Reference [1] gives a brief review of the principal ideas and the history of beam cooling in storage rings; a theoretical dicussion and a few practical examples can be found in [2].

11.1 Damping Rates and Fokker–Planck Equation

In the presence of cooling and in the absence of any excitation by noise, the evolution of a beam distribution function $f(x, x', t)$ for one degree of freedom, here for the horizontal plane, is described by the differential equation [2]

$$\frac{\mathrm{d}f(x, x', t)}{\mathrm{d}t} = \lambda f \,, \tag{11.1}$$

with the solution

$$f(x, x', t) = e^{\lambda t} f_0(x_0, x_0') \,, \tag{11.2}$$

where the subindex 0 characterizes the initial distribution f_0 or the initial phase-space variables. The latter, x_0 and x_0', are related to x and x' by the equation of motion including the damping. Note that the phase space

M. G. Minty et al., *Measurement and Control of Charged Particle Beams*,
https://doi.org/10.1007/978-3-662-08581-3_11

density about each particle increases exponentially. Without a cooling force, the system would be Hamiltonian and the local phase-space density conserved ($\mathrm{d}f/\mathrm{d}t = 0$), so that $\lambda = 0$ in this case.

It is common to introduce action-angle variables I and ψ (where I is proportional to the square of the oscillation amplitude) via the relations

$$\frac{x}{\sqrt{\beta}} = \sqrt{2I}\cos\psi \; , \tag{11.3}$$

$$\sqrt{\beta}\left(x' + \alpha\frac{x}{\beta}\right) = -\sqrt{2I}\;\sin\psi \; , \tag{11.4}$$

where α and β are the usual alpha and beta function describing the linear optics (cf. Chap. 1). The angle variable ψ can be identified with the betatron phase, The distribution in the angle ψ is often uniform and random. In these case, the beam distribution function f only depends on the action variable I and, possibly, on the time, i.e., $f(x, x', t) = f(I, t)$. Indeed, in earlier chapters, e.g., Sect. 4.1, we have implicitly taken the randomness of the betatron phase as the definition of a 'matched' beam. For a mismatched or oscillating beam, the initial beam distribution is not uniform in the angle ψ, but any spread of the betatron frequencies results in a phase 'randomization', after a time roughly equal to the inverse of the frequency spread. A spread in the betatron frequency is always present. It arises, e.g., from a nonzero chromaticity and a finite energy spread, or from a dependence of the betatron tune on oscillation amplitude. Our following treatment assumes that this 'randomization time' is much shorter than the cooling time. The phase randomization and the cooling can then be mathematically decoupled, e.g., by averaging the equations describing the time evolution of the action over the betatron phase.

Cooling in the three phase-space dimensions results in an exponential damping of the 3 action invariants:

$$\left\langle \frac{\partial \dot{I}_i}{\partial I_i} \right\rangle = -\lambda_i \; , \tag{11.5}$$

where $i = (x, y, z)$. The angular brackets in (11.5) denote an average over both the angle variables and the azimuthal position around the storage ring, θ, i.e.,

$$\langle ... \rangle = \int_0^{2\pi} \frac{\mathrm{d}\psi}{2\pi}\,\frac{\mathrm{d}\theta}{2\pi}\,(...) \tag{11.6}$$

and the λ_i are the damping rates in the three planes.

Denoting the physical momenta by $p_i = \gamma m c v_i$ (v_i is the velocity for the ith degree of freedom in units of m/s, m the particle mass, and γ the relativistic factor) and considering a 'cooling force' F which changes the particle momenta according to $\dot{p}_k = F_k$, some algebra yields

$$\begin{aligned}
\sum_i \left\langle \frac{\partial \dot{I}_i}{\partial I_i} \right\rangle &= \sum_i \left\langle \sum_k \frac{\partial}{\partial I_i} \frac{\partial I_i}{\partial p_k} F_k \right\rangle \\
&= \sum_i \left\langle \sum_k \frac{\partial}{\partial I_i} \frac{\partial I_i}{\partial p_k} F_k + \frac{\partial}{\partial \psi_i} \frac{\partial \psi_i}{\partial p_k} F_k \right\rangle \\
&= \left\langle \sum_k \frac{\partial F_k}{\partial p_k} \right\rangle ,
\end{aligned} \tag{11.7}$$

where we have made use of the fact that the average over ψ of any derivative with respect to ψ is zero. The sum of the action damping coefficients is

$$\lambda_x + \lambda_z + \lambda_s = -\left\langle \frac{\partial F_x}{\partial p_x} + \frac{\partial F_z}{\partial p_z} + \frac{\partial F_s}{\partial p_s} \right\rangle = \langle -\boldsymbol{\nabla}_{\boldsymbol{p}} \boldsymbol{F} \rangle , \tag{11.8}$$

independent of any coupling between the three planes of motion.

As an example, cooling due to synchrotron radiation and due to ionization cooling is approximately described by a cooling force that is anti-parallel to the particle velocity $\boldsymbol{v}$ [2],

$$\boldsymbol{F} = -a\boldsymbol{v} , \tag{11.9}$$

where the coefficient a may depend on the particle energy. The cooling is accompanied by a particle energy loss rate W,

$$\frac{\mathrm{d}E}{\mathrm{d}t} = -W = \boldsymbol{F} \cdot \boldsymbol{v} = -av^2 , \tag{11.10}$$

which can be compensated by an rf system. Assuming ultrarelativistic particles ($v \equiv |\boldsymbol{v}| = c$), the cooling force of (11.9) may be rewritten in terms of the energy loss as $\boldsymbol{F} = -\boldsymbol{v}W/c^2$, and direct evaluation then yields:

$$-\boldsymbol{\nabla}_{\boldsymbol{p}} \cdot \boldsymbol{F} = \left(\frac{W}{pc} \right) \left[2 + \frac{\partial \ln W}{\partial \ln p} \right] . \tag{11.11}$$

By inserting this expression into (11.8) the total decrease rate in phase-space volume can be calculated. Equations (11.8) and (11.11) state that the sum of the three damping rates is a constant, only depending on the total rate of energy loss. In the special case of synchrotron radiation, this is known as the 'Robinson theorem'.

One might think it would be possible to produce a beam of nearly zero temperature by cooling for a very long time. However, there is always some noise exciting the beam, which prevents reaching this limit and gives rise to an equilibrium emittance. In the case of synchrotron radiation this noise is due to quantum fluctuations, for ionization cooling it is due to multiple scattering, and in the case of stochastic cooling there is electronic noise in the detector-amplifier chain and Schottky noise arising from the finite number of particles in the beam.

With such noise sources present, the evolution of the distribution function $f(I,t)$ is no longer described by (11.1), but by a Fokker–Planck equation of the form

$$\frac{\partial f(I,t)}{\partial t} = \frac{\partial}{\partial I}\left(-\left\langle\frac{\Delta I}{\Delta t}\right\rangle f(I,t)\right) + \frac{1}{2}\frac{\partial^2}{\partial I^2}\left[\left(\left\langle\frac{(\Delta I)^2}{\Delta t}\right\rangle f(I,t)\right)\right] , \quad (11.12)$$

where now the angular brackets denote an average over the entire beam distribution, including the action variables, and over the noise. For example, if the Fokker–Planck terms $\langle\Delta I\rangle$ and $\langle(\Delta I)^2\rangle$ are linear in I and constant, respectively, the equation reduces to

$$\frac{\partial f}{\partial t} = \frac{\partial}{\partial I}\left(\lambda I f + \frac{D}{2}\frac{\partial f}{\partial I}\right) , \quad (11.13)$$

where $\lambda = \langle\Delta I/\Delta t\rangle/I$, and $D \equiv \langle(\Delta I)^2/\Delta t\rangle$. The beam then asymptotically approaches the distribution, $f_\infty \propto \exp(-I/I_\infty)$, with the equilibrium emittance (for the equality of rms emittance and average action see (1.14) and Ex. 1.1)

$$\epsilon = \langle I\rangle_{t=\infty} = I_\infty = \frac{D}{2\lambda} . \quad (11.14)$$

Using (11.13), this distribution is easily shown to be stationary: $\partial f_\infty/\partial t = 0$.

The cooling of various particles can be coupled, e.g., in stochastic cooling the time resolution may be limited by the amplifier bandwidth, and on each passage through the cooler only the average position of several particles is measured and damped. Under these circumstances, the beam is fully cooled only if the individual particles exchange their positions within the beam, so that on successive turns the measured average position, which is damped, refers to different combinations of particles. This process of particle exchange is called 'mixing'.

11.2 Electron Cooling

Electron cooling was proposed in 1966 by G.I. Budker [3]. The first experiments of electron cooling were performed at the NAP-M storage ring at the INP in Novosibirsk, where a 65-MeV antiproton beam was cooled down to a final momentum spread of 1.4×10^{-6} and to an angular divergence of 12.5 μrad, much smaller than the 3 mrad angular divergence of the 0.3-A 50-keV electron beam. Cooling times of the order of 25 ms were achieved [2].

11.2.1 Basic Description

Electron cooling is based on the heat exchange between a stored hadron beam and an accompanying electron beam via Coulomb collisions. The temperature of the electron beam is held constant and lower than the temperature of the hadron beam to be cooled. This is easily fulfilled since for equal ion and electron velocities, $v_e \approx v_i$, the temperature of the electron beam is

$$T_e \approx \frac{m_e}{M} T_{\text{ion}} , \tag{11.15}$$

where M and m_e denote the ion and electron masses, respectively. Because of their mass ratio, the temperature of the ion beam is much larger than that of the electron beam. The average velocities of the hadron and electron beams should coincide in the cooling interaction region, in order to maximize the Coulomb cross section, which depends on the relative velocity. Viewed in the electron rest frame, moving with the electron beam, the ions are 'stopped' similarly to the slowing down of charged particles traversing matter, because in the Coulomb collisions energy is transferred from the ions to the electrons. The typical layout of an electron cooler and a photo of the electron cooling system at LEAR are depicted in Figs. 11.1 and 11.2, respectively.

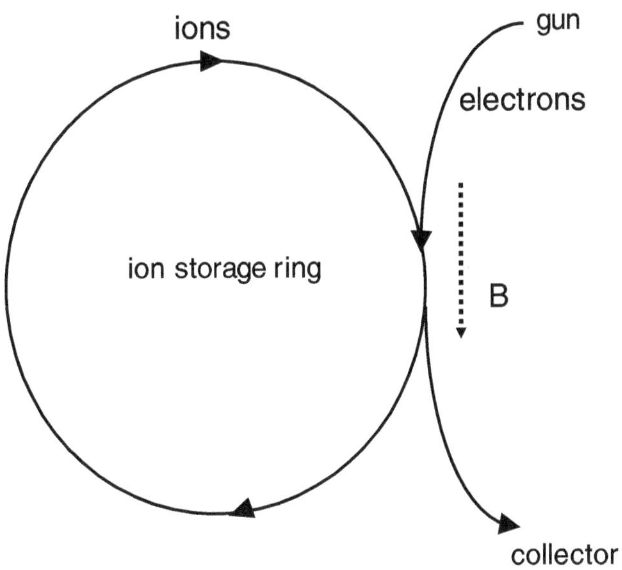

Fig. 11.1. Schematic of electron cooling for an ion storage ring

Transverse and longitudinal temperatures, $T_\perp$ and $T_\parallel$, of the ion beam can be defined by analogy with kinetic gas theory:

$$T_\perp = \frac{M\langle u_\perp^2 \rangle}{k_B} , \tag{11.16}$$

and

$$T_\parallel = \frac{M\langle \Delta u_\parallel^2 \rangle}{k_B} , \tag{11.17}$$

where M is the ion mass, u the ion velocity, and k_B the Boltzmann constant. The velocity components $\langle u_\perp^2 \rangle^{1/2}$ and $\langle \Delta u_\parallel^2 \rangle^{1/2}$ refer to the transverse

Fig. 11.2. Electron cooling system at LEAR (Courtesy M. Chanel, 1999)

and longitudinal rms velocity spread, repectively. The longitudinal velocity is taken to be the difference from the mean velocity of the ion beam, which is indicated by the prefix Δ. The transverse and longitudinal temperatures are usually not the same. Electron-beam temperatures are defined in the same way.

The cooling stops when the temperatures of the electron and ion beam are equal. The velocity of a cooled coasting ion beam (without rf) is equal to that of the electron beam, $v_{\text{ion}} = v_e$. This provides a useful tool for tuning the ion beam energy. For a bunched beam, the rf frequency must be adjusted in order to match the revolution frequency of the ions as determined by the electron beam.

11.2.2 Estimate of the Cooling Rate

The cooling force may be estimated by considering the collision of a single ion with a single electron in a reference frame where the electron is at rest before the collision [2].

To this end, we split the collision into two steps. During the first step, the electron and ion approach each other, and in the second step they are separating again. We assume that during the first part the electron is accelerated by the field of the ion and that it moves in the direction of the impact parameter. The duration of either time step is of the order $\Delta t \approx \rho/u$, where ρ is the impact parameter and u the velocity of the ion. The situation is sketched in Fig. 11.3.

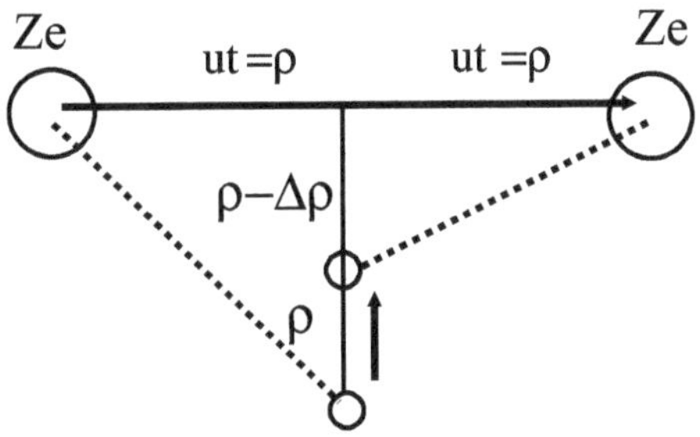

Fig. 11.3. Collision of one ion and one electron during electron cooling [2]

The ion with charge Ze moves from left to right and approaches the electron with an impact parameter ρ. The electron, initially at rest, is accelerated by the ion field. At the end of the first time step, the electron velocity in the direction of the closest approach is

$$\Delta v_e = \frac{Zr_e c^2}{\rho u} \, . \tag{11.18}$$

At this time the electron has moved by about a distance

$$\Delta\rho \approx \Delta v_e \Delta t \approx \frac{Zr_e c^2}{u^2} \, . \tag{11.19}$$

Integrating over the value of the initial impact parameter, the average change of the ion momentum in the direction u is

$$\left\langle \frac{\mathrm{d}p_u}{\mathrm{d}t} \right\rangle = u n_e m_e c^2 \int_{\rho_{\min}}^{\rho_{\max}} \left(\frac{Zr_e c}{\rho u} - \frac{Zr_e c}{(\rho - \Delta\rho)u} \right) 2\pi\rho \; \mathrm{d}\rho \tag{11.20}$$

where n_e denotes the local density of the electron beam, and m_e the electron mass. The limits of integration $\rho_{\max}$ and $\rho_{\min}$ refer to the maximum and minimum impact parameter. Expanding in powers of $\Delta\rho$ and keeping only the leading contribution, one finds

$$\left\langle \frac{\mathrm{d}p_u}{\mathrm{d}t} \right\rangle = \frac{2\pi n_e r_e^2 Z^2 m_e c^4}{u^2} \, L_C \tag{11.21}$$

where we have introduced the Coulomb logarithm $L_C \equiv \ln\left(\rho_{\max}/\rho_{\min}\right)$. As an upper integration limit $\rho_{\max}$ we may take the Debye shielding length of the electron beam:

$$r_D \approx \left[\frac{k_B T}{4\pi m_e c^2 n_e r_e} \right]^{1/2} \, . \tag{11.22}$$

A lower limit $\rho_{\min}$ can be determined, e.g., from the maximum momentum transfer to the electron (classical head-on collision):

$$\rho_{\min} = \frac{Zr_e c^2}{u^2} \, . \tag{11.23}$$

In numerical estimates, L_C is usually taken to be constant, on the order of 10.

Averaging (11.21) over the electron velocity distribution function f_e results in the cooling force

$$F_{\rm el} = \left\langle \frac{{\rm d}\boldsymbol{p}}{{\rm d}t} \right\rangle = 2\pi Z^2 r_e^2 m_e c^4 L_C \int {\rm d}^3 v_e f_e(\boldsymbol{v}_e) \frac{\boldsymbol{v} - \boldsymbol{v}_e}{(\boldsymbol{v} - \boldsymbol{v}_e)^3} \,. \qquad (11.24)$$

The result of a more precise evaluation of the cooling force starting from the Rutherford cross section agrees within a factor of 2 with (11.24) [4].

The cooling time $\tau_{\rm el}$ follows from [5]

$$\frac{1}{\tau_{\rm el}} = \left| \frac{1}{u} \frac{{\rm d}u}{{\rm d}t} \right| = \left| \frac{F_{\rm el}}{Mu} \right| \,. \qquad (11.25)$$

In the laboratory frame the cooling time is larger by a factor γ due to time dilation (there is a further factor of γ due to Lorentz contraction if the distribution function f_e is taken to be that in the laboratory frame). In the limit of large ion velocities, the electron velocity may be replaced by a delta function; in the opposite limit an isotropic Gaussian distribution is assumed. The cooling time in the two limits is [5]:

$$\tau = \frac{\gamma^2}{\eta_{\rm el}} \frac{M}{m_e} \frac{1}{Z^2 r_e^2 c^4} \frac{1}{\rho_L L_C} \begin{cases} \frac{1}{4\pi} u^3 & \text{for } u \gg v_{e,{\rm rms}} \\ \frac{3}{2\sqrt{2\pi}} \left(\frac{\frac{3}{2} k_B T_e}{m_e} \right)^{3/2} & \text{for } u < v_{e,{\rm rms}} \end{cases} \qquad (11.26)$$

where $\eta_{\rm el}$ is the ratio of the cooling section length to the ring circumference, and ρ_L the electron beam density in the laboratory frame. The equation shows that electron cooling becomes inefficient for high energies, $\gamma \gg 1$, and that the cooling time is short for light ions of high charge. The cooling time of hot beams scales as u^3, while the cooling time of cold beams is independent of the ion velocities and only depends on the electron temperature.

Figure 11.4 shows a schematic of the transverse and longitudinal cooling forces, illustrating the two different cooling regimes incurred for high and low ion velocities.

Example parameters for electron cooling are $k_B T_e \approx 0.2$ eV, $n_L = 3 \times 10^8$ cm^{-3}, $L_C = 10$, $\eta = 0.05$, $\gamma = 1$, and $Z = 1$, which results in a cooling time of 40 s.

In reality, there are two additional effects which considerably help to reduce the cooling times: First, the electron velocity distribution is not Gaussian, but Maxwellian, and due to acceleration in the electron gun, the distribution is compressed in the longitudinal direction. This compression of the longitudinal velocity spread leads to shorter longitudinal cooling times. Second, a longitudinal magnetic field is employed to guide and confine the electron beam. This results in a cyclic motion of the electrons. If the cyclotron period is small compared with a typical ion-electron collision time, the cyclotron motion decreases the effective transverse temperature of the electron beam, and can reduce the cooling times, to values below one tenth of a second [2, 6, 5].

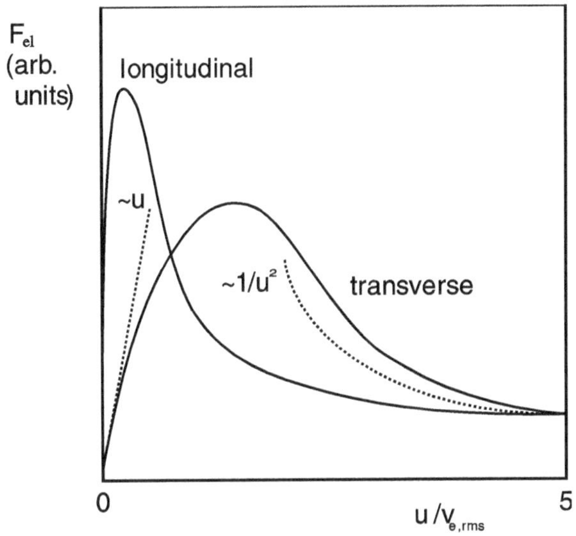

Fig. 11.4. Cooling force $F_{\mathrm{el}} = Mu/\tau_{\mathrm{el}}$ in a flattened electron beam as a function of ion velocity u in units of the rms electron velocity in the beam frame $v_{e,\mathrm{rms}}$. The *dashed curve* corresponds to the asymptotic formulae derived in the text. The difference between the transverse and longitudinal plane is due to the temperature difference, which arises from the longitudinal acceleration. Picture is redrawn from [5]

To relate electron cooling times for different types of particle beams, we note that the cooling rate scales like [1]

$$\frac{1}{\tau} \propto \frac{Z^2}{A}, \tag{11.27}$$

where A is the atomic mass of the ion, and Z the atomic number (i.e., the ion charge in units of the electron charge). We thus expect that cooling is faster for highly charged ions. However, these ions can also more easily capture a cooling electrons, thereby changing their charge, and get lost. The rate of recombination due to radiative electron capture scales approximately as [1]

$$\frac{1}{\tau_r} \propto Z^2 \,. \tag{11.28}$$

It is also worth mentioning that for relativistic energies electron cooling becomes less efficient; see, e.g., (11.26), where the cooling time τ increases as γ^2. In addition, higher electron-beam energies would be required in the cooling system.

11.2.3 Optical Functions at the Electron Cooler

If the electron beam temperature is low compared with that of the ion beam, the electron cooling rate varies as

$$\frac{1}{\tau} \propto \frac{1}{u^3} \sim \frac{1}{\theta_{x,y}^3}, \tag{11.29}$$

where $\theta_{x,y} = \sqrt{\epsilon/\beta_{x,y}}$ is the transverse rms divergence of the ion beam, and $\beta_{x,y}$ here the lattice beta function at the cooler (we assume that $\alpha_x \approx 0$). One might thus imagine that a large value of $\beta_{x,y}$ would give the best cooling results. However, for a large value of $\beta_{x,y}$ also the beam size is large, and the ions sample the nonlinear space-charge field of the electron beam. This space-charge effect complicates the electron-ion velocity matching. In addition, a large ion beam may only incompletely overlap with the electron beam. For this reason, an intermediate beta function turns out to be optimal, where the ion beam is slightly smaller than the electron beam.

One would also expect that the cooling rate increases in proportion to the electron beam current. In practice, however, for larger current one observes a tendency of saturation. Again, the limit arises from the space-charge force in the electron beam.

Let us take a closer look at the electron space-charge effects, and, in doing so, also explore the effect of a nonzero dispersion function at the electron cooler. Consider a cylindrically symmetric electron beam of radius a with a uniform transverse distribution and with a longitudinal density $\lambda = I/(e\beta c)$, where I is the current, e the electron charge, and c the speed of light. For a uniform charge distribution the space charge force is linear for radial positions $r < a$:

$$E_r = \frac{\lambda e}{2\pi\epsilon_0 a^2} r \,. \tag{11.30}$$

Sufficiently far away from the gun, the electron beam reaches an equilibrium state where the sum of kinetic and potential energy is a constant for all electrons and where the electron energy depends on the radial position as [7]

$$m_e c^2 \gamma(r) = m_e c^2 \gamma(0) + e \int_0^r \mathrm{d}r' \; E_r(r') \tag{11.31}$$

or

$$\gamma(r) = \gamma(0) + \lambda r_e \frac{r^2}{a^2} \,, \tag{11.32}$$

where $\gamma(r)$ is the Lorentz factor characterizing the energy of electrons at radius r, and r_e is the classical electron radius. Since

$$\frac{\Delta v(r)}{v} = \frac{1}{\gamma^2 - 1} \frac{\gamma(r) - \gamma(0)}{\gamma} = \frac{1}{\gamma^2 - 1} \frac{\lambda r_e r^2}{\gamma a^2} \,, \tag{11.33}$$

the velocity distribution in the electron beam is roughly parabolic as a function of the radial position. For high currents, the increase in the velocity spread of the electron beam degrades the cooling force. The situation is illustrated in Fig. 11.5, which shows the velocity of electron and ion beams as a function of radial position. From the figure, it is evident that a nonzero dispersion at the electron cooler can reduce the average velocity difference between electrons and a beam which is injected off-momentum, thereby improving the performance. In Fig. 11.5, the part of the beam which has already been cooled – the 'stack' – has zero momentum offset.

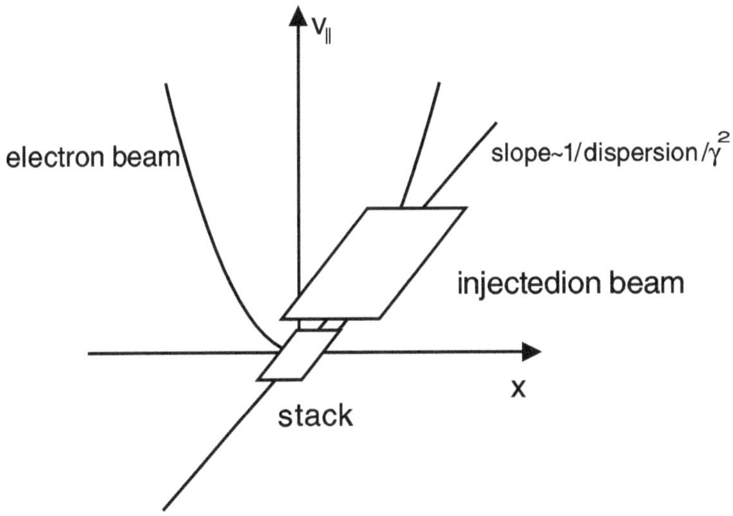

Fig. 11.5. Longitudinal velocity versus horizontal position of the electron and ion beams. Due to space charge the electron velocities lie on a parabola; the ion velocity varies linearly with a slope inversely proportional to the dispersion. Because of betatron oscillations, ions occupy a large area in phase space, as indicated [8, 9, 10] (Courtesy Ch. Carli and M. Chanel, 2002)

The optimum value of the dispersion function scales as [8, 9]

$$D \propto \sqrt{\frac{Ua^2}{I(\Delta p/p)_{\rm rms}}}\,, \tag{11.34}$$

where U is the accelerating voltage of the electron beam, I the electron current, $(\Delta p/p)_{\rm rms}$ the rms momentum spread of the ion beam, and a the electron beam size. The positive effect of a nonzero dispersion was confirmed by observations [8, 9].

11.2.4 Outlook

For the cooling of high-energy beams, it has been proposed to store the electron beam in a storage ring, sharing a common straight section with the ion or proton storage ring, where the cooling takes place [11, 12].

The emittance of the electron beam is then maintained by radiation damping. In such scheme, the bucket spacing of the electron storage ring should be an integral multiple of the bucket spacing of the ion storage ring [12]:

$$\frac{C_e}{h_e} = n\frac{C_i}{h_i} \qquad (n \text{ integer})\,, \tag{11.35}$$

where h_e and h_i denote the harmonic numbers for the electron and ion ring, respectively, and C_e and C_i the ring circumferences.

Recently, a novel scheme of high-energy electron cooling was proposed for the Relativistic Heavy Ion Collider (RHIC). The concept includes acceleration in a superconducting recirculating linac, strong solenoidal fields in the cooling-interaction region, and energy recovery from the electron beam after its passage through the cooling section [13, 14].

11.3 Stochastic Cooling

Excellent rewiews of stochastic cooling are available [1, 5, 15, 16, 17]. Stochastic cooling was conceived in 1968 by van der Meer. Proton beam Schottky noise was first observed at the ISR in 1972, and first experimental demonstration took place in 1975, also at the ISR. In the period 1977–83, cooling tests were performed at CERN, FNAL, Novosibirsk and INS-Tokyo. In the 1977 cooling experiment ICE at CERN the momentum spread of 5×10^7 particles was reduced from 3.5×10^{-3} to 5×10^{-4}. At LEAR, in 1985, the momentum spread of 3×10^9 particles was reduced from 4×10^{-3} to 1.2×10^{-3} in 3 minutes of cooling [1]. At the CERN AA a factor 3×10^8 increase in phase-space density was achieved [1].

11.3.1 Basic Description

Figure 11.6 shows the process of stochastic cooling. A transverse pick up detects the displacement of a particle and feeds a signal related to the measured displacement through an amplifier to a kicker. The kicker applies a deflection which corrects the particle trajectory and reduces its betatron oscillation. The signal pulse induced by an off-axis particle and arriving at the kicker is of length $T_s \sim 1/(2W)$, where W is the bandwidth of the cooling system. The smallest fraction of beam that can be observed, the sample, is

$$N_s = \frac{NT_s}{T_0} = \frac{N}{2WT_0} , \tag{11.36}$$

where T_0 is the revolution time, and N the total number of beam particles.

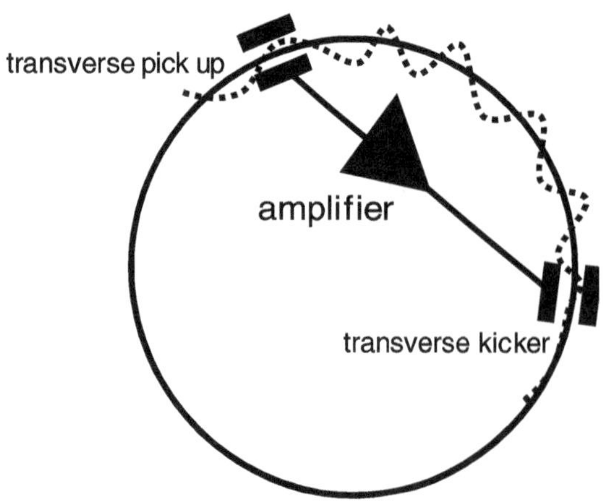

Fig. 11.6. Schematic of stochastic cooling

If the offset of the test particle is x, the applied correction is $-\lambda x$ where λ is related to the sensitivity of the pick up, the strength of the kicker, and the amplification in the cooling loop. The corrected position after the kick is

$$x_c = x - \lambda x - \sum_{\text{sample}'} \lambda x_i \ , \tag{11.37}$$

where the sum with superindex $'$ is over all particles in the sample except for the test particle. This can be rewritten as

$$x_c = x - \lambda N_s \langle x \rangle_s = x - g \langle x \rangle_s \ , \tag{11.38}$$

where $\langle x \rangle_s \equiv \sum_{\text{sample}} x_i / N_s$, and $g \equiv N_s \lambda$ is the fractional correction, also called the *gain*.

If the sample contains only the test particle, and assuming $g = 1$, the cooling time can be roughly estimated as [5]

$$\frac{1}{\tau_x} \equiv -\frac{1}{x}\frac{\mathrm{d}x}{\mathrm{d}t} = \frac{1}{N_s T_0} = 2\frac{W}{N} \ . \tag{11.39}$$

More rigorously, one computes the emittance damping rate

$$\frac{1}{\tau_{x^2}} \equiv \frac{1}{\epsilon_x}\frac{\mathrm{d}\epsilon_x}{\mathrm{d}t} = \frac{W}{N}\left[2g - g^2(M + U)\right] \ , \tag{11.40}$$

where $M \geq 1$ is a 'mixing' term, describing the exchange of particle positions between successive revolutions, and U denotes the noise-to-signal ratio. Simplified one might say that in practice [1]

$$\frac{1}{\tau_{x^2}} \approx \frac{1}{10}\frac{W}{N} \ . \tag{11.41}$$

A typical time constant is $\tau \approx 1$ s for $N \approx 10^7$ and $W \approx 100$ MHz. In (11.40), we have ignored an additional small mixing occurring between the pick up and the kicker.

Comparing (11.41), with the equations for electron cooling we observe that electron cooling works best for cold beams, and stochastic cooling works best for large (hot) beams, where the signal-to-noise ratio is large (U small), and for a small number of particles (small N). Thus stochastic cooling is good for 'halo cleaning', electron cooling for 'core freezing'.

Stochastic cooling for bunched beams has not yet been demonstrated. For this application a much higher bandwidth would be required. In addition, there are large signals at multiples of the revolution frequency, which must be avoided by operating at frequencies well above the $(1/e)$ fall-off frequency of the bunch power spectrum, $f_b \approx \beta c/\sigma_z$. However, at these high frequencies unexpectedly strong coherent signals were observed, which obstruct the observation of Schottky noise and thus the cooling [1, 18].

A promising alternative for bunched beams may be 'optical stochastic cooling', at much higher frequencies and bandwidths [19, 20].

11.3.2 Application: Emittance Growth from a Transverse Damper

It is interesting that the formalism of the stochastic-cooling equations can also be used to estimate the emittance growth induced by the transverse feedback system in a proton storage ring, such as the LHC [21].

We first need to modify (11.40) so as to more accurately include the response of the particle distribution to the cooling ('closing the loop via the beam') and convert the description into the frequency domain. Ignoring the mixing term and only keeping the Schottky noise contribution, the cooling equation becomes [21, 22]

$$\frac{1}{\tau_{x^2}} = \frac{1}{2}\frac{f_0}{N}\sum_{n=-\infty}^{\infty}\left[2\frac{g_n}{1+S_n} - \frac{U_n g_n^2}{(1+S_n)^2}\right] , \tag{11.42}$$

where f_0 is the revolution frequency, N the total number of particles, g_n the 'reduced' feedback gain, S_n the 'feedback via the beam' factor $S_n \approx g_n/(4\delta Q)$ [17], where δQ is the total 'tune spread' (depending on the shape of the distribution approximately equal to 2–3.5 times the rms tune spread), and U_n the ratio of noise and Schottky signal for full mixing. The sum extends over all 'Schottky bands' inside the bandwidth of the system, which, for a bunch-to-bunch damper is $W = 1/(2T_b)$, where T_b is the bunch spacing. In the frequency domain, there are two betatron bands per revolution harmonic and, thus, the total number of beam Schottky bands is $n_b = 2W/f_0$.

The Schottky signal power per band is

$$\left.\frac{\mathrm{d}x^2}{\mathrm{d}f}\right|_{\text{signal}} = \frac{\sigma^2}{N f_0} , \tag{11.43}$$

where σ denotes the rms beam size. Assume that the amplifier noise is dominated by the quantization of the digital processing, and that the least significant bit of the ADC corresponds to a fraction α of the beam size. Then the amplifier noise is

$$\left.\frac{\mathrm{d}x^2}{\mathrm{d}f}\right|_{\text{noise}} = \frac{\alpha^2\sigma^2}{12W} . \tag{11.44}$$

Dividing this by the Schottky signal power gives

$$U_n = \frac{\mathrm{d}x^2/\mathrm{d}f|_{\text{noise}}}{\mathrm{d}x^2/\mathrm{d}f|_{\text{signal}}} = \frac{\alpha^2}{12}N\frac{f_0}{W} . \tag{11.45}$$

If N is very large, we can neglect the damping term in (11.42) – the first term in the square brackets –, and consider the amplifier-noise component alone. In that case, the above equation yields the emittance growth rate [21]

$$\frac{1}{\tau_{x^2}} = -\frac{4}{3}f_0\alpha^2\delta Q^2 , \tag{11.46}$$

where a negative sign of τ_{x^2} indicates growth. As an example, using LHC parameters, $f_0 = 11$ kHz, $\delta Q \approx 10^{-3}$ (due to beam-beam collisions), and $\alpha = 1/512$ (i.e., a 10 bit effective ADC within $\pm\sigma$), one finds $\tau_{x^2} \approx 50$ hours [21]. A similar treatment can be applied to estimate the impact of other noise perturbations, e.g., ground motion.

Without feedback, the beam becomes unstable, if the imaginary tune shift due to an impedance $\Delta Q_\perp$ exceeds the total tune spread divided by π, i.e., if $\Delta Q_\perp > \delta Q/\pi$ [7]. To obtain a stable beam, the gain g_n of the feedback must be larger than about $4(\pi \Delta Q_\perp - \delta Q)$, where $\Delta Q_\perp$ denotes the imaginary tune shift due to an impedance [7, 21].

11.4 Laser Cooling

11.4.1 Ion Beams

Laser cooling of atoms held in electromagnetic traps is well understood and widely used. In 1981 P. Channel suggested to apply laser cooling also to ions circulating in a storage ring [23].

Laser cooling exploits the Doppler shift in frequency such that the laser beam interacts selectively with ions of a certain energy. The Doppler shifted frequency in the ion rest frame is

$$\omega' = \gamma\omega(1 - \beta\cos\theta) , \tag{11.47}$$

where θ is the angle between the ion velocity and the incident laser. We denote by A and B a lower and upper level in the ion electronic state, respectively. Ions with a velocity β so that $\omega = \omega_{AB}$, corresponding to the transition $A \to B$, absorb photons, which are subsequently re-emitted. The emission is isotropic, while the momentum received during absorption is in the direction of the laser. In a single absorption, the ion acquires the recoil velocity:

$$v_r = \frac{\hbar\omega_{AB}}{m_{\rm ion} c} , \tag{11.48}$$

where $m_{\rm ion}$ is the ion mass and $\hbar$ the reduced Planck constant. To avoid isotropic stimulated emission, while yet maintaining a short cooling time, the upper level B of the ion should have a short decay time. The ultimate beam temperature that can be reached is determined either by the energy of a single absorbed photon, or by a balance of cooling and heating due to the randomness in the spontaneous emission recoils,

$$T_{\rm min} = \frac{7}{20}\frac{\hbar\Gamma}{k_B} , \tag{11.49}$$

where Γ is the spontaneous decay rate (inverse lifetime) of the excited ion state. Laser cooling is illustrated schematically in Figs. 11.7 and 11.8.

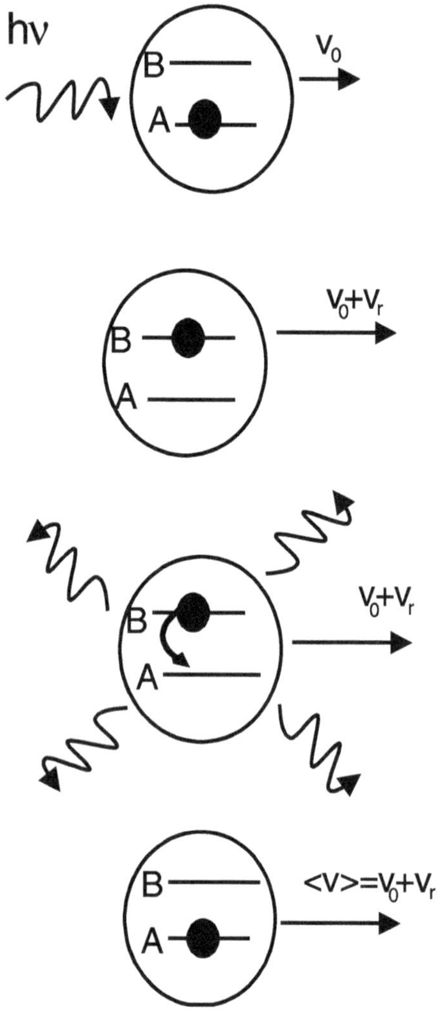

Fig. 11.7. Photon absorption and emission during laser cooling [5]. After each photon absorption the recoil component v_r is added to the initial ion velocity v_0. On the other hand, the emission is isotropic and, thus, on average it does not alter the final ion velocity $\langle v \rangle$

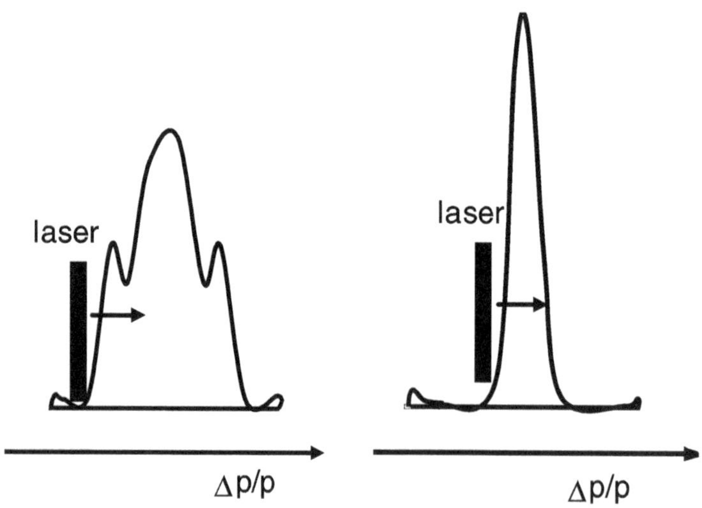

Fig. 11.8. Evolution of ion momentum distribution during laser cooling of a bunched ion beam

As an example [5], consider a 100-keV $^7Li^+$ beam. The ion transition at 548.5 nm is attainable using CW dye lasers. The lifetime of the upper state is 43 ns. The change in energy due to a single absorption is 12 meV. A few mW laser power on a 5-mm spot result in a spontaneous emission of 1.2×10^7 s^{-1}, or about 15 absorptions in an interaction region of 2 m length. This corresponds to a change in energy of 0.2 eV. To cool an ion beam with an energy spread of 1 eV would only require a few revolutions, or a few tenths of microseconds. The ultimate temperature is limited by the recoil momentum acquired in the absorption of a single photon.

Laser cooling requires adequate energy levels and transitions that can be reached by tunable lasers. So far, only 4 ion species fulfill this condition ($^7Li^+$, $^9Be^+$, $^{24}Mg^+$, and $^{166}Er^+$). Laser cooling was demonstrated experimentally in TSR and ASTRID, where energy spreads of less than 10^{-6} were obtained for Li beams [24, 25, 26].

So far laser cooling affects mainly the longitudinal temperature of a beam. However, it is believed that by resonantly coupling the synchrotron and betatron motion, the very fast laser cooling can be extended to the transverse phase space [27]. The coupling between synchrotron motion and horizontal betatron motion may be provided either by a special coupling cavity [28], or, more simply, by momentum dispersion in a regular rf cavity [29]. With such coupling present, the transverse cooling is considerably improved if the tunes are close to a linear resonance:

$$Q_x - Q_s \approx k \,, \tag{11.50}$$

$$Q_x - Q_y \approx l \,. \tag{11.51}$$

where k and l are integers.

11.4.2 Electron Beams

A different type of laser cooling was proposed by Telnov [30] for e^+e^- linear colliders, as a scheme to reduce the transverse emittances and to reach ultimately high luminosities. Collision of an electron beam with a high-power laser beam does not change the beam spot size, nor much the angular divergence. Only the beam energy is decreased, for example, from an inital value E_0 to E. This means that in a laser-cooling stage the two transverse normalized emittances decrease by a factor E/E_0. Telnov estimated that ultimate emittances of $\gamma\epsilon_{x,y} = 2 \times 10^{-7}$ m could be achieved, far better than what can be delivered by conventional damping rings.

More recently, Huang and Ruth studied a laser-electron storage ring (LESR) where radiative laser cooling overcomes the intrabeam scattering effect [31]. The LESR is sketched in Fig. 11.9. It consists of bending magnets, an rf cavity, an injector, and a laser-beam interaction region. A circulating bunch in the ring counterpropagates on each turn through the intense laser pulse. The laser pulse is stored in a high-Q optical resonator, whose path

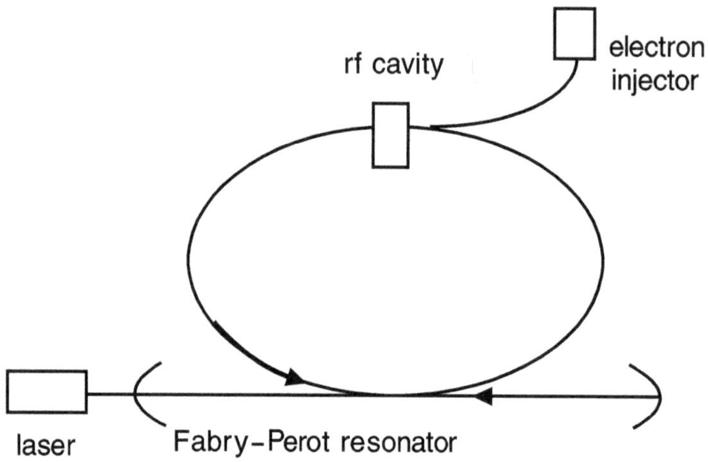

Fig. 11.9. Schematic of a laser-electron storage ring [31]

length is adjusted such that the laser-pulse repetition frequency equals the beam revolution frequency. Thus, a single laser pulse can interact several 10^4 times with the same electron bunch. The LESR can be configured either as a damping ring producing beams with very small transverse emittances, or as a high-intensity X-ray source.

The effect of the laser field is the same as that of a static wiggler with peak field strength [32]

$$B_w = \frac{2}{c}\sqrt{2Z_0 I}\ , \tag{11.52}$$

where I is the laser intensity and Z_R the vacuum impedance (377 Ω). Then the power radiated in the laser field is

$$P_\gamma = \frac{32\pi}{3} r_e^2 \gamma^2 I \tag{11.53}$$

and the energy loss of an electron per turn

$$(\Delta E)_\gamma = \int P_\gamma \frac{\mathrm{d}z}{2c} = \frac{32\pi}{3} r_E^2 \gamma^2 \frac{E_L}{Z_R \lambda_L}\ , \tag{11.54}$$

where Z_R denotes the laser Rayleigh length. The latter characterizes the depth of focus of the laser beam and it is the exact equivalent of a laser-beam 'beta function'. In (11.54), we have assumed that the laser beam is diffraction limited, so that the effective laser emittances are $\epsilon_{L;x,y} \approx \lambda_L/(4\pi)$, in which case its transverse spot area Σ_L at the focal point is given by $\Sigma_L \equiv 2\pi\sigma_{L,x}\sigma_{L,y} = Z_R\lambda_L/2$.

From the energy loss per turn we can compute the longitudinal damping time. It corresponds to a number of turns equal to

$$n_d = \frac{E}{(\Delta E)_\gamma} = \frac{1.6 \times 10^5 \lambda_L[\mu\mathrm{m}] Z_R[\mathrm{mm}]}{E_L[\mathrm{J}]\ E[\mathrm{MeV}]}\ , \tag{11.55}$$

with E the beam energy.

The transverse emittances are damped at the same rate as the energy spread,

$$\Gamma_{x,y}^{\rm RLC} \equiv -\frac{1}{\epsilon_{x,y}} \left\langle \frac{{\rm d}\epsilon_{x,y}}{{\rm d}t} \right\rangle = \frac{1}{n_d T_{\rm rev}} = \frac{\Delta E_\gamma / E}{T_{\rm rev}} \,, \tag{11.56}$$

where $T_{\rm rev}$ is the revolution time, and RLC stands for 'radiative laser cooling'.

The laser field does not only provide damping, but in the same way as regular synchrotron radiation, it also introduces a quantum excitation. The quantum excitation consists of two parts: a dispersive component, which is dominant in conventional storage rings, and a component due to the finite opening angle of photon emission ($\theta \sim 1/\gamma$). The LESR is designed with zero optical dispersion in the laser-beam interaction region. A small amount of dispersion generated by the wiggler field is negligible compared with the effect of the opening angle, since the wiggle angle is much smaller than $1/\gamma$. This is quite different from the situation in a conventional ring, where the dispersive part is always much larger than the opening-angle contribution. Thus, in a conventional ring the emittance is determined by the dispersion (via the 'curly $\mathcal{H}$'; compare (4.67)), while in the LESR it is defined only by the opening angle.

The number of photons scattered into a frequency interval dω is [33]

$$\frac{{\rm d}N_\gamma}{{\rm d}\omega} = \frac{1}{\hbar\omega} \frac{{\rm d}E_\gamma}{{\rm d}\omega} = \frac{3(\Delta E)_\gamma}{\hbar \omega_m^2} \left[1 - 2\left(\frac{\omega}{\omega_m}\right) + 2\left(\frac{\omega}{\omega_m}\right)^2 \right] , \tag{11.57}$$

where the energy loss per turn $(\Delta E)_\gamma$ was given above, and $\omega_m = 4\gamma^2\omega_L = 8\pi\gamma^2 c/\lambda_L$ is the maximum photon frequency. The photon frequency ω and the scattering angle θ are related by

$$\omega = \frac{\omega_m}{1+\gamma^2\theta^2} \,. \tag{11.58}$$

The transverse recoil of the electron is $\delta\psi = \hbar\omega\theta/E$, causing an average change in the normalized transverse emittances of $\Delta\epsilon_{x,y,N} \approx \beta_{x,y}^* \, \delta\psi^2/4$. Here, one factor of 2 is due to the projection onto a transverse plane, the other is due to averaging over the betatron phase.

Integrating over the photon spectrum yields the average emittance excitation per turn

$$\Delta(\epsilon_{x,y,N}) = \frac{\gamma\beta^*}{2} \int_0^{\omega_m} {\rm d}\omega \frac{\delta\psi^2}{2} \frac{{\rm d}N_\gamma}{{\rm d}\omega} = \frac{3}{10} \frac{\lambda_c}{\lambda_L} \frac{(\Delta E)_\gamma}{E} \beta_{x,y}^* \,, \tag{11.59}$$

where $\beta_{x,y}^*$ is the beta function at the laser-electron interaction point, and $\lambda_c = h/(mc) \approx 2.43\times10^{-12}$ m the electron Compton wavelength. The average emittance excitation per unit time reads

$$\left\langle \frac{{\rm d}\epsilon_{x,y,N}}{{\rm d}t} \right\rangle = \frac{(\Delta E)_\gamma}{T_{\rm rev}} \,. \tag{11.60}$$

As usual, the balance of damping, (11.56), and excitation, (11.60), determines the equilibrium emittance:

$$\epsilon_{x,y,N} = \frac{3}{10}\frac{\lambda_c}{\lambda_L}\beta^*_{x,y} \, . \tag{11.61}$$

According to (11.61), small emittances require a small beta function $\beta^*_{x,y}$. Reducing the value of β^* also facilitates the matching of the electron beam to the laser spot size, thus limiting the required laser-pulse energy.

Longitudinally, the energy spread increases due to the energy fluctuation of the emitted photons:

$$\left\langle \frac{\mathrm{d}(\sigma_E)^2}{\mathrm{d}t} \right\rangle = \frac{1}{T_{\mathrm{rev}}} \int_0^{\omega_m} \mathrm{d}\omega \; (\hbar\omega)^2 \; \frac{\mathrm{d}N_\gamma}{\mathrm{d}\omega} = \frac{7}{10} \frac{\hbar\omega_m \; (\Delta E)_\gamma}{T_{\mathrm{rev}}} \, . \tag{11.62}$$

As in a normal storage ring, the longitudinal damping occurs at a rate

$$\frac{1}{\sigma_E^2}\left\langle \frac{\mathrm{d}(\sigma_E)^2}{\mathrm{d}t} \right\rangle = -2\frac{\Delta E_\gamma/E}{T_{\mathrm{rev}}} \equiv -\Gamma_s^{\mathrm{RLC}} \, . \tag{11.63}$$

Equating the excitation and damping terms yields the equilibrium energy spread [31]

$$\sigma_\delta \equiv \frac{\sigma_E}{E} = \sqrt{\frac{7}{5}\frac{\lambda_c}{\lambda_L}\gamma} \, , \tag{11.64}$$

which tends to be much larger than in a conventional storage ring.

The increased energy spread widens the beam size in the arcs of the laser-electron storage ring, where the dispersion function is large. Thereby it both reduces the emittance growth rate due to intrabeam scattering and it keeps the incoherent space-charge tune shift at an acceptable value (for the above parameters, a bunch population of 10^{10}, an average beta function of 0.1 m, and 6 mm rms bunch length, the tune shift is about 0.01) [31]. However, the large energy spread demands a good chromatic correction, and a high-frequency rf system in order to maintain a short bunch length.

The depletion of the laser pulse due to its interaction with the electron beam is negligible. Neither does the laser-pulse energy significantly decrease over several damping times, provided the mirror reflectivity in the optical resonator is sufficiently high (i.e., 99.99% or better).

This scheme has not yet been demonstrated in practice, but several projects have been proposed and proof-of-principle experiments are under way [34, 35].

11.5 Thermal Noise and Crystalline Beams

Laser or electron cooling produce extremely cold beams. These beams have unusual noise spectra [36]. Suppose the azimuthal density of a stored unbunched proton beam is described by a Fourier expansion as

$$\rho(\theta, t) = \sum_{n=-\infty}^{\infty} \frac{A_n(t)}{2\pi} \exp(in\theta) \tag{11.65}$$

and

$$A_n(t) = \sum_{i=1}^{N} e^{-in\theta_i(t)} , \tag{11.66}$$

where i counts the particles and N is the total number of particles in the beam. In an ordinary beam, where the fluctuations arise from so-called Schottky or 'shot' noise, we have $\langle |A_n|^2 \rangle = N$.

Interaction of the particles via the external environment (characterized by the longitudinal impedance) suppresses the density fluctuation at the nth revolution harmonic as [36]

$$\langle |A_n|^2 \rangle = \frac{N}{1 + N/N_{\mathrm{th}}} , \tag{11.67}$$

where the threshold number N_{th} follows from equating the longitudinal coherent frequency shift for the nth revolution harmonic $\Delta\Omega_n$,

$$(\Delta\Omega)_n^2 = n^2 \frac{N r_p m c 4\pi \omega_r (\mathrm{d}\omega_{\mathrm{rev}}/\mathrm{d}p)}{Z_0 C} \left(\frac{Z_n}{n}\right) , \tag{11.68}$$

where p is the momentum, ω_r the angular revolution frequency, Z_0 the vacuum impedance, C the circumference and r_p the classical particle radius, to the spread in the revolution frequency $(n\,\delta\omega_r)$ near $n\omega_r$. This yields [36]:

$$N_{\mathrm{th}} = \frac{C Z_0 \, (\delta\omega_{\mathrm{r}})^2}{4\pi r_p m c \omega_r (\mathrm{d}\omega_r/\mathrm{d}p)} \left(\frac{n}{Z_n}\right) . \tag{11.69}$$

When the beam is cooled, N_{th} becomes smaller than N. Under these conditions the noise power of the beam no longer depends on the number of particles. Instead it is a direct measure of the beam temperature:

$$\langle |A_n|^2 \rangle \approx N_{\mathrm{th}} \propto (\delta\omega_r)^2 \left(\frac{n}{Z_n}\right) . \tag{11.70}$$

The impedance Z_n/n can be determined from the observed shift in coherent frequency as a function of beam current. The remarkable suppression of the noise spectrum for a cold beam was first observed with an electron-cooled proton beam at the NAP-M storage ring in Novosibirsk [36].

The fast cooling techniques open up the exciting possibility to generate a new state of matter: a crystalline beam. Crystalline beams were proposed by Dikanski and Pestrikov [37], motivated by the observation at NAP-M [36]. Theoretical studies of crystal beams were first performed by Schiffer and Rahman [38, 39], and later by Wei, Li, Sessler, Okamoto, and others [40, 41, 42]. A crystalline beam is an ordered state, where the particles forming the beam 'lock' into fixed positions so that the repelling intra-particle Coulomb forces just balance the external focusing force. Crystalline beams might provide a route to obtaining ultra-high luminosity in colliders.

The generation and possible maintenance of the ordered state was investigated with molecular dynamics (MD) methods [38, 39] starting from a Hamiltonian describing the external focusing and the inter-particle forces in the beam frame. For example, and without derivation, in a combined-function cyclotron magnet, this Hamiltonian is [40]

$$H = \frac{1}{2}(P_x^2 + P_y^2 + P_z^2) - \gamma x P_z + \frac{1}{2}(1-n)x^2 + \frac{1}{2}ny^2 + V_c(x,y,z)\,, \quad (11.71)$$

with $n \equiv -\partial B_y/\partial x\ \rho/B_0$ measuring the strength of the quadrupole field, ρ the bending radius associated with the dipole field B_0, and the inter-particle potential

$$V_c = \sum_j [(x_j - x)^2 + (y_j - y)^2 + (z_j - z)^2]^{-1/2}\,, \quad (11.72)$$

where the summation is over all other particles. In the above Hamiltonian all dimensions were scaled by the characteristic distance $\xi = r_p\rho^2/\beta^2\gamma^2$, time is measured in units of $\rho/(\beta\gamma c)$, and energy in units of $\beta^2\gamma^2 Z^2 e^2/\xi$.

The beam-frame is an accelerated frame of reference, and the above Hamiltonian includes, so to speak, the relativistic generalization of centrifugal and Coriolis forces [40]. The effect of shear, given by the term $\gamma x P_z$, can render the Hamiltonian unbounded. This and the time dependent focusing in an alternating gradient focusing lattice may heat and melt the crystal.

Studying the circumstances under which the crystal is stable, one finds that two conditions have to be fulfilled in order to maintain the crystalline state [41]:

1. the storage-ring must be alternating focusing and the beam energy must be below the transition energy, and
2. the ring lattice periodicity should be larger than 2 times the maximum betatron tune.

The first condition arises from the requirement of stable kinematic motion. The second condition ensures that these is no linear resonance between crystal phonon modes and the machine lattice periodicity [42].

Although the crystalline ground state will show a periodic variation of its shape as the beam travels around the storage ring, at low temperatures very little heat is absorbed by the crystal and the crystal can remain stable for a very long time.

When the ion density is very low, the crystalline ground state is a 1-dimensional chain stretching around the ring. As the ion density increases, the 1-dimensional crystal changes into a 2-dimensional crystalline structure. This transition from the 1-dimensional to a 2-dimensional configuration occurs when the nearest-neigbor distance Δ_z (in the scaled units) obeys the equality [40]

$$\min(Q_y^2, Q_x^2 - \gamma^2) = \frac{4.2}{\Delta_z^3} . \tag{11.73}$$

The 2-dimensional structure extends into the transverse plane of weaker focusing. At still larger densities, a 3-dimensional crystal should be formed.

One-dimensional crystal beams have been observed in the ESR and SIS rings at the GSI Darmstadt, where they are generated by electron cooling [43].

11.6 Beam Echoes

An echo is a coherent oscillation which grows out of a quiet beam with some delay after the application of two independent pulse excitations. Echoes can occur in unbunched and in bunched beams, both transversely and longitudinally. The shape and magnitude of the echo signal contains information on diffusion processes in the beam and on the beam temperature (e.g., on the energy spread). Echoes may thus become a useful diagnostics tool for beam cooling.

We first give a simple illustration how an echo signal can arise. We next calculate the echo signal in the transverse plane induced by the successive application of a dipole kick and a quadrupole kick, closely following the pioneering work by Stupakov [44]. Then, we discuss experimental results, addressing both longitudinal echoes in unbunched beams and a different type of transverse echo, which is induced by two dipole kicks.

11.6.1 Illustration

The successive application of a dipole kick (at time $t = 0$) and a quadrupole kick (at a later time $t = \tau$) can generate an echo signal (at time $t \approx 2\tau$), as illustrated schematically in Fig. 11.10.

The first picture shows the dipole kick, which deflects two beam particles to different radial positions in phase space. After the kick, the two particles execute betatron oscillations, which are represented as circular movements about the phase-space origin. If the betatron tune depends on the radial position in phase space (i.e., on the amplitude of the oscillation), the two particles rotate at different angular velocities. This difference in angular velocity is indicated by the different lengths of the dashed arrows. As we shall see, the nonzero tune shift with amplitude is essential for producing the echo signal.

Next, after a certain time interval, a quadrupole kick is applied (the right picture). In this example, at the moment of the kick the particle with the larger betatron amplitude has no transverse offset ($x = 0$), and hence its motion is not affected by the quadrupole kick. On the other hand, the amplitude of the second particle is changed by the kick in such a way that its betatron

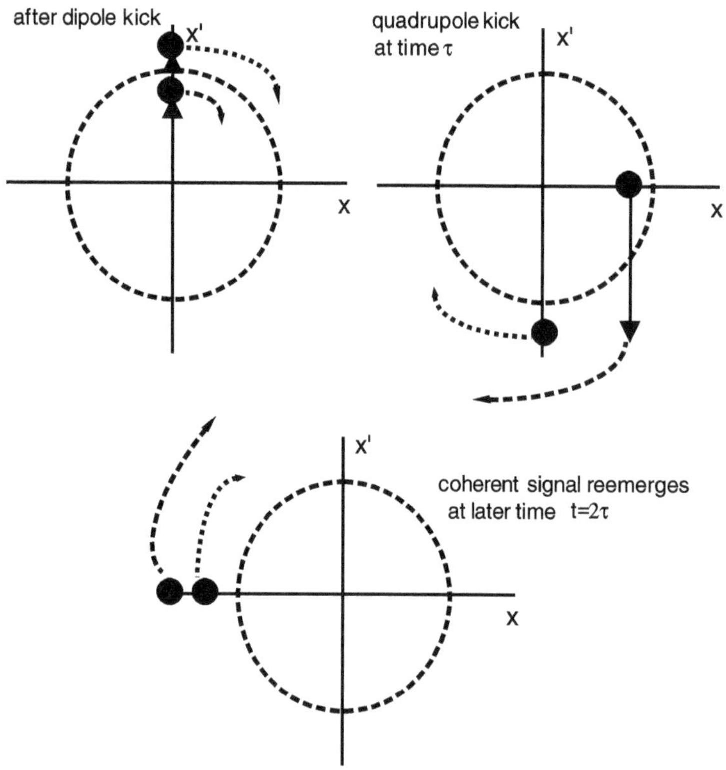

Fig. 11.10. Two-particle model of signal recoherence after applying first a dipole kick and then a quadrupole kick [45]. For the echo generation it is essential that the betatron tune depends on the oscillation amplitude

amplitude increases and now exceeds that of the first particle. Therefore, the quadrupole kick inverts the difference in betatron frequency between the two particles, i.e., the previously more slowly oscillating particle becomes faster and vice versa. After a further time interval, comparable to the time between the two kicks, the particles are again in phase (the last picture). This represents the 'echo'.

By considering the motion of additional particles in phase space, one could also illustrate that, at the moment of the echo, the betatron motion does not 'recohere' for all the particles in the beam, but only for a certain subset.

11.6.2 Calculation of Transverse Echo

We now calculate the response of the beam centroid to the dipole and quadrupole kick and derive an analytical expression for the echo response. To describe the transverse motion of particles in a storage ring, we here employ the normalized coordinates

$$\hat{y} = \frac{y}{\sqrt{\beta}} \quad \text{and} \quad \hat{p} = \frac{1}{Q\omega_r}\frac{\mathrm{d}y}{\mathrm{d}t}\,, \tag{11.74}$$

where β is the beta function, Ω_r the angular revolution frequency, and Q the tune. The beam dynamics can be studied using the distribution function $\rho(\hat{p}, \hat{y}, t)$ which is normalized so that $\int \rho(\hat{p}, \hat{y}, t)\ \mathrm{d}\hat{p}\mathrm{d}\hat{y} = 1$. The initial distribution is assumed to be Gaussian,

$$\rho(\hat{p}, \hat{y}, 0) = \frac{1}{2\pi I_0} \exp\left(-\frac{\hat{p}^2 + \hat{y}^2}{2I_0}\right) , \tag{11.75}$$

where I_0 is a constant equal to the rms beam emittance. It is customary to introduce action-angle coordinates (I, ϕ) via

$$\hat{y} = \sqrt{2I} \cos\phi , \tag{11.76}$$

$$\hat{p} = -\sqrt{2I} \sin\phi . \tag{11.77}$$

The initial distribution function then assumes the form

$$\rho_0(I, \phi) = \rho(I, \phi, 0) = \frac{1}{2\pi I_0} \exp\left(-\frac{I}{I_0}\right) \tag{11.78}$$

and the transformation corresponding to free betatron oscillations conserves the action J:

$$I(t) = I(0) , \tag{11.79}$$

$$\phi(t) = \phi(0) + Q\omega_r t . \tag{11.80}$$

In the original coordinates this oscillation reads

$$\hat{p}(t) = \hat{y}(0)\ \cos(Q\omega_r t) + \hat{p}(0)\ \sin(Q\omega_r t) , \tag{11.81}$$

$$\hat{y}(t) = -\hat{y}(0)\ \sin(Q\omega_r t) + \hat{p}(0)\ \cos(Q\omega_r t) . \tag{11.82}$$

We assume that the tune Q depends on the amplitude of the oscillation as

$$Q(I) = Q_0 + \Delta Q \frac{I}{I_0} , \tag{11.83}$$

where ΔQ has the meaning of a tune shift with amplitude, which is crucial for the echo effect.

From the distribution function $\rho(I, \phi, t)$ we can calculate the evolution of the averaged (centroid) displacement, by means of a simple integration:

$$\langle\hat{y}\rangle = \int_{-\infty}^{\infty} \mathrm{d}\hat{p} \int_{-\infty}^{\infty} \hat{y}\rho(\hat{p}, \hat{y}, t)\,\mathrm{d}\hat{y} = \sqrt{2} \int_0^{\infty} \sqrt{I}\ \mathrm{d}I \int_0^{2\pi} \cos\phi\ \rho(I, \phi, t)\ \mathrm{d}\phi . \tag{11.84}$$

Our strategy is to compute $\rho(I, \phi, t)$ after applying the two transverse excitations, and then to obtain the echo signal in the motion of the beam centroid from (11.84). The evolution of the distribution function is governed by the Vlasov equation:

$$\frac{\partial\rho}{\partial t} + \frac{\partial\rho}{\partial\hat{y}}\frac{\mathrm{d}\hat{y}}{\mathrm{d}t} + \frac{\partial\rho}{\partial\hat{p}}\frac{\mathrm{d}\hat{p}}{\mathrm{d}t} = 0 . \tag{11.85}$$

Alternatively and equivalently, the distribution function at time t can be obtained from that at time 0 by simply expressing the coordinates $\hat{p}(t)$ and $\hat{y}(t)$, or the corresponding action-angle variables, in terms of those at time 0. In other words, the Hamiltonian mapping induces the following transformation of the distribution function:

$$\rho(\hat{p}(0), \hat{y}(0), 0) \to \rho(\hat{p}(t), \hat{y}(t), t) = \rho(\hat{p}(\hat{p}(t), \hat{y}(t), 0), \hat{y}(\hat{p}(t), \hat{y}(t), 0), 0) ,$$

which links the distributions at times 0 and t. We will use this second method for computing $\rho(\hat{p}, \hat{y}, t)$.

Suppose that at time $t = 0$, the beam is displaced from the closed orbit by a transverse dipole kick of size $\Delta\hat{p} = \epsilon$. This dipole kick gives rise to the new distribution function:

$$\rho_1(\hat{p}, \hat{y}) = \rho_0(\hat{p} - \epsilon, \hat{y}) , \tag{11.86}$$

where ρ_0 is the initial distribution function at time $t = 0$, which we assume to be Gaussian. Assuming that the kick ϵ is small, we can expand the above equation to first order in ϵ:

$$\rho_1(\hat{p}, \hat{y}) \approx \rho_0(\hat{p}, \hat{y}) - \epsilon \frac{\partial \rho_0}{\partial \hat{p}} = \rho_0(I) + \epsilon\sqrt{2I}\sin\phi \, \frac{\mathrm{d}\rho(I)}{\mathrm{d}I} . \tag{11.87}$$

The first kick is followed by a free betatron oscillation over a time τ. This changes the distribution function as

$$\rho_2(I, \phi, \tau) = \rho_1(I, \phi - Q\omega_r\tau) . \tag{11.88}$$

Inserting the previous expression for ρ_1 we find

$$\rho_2 = \rho_0(I) + \epsilon\sqrt{2I}\sin(\phi - Q(I)\omega_r\tau)\frac{\mathrm{d}\rho_0(I)}{\mathrm{d}I} . \tag{11.89}$$

Using (11.83), (11.84), and (11.89), and performing the integration, we calculate the centroid motion after the dipole kick:

$$\langle \hat{y} \rangle = \epsilon \left[\frac{1 - \Delta Q^2 \omega_r^2 \tau^2}{(1 + \Delta Q^2 \omega_r^2 \tau^2)^2} \sin(Q_0 \omega_r \tau) + \frac{2\Delta Q \; \omega_r \tau}{(1 + \Delta Q^2 \omega_r^2 \tau^2)^2} \; (\cos Q_0 \omega_r \tau) \right] .$$

It is illustrated in Fig. 11.11 and clearly shows the decoherence of the signal. For large τ, the average displacement $\langle \hat{y} \rangle$ decreases as τ^{-2}.

At a later time, $t = \tau$ we apply a quadrupole kick of strength q:

$$\hat{p}_{\text{new}} = \hat{p}_{\text{old}} + \Delta\hat{p}_{\text{quad}} = \hat{p}_{\text{old}} - q\hat{y} . \tag{11.90}$$

The new distribution function is

$$\rho_3(\hat{p}, \hat{y}) = \rho_2(\hat{p} - \Delta\hat{p}_{\text{quad}}, \hat{y}) . \tag{11.91}$$

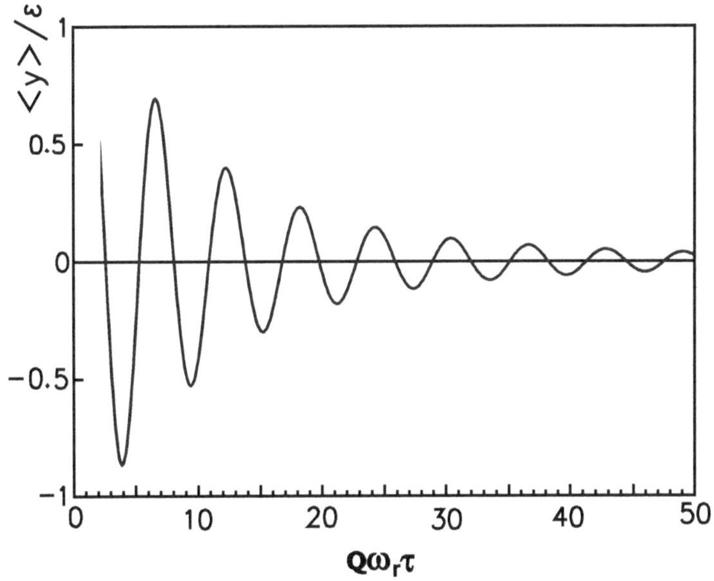

Fig. 11.11. Average displacement of the beam as a function of time following a dipole kick, for a tune spread $\Delta Q \approx 10\%$ [44]

We again perform a Taylor expansion, also assuming that the quadrupole kick is small, or, more precisely, that

$$qQ\omega_r\tau \ll 1 . \tag{11.92}$$

Inserting all the terms from above we have

$$\begin{aligned}\rho_3(\hat{p},\hat{y}) \approx\ & \rho_0(I) + \epsilon\sqrt{2I}\sin(\phi - Q(I)\omega_r\tau)\ \frac{\mathrm{d}\rho_0(I)}{\mathrm{d}I} \\ & + q\hat{y}\frac{\partial}{\partial\hat{p}}\left[\rho_0(I) + \epsilon\sqrt{2I}\sin(\phi - Q(I)\omega_r\tau)\frac{\mathrm{d}\rho_0(I)}{\mathrm{d}I}\right] .\end{aligned} \tag{11.93}$$

The echo effect is contained in the last term on the right-hand side of this equation [44]. Using the relation

$$\frac{\partial}{\partial\hat{p}} = -\sqrt{2I}\sin\phi\frac{\partial}{\partial I} - \frac{1}{\sqrt{2I}}\cos\phi\frac{\partial}{\partial\phi} , \tag{11.94}$$

the largest term that contributes to the echo comes from the derivative of $\sin(\phi - Q(I)\omega_r\tau)$ with respect to I. Denoting this term by ρ_3^{echo} one has

$$\rho_3^{\mathrm{echo}} \approx 2\epsilon q\Delta Q\ \omega_r\tau\sin(\phi)\cos(\phi - Q(I)\omega_r\tau)\frac{I}{I_0}\frac{\mathrm{d}\rho_0(I)}{\mathrm{d}I} . \tag{11.95}$$

Following the quadrupole kick, there is another free betatron oscillation of duration s (here the variable s is in units of time), with

$$\rho_4(I,\phi) = \rho_3^{\mathrm{echo}}(I, \phi - Q\omega_r s) . \tag{11.96}$$

Putting this into (11.84) and integrating, we finally obtain the equation for the echo response:

$$\langle \hat{y}^{\mathrm{echo}} \rangle \approx q\epsilon \Delta Q \omega_r \tau \left[\frac{A(A^2-3)}{(1+A^2)^3} \cos(Q_0 \omega_r (\tau - s)) + \frac{3A^2-1}{(1+A^2)^3} \sin(Q_0 \omega_r (\tau - s)) \right] , \tag{11.97}$$

where $A \equiv \Delta Q\ \omega_r(\tau - s)$. The echo is illustrated in Fig. 11.12 as a function of the time difference $(\tau - s)$, for a tune spread of 10% $(\Delta Q/Q_0 = 0.1)$. The peak of the echo signal is proportional to the strengths of the two kicks. It does not depend on the time interval between them. However, the time of the echo occurrence around $s = \tau$ (or $t = 2\tau$) of course does.

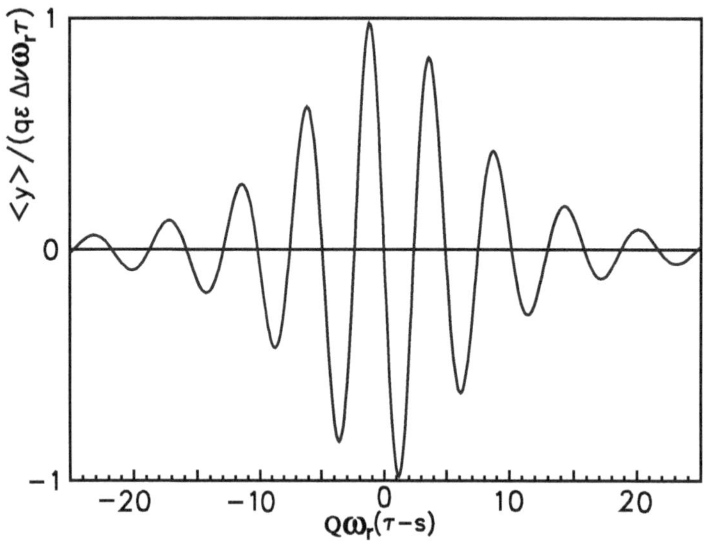

Fig. 11.12. Echo signal of the beam after a second (quadrupole) kick was applied [44]

11.6.3 Measurements of Longitudinal Echoes

Experimental results were first obtained for longitudinal echoes in unbunched beams. Such experiments were performed at the Fermilab Accumulator [46] and at the CERN SPS [47]. In these studies, two rf kicks were applied at frequencies f_{kick1} and f_{kick2}. The response was observed at the difference frequency [46]

$$f_{\mathrm{echo}} = f_{\mathrm{kick2}} - f_{\mathrm{kick1}} . \tag{11.98}$$

For example if $h_{\mathrm{kick1}} = 10$ and $h_{\mathrm{kick2}} = 9$ (h is the harmonic number), the response occurred near the fundamental frequency $h_{\mathrm{echo}} = 1$. Generalizing the previous discussion, now the time of the echo, counted from the first kick, is

$$t_{\text{echo}} = \frac{f_{\text{kick2}}}{f_{\text{kick2}} - f_{\text{kick1}}} \tau , \tag{11.99}$$

where τ as before denotes the time separation between the two kicks.

The presence of diffusion destroys the reversibility of the decoherence. Diffusion thus reduces the response of the echo signal, especially for echoes at large times t_{echo}. The amplitude of the echo is of the form [46]

$$I_{\text{echo}} \propto J_1(\delta k_1 \tau) \exp(-D k_2 t_{\text{echo}}^3) , \tag{11.100}$$

where δ is proportional to the kick strength, k_2 is a constant which depends on the two kick harmonics (and on the echo harmonic), D is the diffusion rate (or collision rate), and t_{echo} the time interval from the first kick to the center of the echo. The decorrelation due to diffusion results in an exponential decay of the echo signal as t^3. By comparing the echo responses for different sets of harmonics, the contributions from the Bessel function and from the diffusion can be distinguished. In the Tevatron Accumulator, a diffusion rate of $d \approx 3 \times 10^{-4}$ Hz was measured [46].

A few further points should be mentioned. Exactly at the center of the echo the measured signal is zero. We have seen the same behavior in our above analysis for the transverse echo (compare Fig. 11.12). It is related to the fact that the echo signal is proportional to the slope of the distribution function, which is zero at the center of Gaussian a bunch. The separation of the two peaks, on either side of this zero, is inversely proportional to the energy spread within the bunch as [46]

$$\Delta t_{\text{peak}} = \frac{\beta^2}{h_{\text{echo}} \pi f_{\text{rev}} |\eta| \frac{\sigma_E}{E}} , \tag{11.101}$$

where η is the slippage factor, f_{rev} the revolution frequency, and β the velocity divided by the speed of light ($\beta = v/c$). This equation was confirmed experimentally.

If the distribution function is not Gaussian, the shape of the echo response changes. The echo signal thus permits a reconstruction of the actual beam distribution. Care has to be taken, as the echo shape may also be modified by longitudinal wake fields.

Another interesting observation is that for sufficiently large energy spread the notch at the center of the echo signal disappears. A possible explanation is the contribution from higher-order momentum compaction (or slippage) to the spread in revolution frequencies:

$$\frac{\Delta f_{\text{rev}}}{f_{\text{rev}}} = -\frac{\eta}{\beta^2} \frac{\Delta E}{E_0} = -\frac{1}{\beta^2} \frac{\Delta E}{E_0} \left(\eta_0 + \eta_1 \frac{1}{\beta^2} \frac{\Delta E}{E_0} + \ldots \right) . \tag{11.102}$$

For a larger energy spread, the nonlinear contributions destroy the linear correlation between particle energy and phase.

In 2000, the first longitudinal echoes for a bunched beam were observed at HERA [48]. In the HERA experiments, first an rf phase kick was applied, which was then followed by an rf amplitude kick. This excitation pattern is the exact longitudinal analogue to the combination of a dipole kick and a quadrupole kick in the transverse plane, which we have analysed in Sect. 11.6.2.

11.6.4 Measurements of Transverse Echoes

Recently, it was discovered by F. Ruggiero that a sequence of two dipole kicks of largely different amplitude can result in a transverse echo [49]. Since all rings are equipped with (injection) kicker magnets, this scheme overcomes the difficulties that had been associated with transverse echo measurements, e.g., the assumed necessity of a quadrupole exciter.

As in Sect. 2.7.4, we assume that the betatron tune changes quadratically with amplitude

$$Q = Q_0 - \mu a^2 , \tag{11.103}$$

where a denotes the oscillation amplitude in units of σ and μ characterizes the strength of the nonlinear detuning. We denote the turn number by N and the magnitude of a dipole kick in units of $\sigma_{x'}$ by $Z = \beta \Delta x'/\sigma_x$ (for simplicity, we here assume that $\alpha_x = 0$ at the kicker).

During filamentation following a large kick Z_1, the beam distribution in betatron phase space will evolve into a spiral-like shape with an increasing number of closely spaced filaments. After N_t turns the distance between two adjacent filaments (i.e., occupied circular regions of phase space) is $\Delta a \approx 1/(2\pi\mu Z_1 N_t)$.

Numerical simulations suggest that for a maximum echo signal the amplitude of the second kick, Z_2, also normalized to $\sigma_{x'}$, should be chosen as half this distance, or

$$Z_{2,\mathrm{opt}} \approx \frac{N_{t,1/2}}{N_t Z_1} . \tag{11.104}$$

where we have introduced the number of turns, $N_{t,1/2}$, after which the initial signal amplitude has decreased by a factor of two (compare the discussion of filamentation in Chap. 2)

$$N_{t,1/2} \approx \frac{1}{4\pi\mu} . \tag{11.105}$$

Figure 11.13 presents a simulation result of the beam centroid response to two subsequent kicks of strength 5σ and 0.25σ, respectively. In the simulation a clear echo is observed if, as here, the first kick is several times the rms divergence and the second kick is so small that the displacement in phase space caused by the latter corresponds to roughly half the distance between filaments. Note also that the normalised rms beam size after applying a kick of strength Z and subsequent filamentation is given by

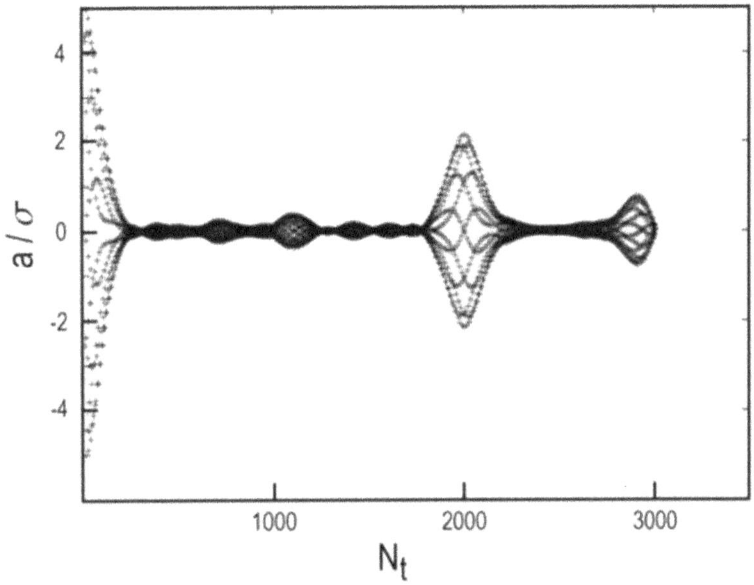

Fig. 11.13. Simulated centroid position in units of σ as a function of turn number for a normalized detuning $\mu = -2 \times 10^{-4}$ [49]. A first kick of 5σ applied at turn 0 is followed by second kick of strength 0.25σ at turn 1000. A clear echo signal is seen around turn 2000, and a second echo at turn 3000

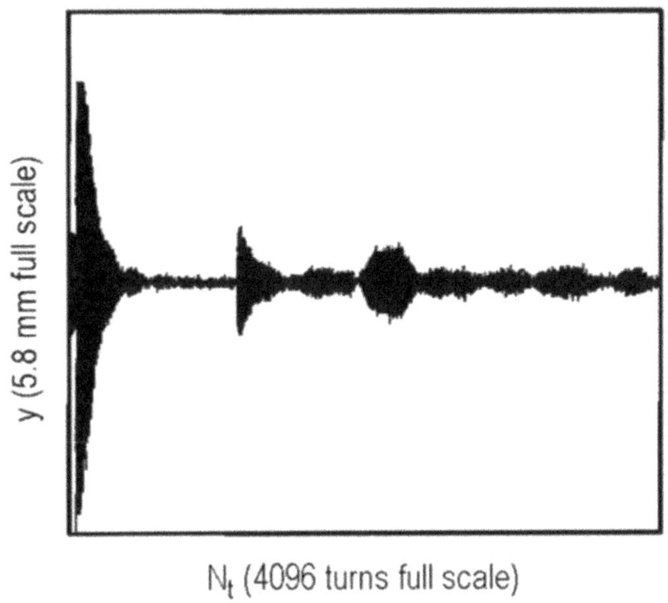

Fig. 11.14. Measured echo signal over 4096 turns [49]. The vertical centroid position is shown as a function of turn number. A first kick at 12884 ms (turn 0) of strength 5 kV, about 0.9σ, is followed at 12910 ms (turn 1128) by a second kick of strength 1 kV, approximately 0.2σ. An echo signal is visible around turn 2000. Octupole magnets were used to adjust the detuning with amplitude, μ, while the chromaticity was corrected by sextupoles. The bunch population was about 8×10^9 protons

$$
\begin{aligned}
\langle a^2(Z)\rangle_\phi = \frac{1}{(2\pi)^2}\int_0^{2\pi} \mathrm{d}\phi \int_0^{2\pi} \mathrm{d}\phi_0 \int_0^{\infty} \mathrm{d}a\ a^3 \sin(\phi+\phi_0)^2 \\
\times \exp\left[-(a^2+Z^2-2aZ\cos\phi_0)/2\right] = \left(1+\frac{Z^2}{2}\right) ,
\end{aligned}
$$

where ϕ_0 is the inital betatron phase of a particle prior to the kick, and ϕ is the phase advance over many turns, which becomes a random variable due to filamentation.

Figure 11.14 shows an echo signal measured during an exploratory experiment at the SPS [49]. In the vertical plane we first see the strong kick. Then the centroid motion is damped due to the decoherence induced by the nonlinearities. Next a second kick is applied after roughly 26 ms. And finally the echo signal appears about 52 ms (2256 turns) after the first kick.

A simulation was performed for parameters approximating those of the experiment, i.e., for a first kick of 0.9σ followed by a second kick of 0.2σ, separated by about 1000 turns. The result shown in Fig. 11.15 resembles the measurement in Fig. 11.14.

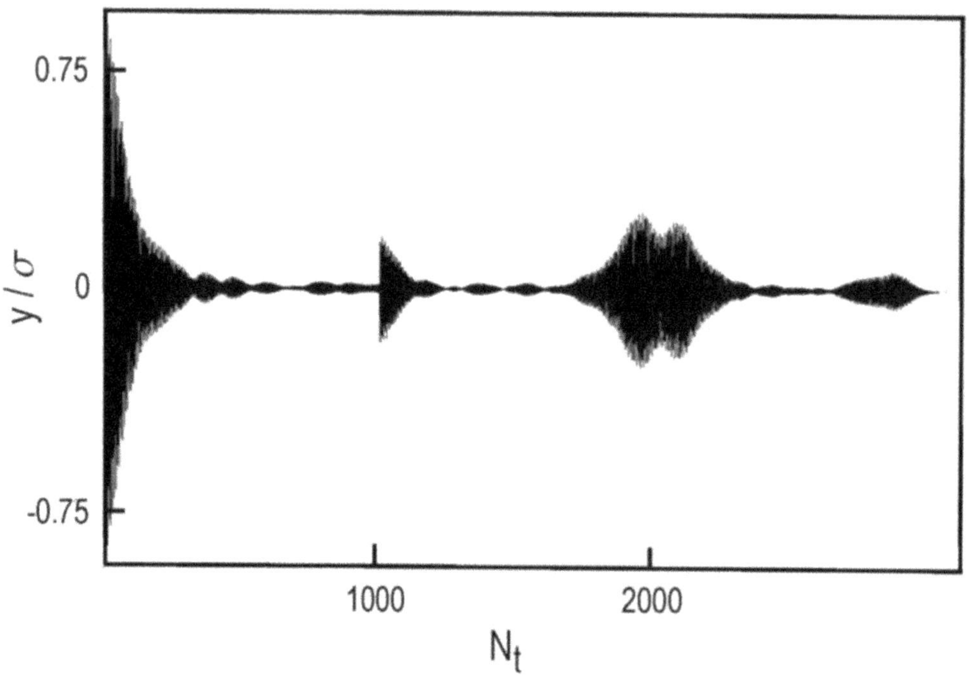

Fig. 11.15. Simulated centroid position in units of σ as a function of turn number for conditions similar to those in the previous figure [49]. A first kick of 0.9σ applied at turn 0 is followed by second kick of strength 0.2σ at turn 1000. Fig. 11.14 The simulation assumed a vertical tune of 0.5785, and a detuning with amplitude equal to $\mathrm{d}Q/\mathrm{d}I = 4000\ \mathrm{m}^{-1}$ corresponding to a normalized detuning $\mu = (-1)/(4\pi N_{t,1/2}) \approx -8\times10^{-4}$, with $N_{t,1/2}\approx 100$

11.7 Ionization Cooling

The successful operation of a future muon collider requires a reduction of the 6-dimensional beam phase space by about a factor of 10^{-6} [50]. The approach proposed to achieving this reduction is ionization cooling. Ionization cooling is similar to electron cooling, but the electron beam is replaced by a solid or liquid.

In ionization cooling the muon beam is passed through some material, in which the muons lose energy, experiencing an average force opposite to their momentum, as in (11.9). The average energy loss is described by the Bethe–Bloch formula

$$-\frac{\mathrm{d}E_\mu}{\mathrm{d}s} = 4\pi N_A r_e^2 m_e c^2 \rho \frac{Z}{A}\frac{1}{\beta^2}\left[\ln\left(\frac{2m_e c^2\beta^2\gamma^2}{I_{\mathrm{ion}}}\right) - \beta^2 - \frac{\delta(\gamma)}{2}\right], \quad (11.106)$$

where N_A is Avogadro's number, the product $4\pi N_A r_e^2 m_e c^2$ equals 0.3071 MeV cm^2 g^{-1}, ρ is the material density, A and Z are mass number and atomic number, respectively, I_{ion} the average ionization energy, and $\delta(\gamma)$ in this expression represents a *density effect* (shielding by the atomic electrons), which at high energies approaches the value $2\ln\gamma$. The energy loss per length for Beryllium is shown in Fig. 11.16 as a function of the momentum of the incident muon. The muons lose kinetic energy in the direction of their motion. Only the longitudinal momentum is restored by subsequent rf sections, resulting in a transverse emittance reduction. This cooling effect is similar to the radiation damping arising from the energy loss due to synchrotron radiation in an electron storage ring. The ionization-cooling process must be repeated many times to achieve a significant emittance reduction. Figure 11.17 illustrates the concept of transverse ionization cooling.

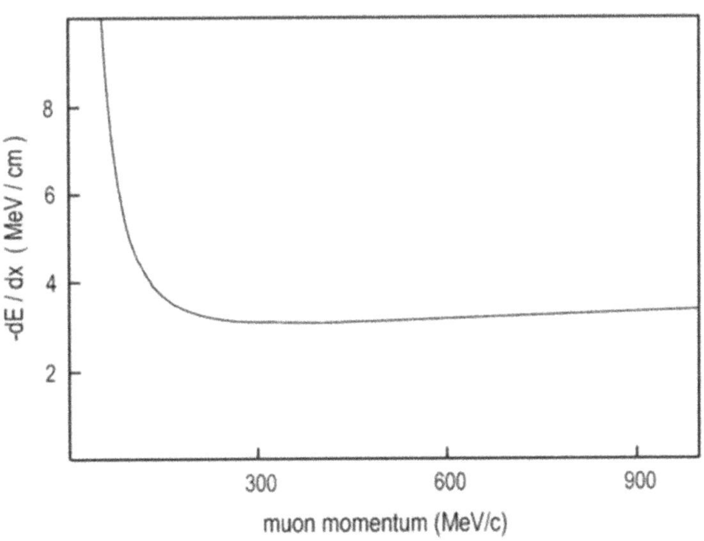

Fig. 11.16. Average muon energy loss per length in beryllium [51]

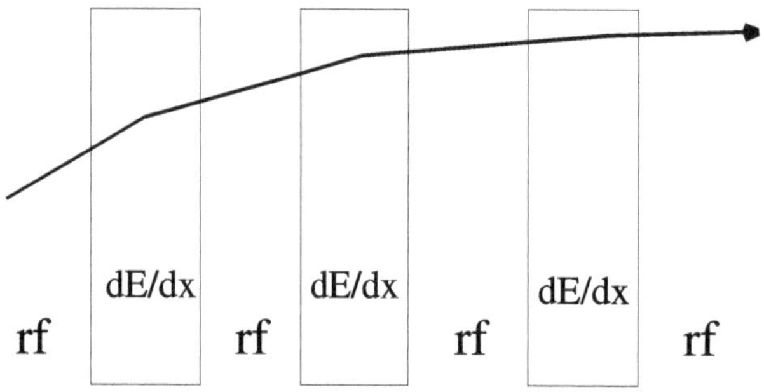

Fig. 11.17. Schematic of ionization cooling in the transverse phase space using a series of low-Z energy absorbers and reacceleration [52]

The equation describing the transverse cooling is [53]

$$\frac{\mathrm{d}\epsilon_N}{\mathrm{d}s} = -\frac{1}{\beta^2}\frac{\mathrm{d}E_\mu}{\mathrm{d}s}\epsilon_N + \frac{1}{\beta^3}\frac{\beta_\perp}{2}\frac{(14\mathrm{MeV})^2}{E_\mu m_\mu c^2 L_R}, \tag{11.107}$$

where ϵ_N is the normalized transverse emittance, E_μ the total muon energy, $\beta_\perp$ the beta function at the absorbing material, $\mathrm{d}E_\mu/\mathrm{d}s$ the energy loss per unit length, and L_R the radiation length. The first term in this equation describes the cooling, and the second the heating term due to multiple scattering. The heating is minimized if $\beta_\perp$ is small, and L_R large (low Z material).

If no further action is taken, the energy spread σ_E evolves according to

$$\frac{\mathrm{d}\sigma_E^2}{\mathrm{d}s} = -2\frac{\mathrm{d}\left(\frac{\mathrm{d}E_\mu}{\mathrm{d}s}\right)}{\mathrm{d}E_\mu}\langle\sigma_E^2\rangle + \frac{\mathrm{d}\left\langle(\Delta E_\mu)^2_{\mathrm{straggling}}\right\rangle}{\mathrm{d}s}, \tag{11.108}$$

where the first term is the cooling (or heating) due to the average energy loss and the second is the "energy-straggling" term given by [54]

$$\frac{\mathrm{d}\left\langle(\Delta E_\mu)^2_{\mathrm{straggling}}\right\rangle}{\mathrm{d}s} \approx 2\pi(r_e m_e c^2)^2 N_A \frac{Z}{A}\rho\gamma^2, \tag{11.109}$$

where N_A is Avogadro's number and ρ the density.

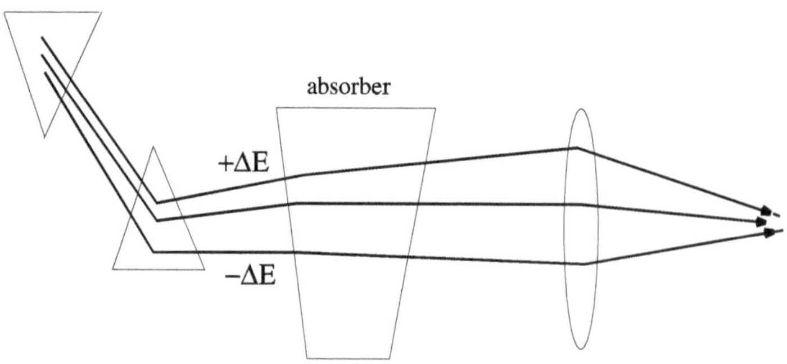

Fig. 11.18. Schematic of ionization cooling in the longitudinal phase space using a wedge [52]

The energy spread can be reduced by a transverse variation in the absorber thickness at a location with dispersion, as shown in Fig. 11.18. The use of such wedges reduces the longitudinal emittance, and it increases the transverse emittance. In other words, the longitudinal cooling is based on emittance exchange with the transverse plane.

Experiments to demonstrate the feasibility of ionization cooling are under study [52].

11.8 Comparison of Cooling Techniques

Table 11.1 compares the different cooling methods. Synchrotron radiation is most suitable for electrons and positrons, ionization cooling can be used for muons, laser cooling for ions, and, possibly, laser cooling of a different kind for electrons and positrons. Stochastic cooling and electron cooling are rather universal, and complementary. Stochastic cooling functions best for a hot beam and the beam halo. Electron cooling times decrease with decreasing beam temperature and electron cooling tends to damp the beam core. Both electron and stochastic cooling are by now well established and used at various storage and accumulation rings. Because of their strong complementarity often these two cooling schemes are employed together in the same ring.

Table 11.1. Comparison of cooling techniques, modified from [5]; N is the number of particles in the beam.

Technique	Stoch.	Electron	Synchr. rad.	Laser (ions)	Laser ($e^{\pm}$)	Ioniz.
Species	All	Ions	e^+, e^-	Some ions	e^+, e^-	Muons
favored beam velocity	high	medium $0.01 < \beta < 0.1$	very high $\gamma > 100$	any $\gamma < 5$	very high $\gamma > 100$	medium
beam intensity	low	any	any	any	any	any
cooling time	$N \cdot 10^{-8}$ s	1–10^{-2} s	$\sim 10^{-3}$ s	$\sim 10^{-4}$–10^{-5} s	$\sim 10^{-2}$–10^{-5} s	$< 10^{-7}$ s
favored temp.	high	low	any	low	any	any

It is predicted that stochastic cooling could be improved by orders of magnitude and possibly extended to bunched beams, if it is applied at optical frequencies. Like optical stochastic cooling, also the laser cooling of electron or positron beams is still waiting for experimental verification. Laser cooling of ion beams in a synchrotron has already been demonstrated, however. This scheme is presently applicable to 4 types of ions, Mg^+, Li^+, Be^+ and Er^+, for which impressively small momentum spreads of less than 10^{-6} have been achieved. Laser cooling has opened the path to a new regime of low temperatures and to the generation of *crystalline beams*. The cooling times for laser cooling are of the order of 10s or 100s of microseconds. They are surpassed only by the even shorter time scale projected for ionization cooling.

Exercises

11.1 Longitudinal Damping Rate with Beam Cooling

Consider two particles which interact simultaneously with the cooling system [2]. Let the cooling act on the momentum variable only. The equations of motion then read

$$\frac{\mathrm{d}p_1}{\mathrm{d}t} = -\lambda(p_1 + p_2) \tag{11.110}$$

and

$$\frac{\mathrm{d}p_2}{\mathrm{d}t} = -\lambda(p_1 + p_2) \; . \tag{11.111}$$

Calculate the damping rate of the centroid motion and the momentum spread.

11.2 Temperature of a Cooled Beam

For each plane of motion a beam temperature can be defined by analogy with kinetic gas theory:

$$\frac{\langle p_x^2 \rangle}{2m} = \frac{1}{2} k_B T_x \; , \qquad \frac{\langle p_y^2 \rangle}{2m} = \frac{1}{2} k_B T_y \; , \qquad \frac{\langle \Delta p_{\parallel}^2 \rangle}{2m} = \frac{1}{2} k_B T_{\parallel} \; , \tag{11.112}$$

where all quantities refer to the beam rest frame.

a) Show that [55] $T_x = \frac{mc^2}{k_B} \beta \, \gamma \frac{\epsilon_{x,N}}{\beta_x}$ and $T_{\parallel} = \frac{mc^2}{k_B} \beta^2 \sigma_p^2$ where $\epsilon_{x,N} \equiv (\gamma\beta)\epsilon_{x,y}$ is the normalized emittance (assumed to be equal in horizontal and vertical plane), $\sigma_p = (\Delta p/p)_{\mathrm{rms}}$ the rms momentum spread in the laboratory frame, and β_x the horizontal beta function (so the temperature is position dependent).

b) Calculate the horizontal and longitudinal temperature for the beam from a proton linac at injection into a cyclotron, with $\epsilon_{x,N} = 0.5$ mm mrad, $\beta\gamma \approx 0.7$, $\beta_x = 10$ m, and $\Delta p/p \approx 10^{-3}$. Compare this with the transverse and longitudinal temperatures of an electron, which is generated at the cathode with $k_B T^c = 0.1$ eV in all directions and then accelerated by a voltage $U_0 = 100$ kV.

c) What is the transverse Debye shielding length of the electron beam at this temperature? Assume a typical electron-beam density of 3×10^8 cm^{-3} in the laboratory frame.

d) For a longitudinal solenoidal guide field of strength 500 Gauss, calculate the electron cyclotron period and compare it with a typical impact time of $\sim r_D/u_\perp$, where $u_\perp$ is the relative transverse velocity (since the electron beam temperature is much smaller, this is determined by the temperature of the proton beam).

11.3 Recombination of Ion Beams during Electron Cooling

Assume an electron cooler for protons provides a cooling time of 10 ms, with a recombination time of 10^5 s. Suppose the same cooling system is used for a beam of fully stripped lead ions ($A = 207$, $Z = 82$). What is the fraction of lead ions that would be lost by recombination during one cooling time?

11.4 Electron-Beam Energy for Electron Cooling

What would be the electron-beam energy required to cool the 7-TeV LHC proton beam?

11.5 Derivation of the Debye Length

Derive the formula for the Debye length, (11.22), by calculating the electron density distribution in the potential of the ion charge and assuming the electrons are in thermal equilibrium. Make appropriate approximations.

11.6 Interaction Probabilities with Electron Cooling

Compare the minimum ion-electron interaction time $\Delta t = \rho_{\min}/u$ (in the beam frame), with the time of traversal through a 10 m cooling section. Estimate the ion velocity u assuming a normalized emittance 10 μm and a 5 m beta function. Can the two times become equal?

11.7 Beam Temperature with Ion-Beam Laser Cooling

Consider laser cooling for 100 keV Mg^+ ions ($A = 24$). Suppose the laser operates at a wavelength of 280 nm, equal to a short-lived transition with a natural linewidth of 46 MHz. (a) Which relative ion velocity $\Delta\beta/\beta$ corresponds to the laser tuning range of 20 GHz? (b) What is the ultimate temperature one might hope to achieve?

11.8 Damping Times with Electron-Beam Laser Cooling

Calculate n_d for the parameters $E_L \approx 1$ J, $\lambda_L \approx 1$ μm, $Z_R \approx 1$ mm, and $E \approx 100$ MeV. What is the equivalent damping time for an average ring radius of 1 m, assuming that electron and laser beams collide on each turn?

11.9 Equilibrium Emittances with Electron-Beam Laser Cooling

As an example, consider a ring with $E = 100$ MeV and $\beta^*_{x,y} = 1$ cm, and a laser with wavelength $\lambda_L = 1$ μm. Calculate the equilibrium emittance and relative energy spread.

11.10 Damping Rates and Equilibrium Emittances with Ionization Cooling

The emittance evolution in an ionization cooling system is described by the equation

$$\frac{\mathrm{d}\epsilon_N}{\mathrm{d}s} = -\frac{1}{\beta^2}\frac{dE_\mu}{ds}\frac{\epsilon_N}{E_\mu} + \frac{1}{\beta^3}\frac{\beta_\perp}{2}\frac{(14\mathrm{MeV})^2}{E_\mu m_\mu c^2 L_R}\,, \tag{11.113}$$

where ϵ_N is the normalized transverse emittance, $\beta_\perp$ the beta function at the absorbing material, $\mathrm{d}E_\mu/\mathrm{d}s$ the energy loss per unit length, and L_R the radiation length. Consider a muon beam with an initial normalized emittance of 0.01 m-rad and a kinetic energy E_k of 150 MeV. The muon mass is about 105.7 MeV. Assume that the beta function at the absorber is 10 cm, and that the minimum energy loss per length, $\mathrm{d}E_\mu/\mathrm{d}s$ is 0.29 MeV/cm.

(a) Calculate the average cooling rate λ (in units of m^{-1}) and the emittance reduction in a 10-m long cooling section containing 320 cm of liquid H_2 (radiation length L_R equal to 890 cm).

(b) Ignoring the second (heating) term, how many such stages and which total length would be required to damp the transverse emittance by a factor 10? In reality the complete cooling system might have a length of 500 m. Which fraction of muons is left after traversing this distance at 150 MeV? Note that the muon lifetime at rest is 2.2 μs.

(c) What is the minimum normalized emittance that can be achieved in such a cooling system of arbitrary length?

(d) Can you derive (11.113)? Note that the projected angular distribution due to multiple scattering is approximately Gaussian with an rms width after distance s equal to

$$\theta \approx \frac{14\ \mathrm{MeV}}{\beta c p}\sqrt{\frac{s}{L_R}}\,. \tag{11.114}$$

12 Solutions to Exercises

1.1 Beam Emittance in terms of Action Angle Variables

From (1.12) at a fixed location s we can write $x = \sqrt{2I_x\beta_x}\cos\phi_x$, where ϕ_x includes the initial phase ϕ_0. We then have

$$\begin{aligned}\epsilon_x &= \frac{\langle x^2(s)\rangle}{\beta_x(s)}\\ &= \int \mathrm{d}\phi_x\,\mathrm{d}I_x 2I_x\cos^2\phi_x\rho(I_x,\phi_x)\\ &= \int \mathrm{d}\phi_x\,\mathrm{d}I_x 2I_x\cos^2\phi_x\rho(I_x)\frac{1}{2\pi}\\ &= \int \mathrm{d}I_x I_x\rho(I_x) = \langle I_x\rangle\,. \end{aligned} \tag{12.1}$$

1.2 Projected Beam Emittances

a) The beam matrix after the skew quadrupole is

$$\Sigma^{xy}_{\mathrm{beam}} = \begin{pmatrix} \beta_x\epsilon_{x0} & 0 & 0 & K_s\beta_x\epsilon_{x0}\\ 0 & \epsilon_{x0}/\beta_x + K_s^2\epsilon_{y0}\beta_y & K_s\beta_y\epsilon_{y0} & 0\\ 0 & K_s\beta_y\epsilon_{y0} & \beta_y\epsilon_{y0} & 0\\ K_s\beta_x\epsilon_{x0} & 0 & 0 & \epsilon_{y0}/\beta_y + K_s^2\epsilon_{x0}\beta_x \end{pmatrix}. \tag{12.2}$$

b) The projected emittances are

$$\epsilon_x = \epsilon_{x0}\sqrt{1+\beta_x\beta_y K_s^2\frac{\epsilon_{y0}}{\epsilon_{x0}}}\,, \tag{12.3}$$

$$\epsilon_y = \epsilon_{y0}\sqrt{1+\beta_x\beta_y K_s^2\frac{\epsilon_{x0}}{\epsilon_{y0}}}\,. \tag{12.4}$$

2.1 Schottky Signals

a) The spectrum corresponds to lines of equal amplitude spaced either by $2\pi/\omega_{\mathrm{rev}}$, in time domain, or by $\omega_{\mathrm{rev}}/(2\pi)$ in frequency domain.

b) Since $\langle\cos n\omega_{\mathrm{rev},k}t\rangle_t = 0$, the average current is given by $e\sum_{k=1}^N f_{\mathrm{rev},k} \approx eNf_{\mathrm{rev}}$, where f_{rev} is the average revolution frequency of the particles.

c) The time average of the mixed terms $\cos(n\omega_{\mathrm{rev},k}t+\phi_k)\cos(n\omega_{\mathrm{rev},l}t+\phi_l)$ with $k \neq l$ is zero. The only terms with nonzero average in $(\sum_k i_k)^2$ are

M. G. Minty et al., *Measurement and Control of Charged Particle Beams*,
https://doi.org/10.1007/978-3-662-08581-3_12

$\sum_{k=1}^{N} \langle \cos^2(n\omega_{\mathrm{rev},k} t + \phi_k) \rangle_t = N/2$. After taking the square root, we obtain the desired result.

2.2 Betatron Tunes

a) The synchrotron tune is about $Q_s = 0.125$. (This is much higher than typical for lower-energy storage rings.)

b) The horizontal betatron tune is about $Q_x = 0.266$. The fact that the tune moves to the right by increasing the horizontally focusing quadrupoles shows that the tune lies between 0 and 0.5.

c) A particles would return to the same place in longitudinal phase space after $1/Q_s \approx 8$ turns, and to the same place in horizontal phase space after about $4/Q_x \approx 15$ turns.

2.3 Application of Multipole Field Expansion

a) For $b_2 \neq 0$ and $a_n = 0$ we have $B_y = B_0 b_2 (x^2 - y^2)$ and $B_x = 2B_0 b_2 xy$. Assuming that the particle is relativistic and moves longitudinally at the speed of light, the Lorentz force is $F_x = -cqB_y$ and $F_y = cqB_x$, where q is the particle charge.

b) Inserting a horizontal and vertical orbit offset, we find the aditional field components $\Delta B_y = 2B_0 b_2 (x_{\mathrm{co}} x_\beta - y_{\mathrm{co}} y_\beta)$ and $\Delta B_x = 2B_0 b_2 (y_{\mathrm{co}} x_\beta + x_{\mathrm{co}} y_\beta)$. It is easily verified that the field components proportional to x_{co} have the same dependence on x_β and y_β as one obtains for a normal quadrupole b_1, while those proportional to $y_{\mathrm{co}} \neq 0$ equal those for a skew quadrupole a_1.

c) If the dispersive contributions to the horizontal and vertical orbit ($x_\delta = D_x \delta$ and $y_\delta = D_y \delta$) are also included, the sextupole b_2 produces additional coupling terms $F_y \propto (x_\beta D_y \delta + y_\beta D_x \delta)$, and $F_x \propto (x_\beta D_x \delta + y_\beta D_y \delta)$.

2.4 Beta-Beat

a) The trace of the matrix product $R = R_Q R R_q$ is $\mathrm{Tr} R = 2\cos\phi_0 - \frac{\beta}{f} \sin\phi_0$. This must also be equal to $2\cos\phi$, from which follows that

$$\cos\phi = \cos\phi_0 - \frac{\beta_0}{2f} \sin\phi_0 \,. \tag{12.5}$$

b) In Fig. 2.9, the maximum phase advance error $\pm\Delta\phi$ is about $\pm 5^\circ$ or ± 0.087 rad. The beta beat oscillates at twice the betatron frequency. The design phase advance per arc cell is $\pi/2$. Thus, the length of an arc cell corresponds to the distance between a location at which the phase equals the (local) average value and the next maximum in $\Delta\phi$. If the optics error is not introduced in this region of the ring, the perturbed R_{12} optical transport matrix element $\sqrt{\hat{\beta}\beta_0} \sin(\pi/2 + \Delta\phi)$ between these two locations is equal to the design matrix element $R_{12,0} = \beta_0 \sin\pi/2 = \beta_0$. From this equality, we can infer that

$$\frac{\hat{\beta}}{\beta_0} = \frac{\sin^2 \pi/2}{\sin^2(\pi/2 + \Delta\phi)} \,, \tag{12.6}$$

or $(\hat{\beta} - \beta_0)/\beta_0 \approx 0.8\%$.

2.5 Quadrupole with a Shorted Coil

a) The upper inboard coil is suspected of a short.

b) The faulted coil gives rise to an additional dipole-like deflection of angle $\Delta\theta \approx \Delta B l_q/(B\rho) = (\Delta B/B)aK$, where ΔB is proportional to the field change ΔI through this coil. The deflection is measured a distance l downstream at a BPM with resolution Δx. The relative change in the coil current I can then be determined with a resolution of $\Delta I/I \propto \Delta B/B = \Delta x/(Kla) \approx 10^{-3}$.

2.6 Quadrupole Gradient Errors

a) The π bump is made from two correctors separated by a total betatron phase advance of π. The first corrector applies a deflection of angle θ_1. This results in an offset $\Delta x = \sqrt{\beta_1\beta_q}\sin\phi_{1q}$ at the quadrupole, where β_1 is the beta function at the corrector, β_q that at the quadrupole and ϕ_{1q} the phase advance from the corrector to the quadrupole. The second corrector at a location with beta function β_2 and a phase advance ϕ_{q2} behind the quadrupole gives rise to a deflection angle θ_2:

$$\theta_2 = -\theta_1\sqrt{\frac{\beta_1}{\beta_2}}(\cos\phi_{12} - \alpha_2\sin\phi_{12}) = \theta_1\sqrt{\frac{\beta_1}{\beta_2}} \,. \tag{12.7}$$

Since the two correctors are located exactly π apart there is no residual oscillation.

b) The gradient error ΔK gives rise to an additional deflection $\Delta\theta = \Delta K\,\Delta x = \Delta K\sqrt{\beta_1\beta_q}\sin\phi_{1q}\theta_1$, at the quadrupole, which translates into an offset

$$\Delta x_2 = \Delta\theta\sqrt{\beta_2\beta_q}\sin\phi_{q2} = \Delta K\beta_q\sqrt{\beta_2\beta_1}\sin\phi_{q2}\sin\phi_{1q}\theta_1 \tag{12.8}$$

at the location of the second corrector. Denoting the normalized residual amplitude by $A = \Delta x_2/(\sqrt{\beta_2}\sin\phi_{q2})$ and find

$$A = \Delta K(\beta_q\sqrt{\beta_1}\sin\phi_{1q})\theta_1 \,, \tag{12.9}$$

which relates the measured leakage A to the gradient error ΔK.

2.7 Multiknobs

We compute or measure the 2×2 sensitivity matrix S relating the strengths of the two quadrupole families (ΔK_1 and ΔK_2) and the changes in the two tunes:

$$\begin{pmatrix}\Delta Q_x\\ \Delta Q_y\end{pmatrix} = \begin{pmatrix}S_{11} & S_{12}\\ S_{21} & S_{22}\end{pmatrix}\begin{pmatrix}\Delta K_1\\ \Delta K_2\end{pmatrix} \,. \tag{12.10}$$

Next, we invert the matrix S,

$$\begin{pmatrix}\Delta K_1\\ \Delta K_2\end{pmatrix} = \frac{1}{\det S}\begin{pmatrix}S_{22} & -S_{12}\\ -S_{21} & S_{11}\end{pmatrix}\begin{pmatrix}\Delta Q_x\\ \Delta Q_y\end{pmatrix} \,. \tag{12.11}$$

From this equation, we obtain the linear combinations of ΔK_1 and ΔK_2 for which $\Delta Q_x \neq 0$ and $\Delta Q_y = 0$, or vice versa.

3.1 Design of an Orbit Feedback Loop

a) The two correctors are placed at two different locations upstream of the BPMs. We design a feedback loop which adjusts the strength of the two correctors so that the beam position is zero. Denoting the beam positions measured at the two BPMs without feedback correction by x_1 and x_2, the equations for the feedback loop are

$$x_1 + C_{k1,1}\theta_1 + C_{k2,1}\theta_2 = 0\,, \tag{12.12}$$

$$x_2 + C_{k1,2}\theta_1 + C_{k2,2}\theta_2 = 0\,, \tag{12.13}$$

where θ_1 and θ_2 are the deflection angles applied by the two correctors, and the coefficients

$$C_{ki,j} = \sqrt{\beta_{ki}\beta_j}\sin\phi_{ki,j} \quad \text{for} \quad i,j = 1,2\,, \tag{12.14}$$

are the (1,2) transfer matrix elements between the ith corrector and the jth BPM, and $\phi_{ki,j}$ the associated betatron phase advance. Combining (12.12) and (12.13) and solving for either θ_1 or θ_2, we obtain

$$\theta_2 = \frac{C_{k1,2}x_1 - C_{k1,1}x_2}{C_{k2,1},C_{k1,2} - C_{k2,2}C_{k1,1}} \tag{12.15}$$

$$= \frac{1}{\sqrt{\beta_{k2}}}\frac{x_1\sin\phi_{k1,2}/\sqrt{\beta_1} - x_2\sin\phi_{k1,1}/\sqrt{\beta_2}}{\sin\phi_{k2,1}\sin\phi_{k1,2} - \sin\phi_{k2,2}\sin\phi_{k1,1}}\,, \tag{12.16}$$

$$\theta_1 = \frac{C_{k2,2}x_1 - C_{k2,1}x_2}{C_{k1,1},C_{k2,2} - C_{k1,2}C_{k2,1}} \tag{12.17}$$

$$= \frac{1}{\sqrt{\beta_{k1}}}\frac{x_1\sin\phi_{k2,2}/\sqrt{\beta_1} - x_2\sin\phi_{k2,1}/\sqrt{\beta_2}}{\sin\phi_{k1,1}\sin\phi_{k2,2} - \sin\phi_{k1,2}\sin\phi_{k2,1}}\,. \tag{12.18}$$

The phase advances between the correctors and the BPMs should not be all equal to 0 or π. In particular, $\sin\phi_{ki,1}$ and $\sin\phi_{ki,2}$ should not be both equal to 0, for $i = 1$ or 2, and $\sin\phi_{k1,j}$ and $\sin\phi_{k2,j}$ shoud not be both equal to zero, as otherwise neither corrector would affect the orbit reading at BPM j. Moreover, the ratios of coefficients $C_{k1,2}/C_{k1,1}$ and $C_{k2,2}/C_{k2,1}$ should not be equal, to avoid a degeneracy and an identical effect of the two correctors. Note that in the extreme case of $C_{k1,2} = 0$ and $C_{k2,1} = 0$, corrector 1 only interacts with BPM 1 and corrector 2 only with BPM 2. The beta functions should be large at the correctors, which minimizes the corrector strength required, and they should also large at the BPMs, which maximizes the sensitivity to orbit changes.

b) In the case of a storage ring, (12.15) and (12.17) still apply, but the coefficients $C_{ki,j}$ now follow from the formula for the closed-orbit distortion (2.34):

$$C_{ki,j} = \frac{\sqrt{\beta_{ki}\beta_j}\cos(|\phi_{ki,j}| - \pi Q_x)}{2\sin\pi Q_x} \quad \text{for} \quad i,j = 1,2\,, \tag{12.19}$$

where Q is the betatron tune, and $\phi_{ki,j}$ as before denotes the betatron phase advance from corrector i to BPM j. The dependence on the beta functions is the same as for a transport line, but the optimum phase advance between the correctors and BPMs now depends on the betatron tune. Again, the ratios of the coefficients $C_{k1,2}/C_{k1,1}$ and $C_{k2,2}/C_{k2,1}$ should be different, to avoid degeneracy.

3.2 Linac Dispersion and Orbit Correction

a) Equation (3.48) describes a harmonic oscillator. The solution is [1]

$$x_1(s) = \frac{\theta}{k_\beta\sqrt{1+\delta_1}} \sin \frac{k_\beta s}{\sqrt{1+\delta_1}} \approx \frac{\theta}{k_\beta\sqrt{1+\delta_1}} \left[\sin k_\beta s - \frac{1}{2} k_\beta s \delta_1 \cos k_\beta s\right] \tag{12.20}$$

or

$$x_1(s) \approx \frac{\theta}{k_\beta} \sin k_\beta s - \frac{1}{2}\left[\theta s \cos k_\beta s + \frac{\theta}{k_\beta} \sin k_\beta s\right] \delta_1 + \mathcal{O}(\delta_1^2) . \tag{12.21}$$

From the term linear in δ_1 we infer the dispersion at the bunch head,

$$D_1(s) = -\frac{1}{2}\left[\theta s \cos k_\beta s + \frac{\theta}{k_\beta} \sin k_\beta s\right] . \tag{12.22}$$

The solution is illustrated in Fig. 12.1. The linear increase with s reflects that the dispersion is resonantly driven [1].

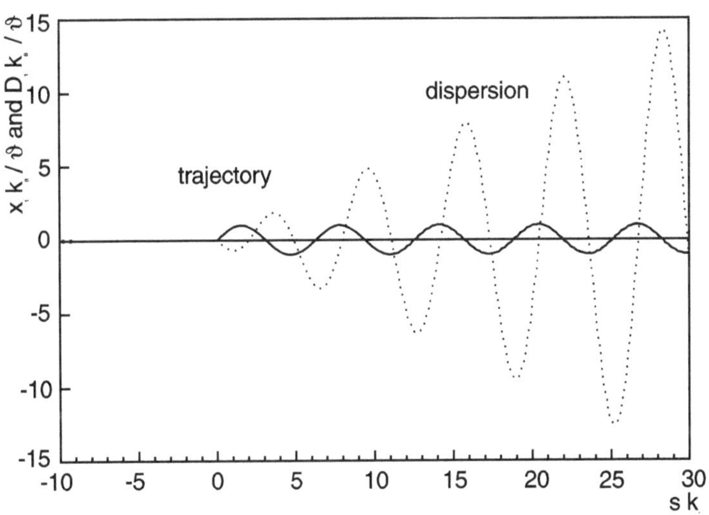

Fig. 12.1. Trajectory oscillation, $x_1 k_\beta/\theta$ for $\delta_1 = 0$, and resonantly growing dispersion at the bunch head, $D_1 k_\beta/\theta$, induced by a deflection at $s = 0$, according to (12.21) and (12.22)

b) The dispersion generated by a single kick at $s = 0$, $D_1(s)$, was computed in (12.22). The dispersion generated by the second kick is obtained by simply shifting the argument by s_2, i.e., it is given by $D_1(s - \pi/k_\beta)$. The dispersion arising from the π bump is then the sum of the terms generated by the two kicks [1]:

$$D_\pi = D_1(s) + D_1(s - \pi/k_\beta) = -\frac{\theta\pi}{2k_\beta}\cos k_\beta s\,. \tag{12.23}$$

The solution is illustrated in Fig. 12.2. While the orbit after the π bump is zero, the dispersion propagates at a constant amplitude. A perfectly centered orbit in the downstream linac section does not imply that the dispersion is zero as well.

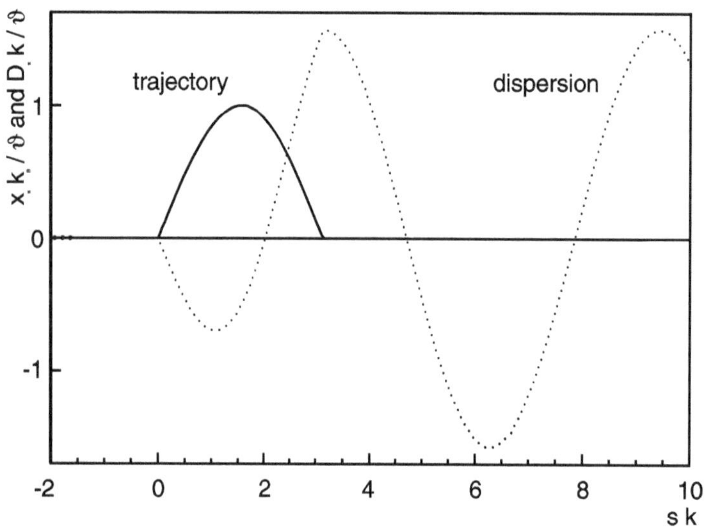

Fig. 12.2. Trajectory perturbation, $x_\pi k_\beta/\theta$, and subsequent constant dispersion, $D_\pi k_\beta/\theta$, induced by a π bump, according to (12.23)

4.1 Beta Mismatch

From (1.15), the action variable of a particle with respect to the matched design optics (subindex 'D') is

$$I = \frac{x^2 + (\beta_D x' + \alpha_D x)^2}{2\beta_D}\,. \tag{12.24}$$

After filamentation and phase randomization the average action is equal to the rms emittance (see (1.14)), i.e., $\epsilon = \langle I\rangle_\mathrm{f} = \langle x^2\rangle_\mathrm{f}/\beta_D$ where the subindex f refers to averaging after filamentation. In a linear system I is conserved, and, hence, its initial average value $\langle I\rangle$ does not change, or $\langle I\rangle = \langle I\rangle_\mathrm{f}$. Averaging I over the initial distribution and using the relations $\langle x^2\rangle = \beta\epsilon_0$, $\langle x'^2\rangle = \gamma\epsilon_0$, and $\langle xx'\rangle = -\alpha\epsilon_0$, we then obtain the final emittance

$$\epsilon = \langle I\rangle = \frac{\gamma_D\beta - 2\alpha_D\alpha + \beta_D\gamma}{2}\,\epsilon_0 = B_\mathrm{mag}\epsilon_0\,. \tag{12.25}$$

4.2 Propagation of Twiss Parameters

Let the initial phase-space ellipse be

$$\gamma_0 x(0)^2 + 2\alpha_0 x(0)x'(0) + \beta_0 {x'(0)}^2 = \epsilon \,. \tag{12.26}$$

In this case ϵ is not the rms beam emittance, but it describes the phase-space area enclosed by the ellipse. Except for a factor 2 this area is equal to the action variable of a particle located on the ellipse, and, in particular, it is a conserved quantity under linear beam transport.

We can evaluate the phase space ellipse at a later location s, namely

$$\gamma_s x(s)^2 + 2\alpha_s x(s)x'(s) + \beta_s {x'(s)}^2 = \epsilon \,. \tag{12.27}$$

Now the trick is to express the initial parameters $x(0)^2$, $x(0)x'(0)$, and $x'(0)^2$ in terms of the final quantities $x(s)^2$, $x(s)x'(s)$, and $x'(s)^2$ using the inverse transport matrix between the two locations:

$$\begin{aligned}
x(0)^2 &= {S'}^2 x(s)^2 - 2S'Sx(s)x'(s) + S^2x'(s)^2 \\
x(0)x'(0) &= -S'C'x(s)^2 - SCx'(s)^2 - (SC + S'C')x(s)x'(s) \\
x'(0)^2 &= {C'}^2 x(s)^2 - 2C'Cx'(s)x'(s) + S^2x'(s)^2 \,.
\end{aligned}$$

Inserting the last expressions into (12.26), expanding the products, and comparing coefficients of x_0^2, $x_0x'(0)$, and $x'(0)^2$ with those in (12.27), we arrive at the desired result:

$$\begin{pmatrix} \gamma_s \\ \alpha_s \\ \beta_s \end{pmatrix} = \begin{pmatrix} S^2 & -2S'C' & {C'}^2 \\ -SS' & SC' + S'C & -CC' \\ S^2 & -2SC & C^2 \end{pmatrix} \begin{pmatrix} \gamma_0 \\ \alpha_0 \\ \beta_0 \end{pmatrix} \,. \tag{12.28}$$

4.3 Static and Dynamic change of Partition Numbers

a) We apply (4.82)

$$\Delta\mathcal{D} \approx -\left(\sum_q k_q^2 D_{x,q} L_q\right) \frac{2\rho^2}{C} \, \Delta x^{\mathrm{mag}} \,, \tag{12.29}$$

which gives $\Delta\mathcal{D} = -0.16$ (16% change) for $\Delta x^{\mathrm{mag}} = 1.5$ mm.

b) From (4.83) we estimate the equivalent change in the rf frequency as

$$\Delta f_{\mathrm{rf}} \approx f_{\mathrm{rf}} \frac{2\pi \Delta x^{\mathrm{mag}}}{C} \,, \tag{12.30}$$

and obtain $\Delta f_{\mathrm{rf}} \approx 192$ kHz.

4.4 Effect of Wiggler on Equilibrium Emittance

Rewriting (4.103) und using $\theta_w = \lambda_p/(\rho_w 2\pi)$, we have

$$\begin{aligned}\gamma\epsilon_{x,w} &\approx \frac{16}{30\pi}\frac{C_q\beta_x}{\rho_w}\gamma^3\theta_w^2 \\ &= \frac{16}{30\pi}\frac{C_q\beta_x}{\rho_w^3}\gamma^3\frac{\lambda_p^2}{(2\pi)^2} \\ &\approx 3.3\times 10^{-16}\ \mathrm{m}^4\ \frac{\gamma^3 B_w^3}{(B\rho)^3} \\ &\approx 4.2\ \mu\mathrm{m}\ ,\end{aligned} \tag{12.31}$$

where $(B\rho)$ is the magnetic rigidity. The normalized emittance is independent of the beam energy. Applying (4.105),

$$\tau_{x,w} \approx \frac{2\rho_w^2}{C_d J_x E^3}\ , \tag{12.32}$$

with $C_d \approx 2.1\times 10^3\ \mathrm{m^2GeV^{-3}s^{-1}}$ and $J_x = 1$, yields a damping time of 670 μs at 1 GeV and about 130 μs at 5 GeV.

The normalized emittance is about the same as in a typical damping ring design for a future linear collider, but the damping time is 5–20 times shorter.

4.5 BNS Damping at the SLC

From (4.112), we have $\xi \approx -1.27$. Combining the generalization of (4.110) to accelerated beams and (4.111) yields

$$\delta_{\mathrm{BNS}} = \frac{N_b r_e \beta^2 W_1(z)}{4L\xi}\ \frac{\ln(\gamma_f/\gamma_i)}{\gamma_f}\ , \tag{12.33}$$

where γ_f and γ_i refer to the final and initial beam energy, respectively. Inserting numbers, we find $\delta_{\mathrm{BNS}} \approx -0.1$, or a 10% energy spread across the bunch.

5.1 Solenoidal Focusing

The phase-space coordinates after the distance z_f are

$$r'(z_f) = \Lambda z_f \tag{12.34}$$

$$r(z_f) = \frac{1}{2}\Lambda z_f^2 + r_0\ , \tag{12.35}$$

where, for simplicitiy, we have dropped the arguments ρ and ξ of r, r' and Λ. After passing through the lens of focal length f and traversing a further distance z_d, we have

$$r'(z_f + z_d) = \Lambda z_f - \frac{1}{f}\left(\frac{1}{2}\Lambda z_f^2 + r_0\right) + \Lambda z_d \tag{12.36}$$

$$\begin{aligned} r(z_f) &= \frac{1}{2}\Lambda z_f^2 + r_0 + z_d z_f \Lambda \\ &\quad -\frac{1}{f} z_d\left(\frac{1}{2}\Lambda z_f^2 + r_0\right) + \frac{1}{2}\Lambda z_d^2\ . \end{aligned} \tag{12.37}$$

Inserting $f = z_d^2/(2(z_f + z_d))$ and rearranging terms gives

$$r'(z_f + z_d) = \frac{2(z_f + z_d)}{z_d^2}\left[-\frac{1}{2}\Lambda z_f\left(z_f - \frac{z_d^2}{z_f}\right) - r_0\right] \quad (12.38)$$

$$r(z_f + z_d) = \left(1 + 2\frac{z_f}{z_d}\right)\left[-\frac{1}{2}\Lambda z_f\left(z_f - \frac{z_d^2}{z_f}\right) - r_0\right]. \quad (12.39)$$

Dividing (12.38) by (12.39) yields the desired result.

5.2 Flat-Beam Transformer

a) For $\mu = 2\pi$, $\Delta = -\pi/2$, and $\alpha = 0$, the matrices A and B are

$$A = I\cos 2\pi + J\sin 2\pi = I \quad (12.40)$$

$$B = I\cos(3\pi/2) + J\sin(3\pi/2) = -J \quad (12.41)$$

The matrix M in (5.18) becomes

$$M = \frac{1}{2}\begin{pmatrix} 1 & -\beta & 1 & \beta \\ 1/\beta & 1 & -1/\beta & 1 \\ 1 & \beta & 1 & -\beta \\ -1/\beta & 1 & 1/\beta & 1 \end{pmatrix} = M = \frac{1}{2}\begin{pmatrix} 1 & -1/k & 1 & 1/k \\ k & 1 & -k & 1 \\ 1 & 1/k & 1 & -1/k \\ -k & 1 & k & 1 \end{pmatrix}, \quad (12.42)$$

where we have used $\beta = 1/k$.

b) Multiplying the matrix M and the vector (5.24), we obtain the final coordinates

$$\begin{pmatrix} x \\ x' \\ y \\ y' \end{pmatrix}_1 = \frac{1}{2}\begin{pmatrix} 1 & -1/k & 1 & 1/k \\ k & 1 & -k & 1 \\ 1 & 1/k & 1 & -1/k \\ -k & 1 & k & 1 \end{pmatrix}\begin{pmatrix} x_0 \\ -ky_0 + x'_0 \\ y_0 \\ kx_0 + y'_0 \end{pmatrix}. \quad (12.43)$$

The equations for y_1 and y'_1 are

$$y_1 = \frac{1}{2k}(x'_0 - y'_0) \quad (12.44)$$

$$y'_1 = \frac{1}{2}(x'_0 + y'_0), \quad (12.45)$$

from which we obtain the second moments

$$\langle y_1^2\rangle = \frac{1}{4k^2}\left(\sigma'^2_{x0} + \sigma'^2_{y0}\right) \quad (12.46)$$

$$\langle y_1 y'_1\rangle = \frac{1}{4}\left(\sigma'^2_{x0} - \sigma'^2_{y0}\right) = 0 \quad (12.47)$$

$$\langle y'^2_1\rangle = \frac{1}{4}\left(\sigma'^2_{x0} + \sigma'^2_{y0}\right). \quad (12.48)$$

The final vertical rms emittance is

$$\epsilon_{y,1} = \sqrt{\langle y_1^2\rangle\langle y'^2_1\rangle - \langle y_1 y'_1\rangle^2} = \frac{\sigma'^2_{y0}}{2k}, \tag{12.49}$$

which demonstrates (5.25). The equations for x_1 and x'_1 are

$$x_1 = x_0 + y_0 - \frac{x'_0}{2k} + \frac{y'_0}{2k} \tag{12.50}$$

$$x'_1 = k(x_0 - y_0) + \frac{1}{2}(x'_0 + y'_0). \tag{12.51}$$

In this case, the second moments are

$$\langle x_1^2\rangle = 2\sigma_{x0}^2 + \frac{\sigma'^2_{x0}}{2k^2} \tag{12.52}$$

$$\langle x_1 x'_1\rangle = 0 \tag{12.53}$$

$$\langle x'^2_1\rangle = 2k^2\sigma_{x0}^2 + \frac{1}{2}\sigma'^2_{x0}. \tag{12.54}$$

The final horizontal rms emittance is

$$\begin{aligned}\epsilon_{x,1} &= \sqrt{\langle x_1^2\rangle\langle x'^2_1\rangle - \langle x_1 x'_1\rangle^2} \\ &= \sqrt{4k^2\sigma_{x0}^2 + 2\sigma_{x0}^2\sigma'^2_{x0} + \frac{1}{4}\sigma'^4_{x0}\frac{1}{k^2}} \\ &= \sqrt{4k^2\sigma_{x0}{}^4 + 4k\epsilon_{y,1}\sigma_x^2 + \epsilon_{y,1}^2} \\ &= 2k\sigma_{x0}^2 + \epsilon_{y,1}.\end{aligned} \tag{12.55}$$

The last equation can also be written as

$$\frac{\epsilon_{x,1}}{\epsilon_{y,1}} = 1 + \frac{4k^2\sigma_{x0}^2}{\sigma'^2_{x0}}, \tag{12.56}$$

which confirms (5.26).

6.1 Scattering off Thermal Photons

a) The beam lifetime due to scattering off thermal photons is

$$\tau \approx \frac{1}{\rho_\gamma c \sigma_T}, \tag{12.57}$$

where $\rho_\gamma \approx 5 \times 10^{14}$ m^{-3} denotes the photon density at 300 K, c the speed of light, and $\sigma_T \approx 0.67$ barn the Thomson cross section. This yields a beam lifetime of 28 hr.

b) The photon density varies with the third power of the temperature. If the vacuum chamber is cooled to 4 K, the beam lifetime increases to about 1400 years.

c) The number of particles lost per train is

$$\Delta N = L\sigma_T \rho_\gamma N\,, \tag{12.58}$$

where L is the length of the linac, ρ_γ the photon density, and N the total number of particles. Inserting numbers, for a chamber temperature of 300 K we find $\Delta N \approx 335$ lost particles.

7.1 Review of Fourier Transformations and an Application

a) For simplicity we initially set $\Delta t = 0$. The Fourier spectrum of the current signal is

$$\begin{aligned} I(\omega) &= \frac{1}{2\pi}\int_{-\infty}^{\infty} i(t)e^{-\mathrm{i}\omega t}\,\mathrm{d}t \qquad (12.59)\\ &= \frac{Q}{2\pi}\int_{-\infty}^{\infty}\sum_{n=-\infty}^{\infty}\left[\delta\left[t-nT-\tau_a\cos(\omega_s nT)\right]e^{-\mathrm{i}\omega t}\right.\\ &\left.+\,\delta\left[t-nT-\frac{T}{2}-\tau_a\cos(\omega_s nT+\phi)\right]e^{-\mathrm{i}\omega t}\,\mathrm{d}t.\right. \end{aligned}$$

Application of the given property of Delta-functions yields

$$\begin{aligned} I(\omega) &= \frac{Q}{2\pi}\sum_{n=-\infty}^{\infty} e^{-\mathrm{i}\omega[nT+\tau_a\cos(\omega_s nT)]}\\ &+\frac{Q}{2\pi}\sum_{n=-\infty}^{\infty} e^{-\mathrm{i}\omega[nT+\frac{T}{2}+\tau_a\cos(\omega_s nT+\phi)]}\,. \end{aligned} \tag{12.60}$$

Using the Bessel function sum rule,

$$\begin{aligned} I(\omega) &= \frac{Q}{2\pi}\sum_{n,k=-\infty}^{\infty} e^{-\mathrm{i}\omega nT}(-i)^k J_k(\tau_a\omega)e^{\mathrm{i}k\omega_s nT}\\ &+\frac{Q}{2\pi}\sum_{n,k=-\infty}^{\infty} e^{-\mathrm{i}\omega(nT+\frac{T}{2})}(-i)^k J_k(\tau_a\omega)e^{\mathrm{i}k(\omega_s nT+\phi)}\,. \end{aligned} \tag{12.61}$$

With $\omega_r T = 2\pi$, this becomes

$$\begin{aligned} I(\omega) = \frac{Q}{2\pi}&\sum_{n,k=-\infty}^{\infty} e^{-\mathrm{i}2\pi n(\frac{\omega-k\omega_s}{\omega_r})}(-i)^k J_k(\tau_a\omega)\\ &\left[1+e^{-\mathrm{i}(\frac{\pi\omega}{\omega_r}-k\phi)}\right], \end{aligned} \tag{12.62}$$

and using the Poisson sum rule, it further simplifies to

$$I(\omega) = Q\omega_r\sum_{n,k=-\infty}^{\infty}(-i)^k J_k(\tau_a\omega)\left[1+e^{-\mathrm{i}(\pi\frac{\omega}{\omega_r}-k\phi)}\right]\delta(\omega-n\omega_r-k\omega_s)\,. \tag{12.63}$$

This expression shows that the spectrum contains the usual rotation harmonics ($k = 0$), and synchrotron sidebands ($k \neq 0$) where the height of the sidebands is given by the Bessel function of appropriate order k.

b) Dipole-mode oscillations correspond to $k = \pm 1$. Consider the phase factor, given in square brackets in (12.63), and the spectrum of sidebands with $k = 1$,

$$\left[1 + e^{-\mathrm{i}(\pi \frac{\omega}{\omega_r} - \phi)}\right], \tag{12.64}$$

which is to be evaluated at $\omega = n\omega_r \pm \omega_s$. Assuming that $T\omega_s \ll 1$, we can neglect the imaginary part of this expression, and for the in-phase oscillations ($\phi = 0$) the real part of the phase factor approximately becomes

$$1 + \cos n\pi \approx \begin{cases} 2 & \text{for even } n \\ 0 & \text{for odd } n \end{cases}, \tag{12.65}$$

while for the out-of-phase oscillations ($\phi = \pi$),

$$1 + \cos n\pi \approx \begin{cases} 0 \text{ for even } n \\ 2 \text{ for odd } n \end{cases}. \tag{12.66}$$

Therefore, one can determine experimentally which of the two normal modes dominates by determining whether the sidebands are located around the even or odd revolution harmonics. The same results apply for $k = -1$.

c) Considering now unequal bunch spacings ($\Delta t \neq 0$), the phase factors become

$$1 + \cos n\pi \cos n\omega_r \Delta t \quad \text{for} \quad \phi = 0 \tag{12.67}$$

$$1 - \cos n\pi \cos n\omega_r \Delta t \quad \text{for} \quad \phi = \pi\,. \tag{12.68}$$

The location of a π-mode sideband for unequally spaced bunches may coincide with the location of a 0-mode sideband for equally spaced bunches, namely if

$$1 - \cos n\pi \cos n\omega_r \Delta t = 1 + \cos n\pi\,. \tag{12.69}$$

This condition is satisfied for

$$\Delta t = \frac{T}{2n}\,. \tag{12.70}$$

The SLC had two damping rings one with equally spaced bunches and the other with unequally spaced bunches (the bunches in this case were separated by 40 buckets and 44 buckets with harmonic number 84). To achieve a common design of "π-mode" cavity for both rings a compromise was made in selecting the cavity resonance frequency such that each ring had partial (though not fully efficient) damping of the π-modes.

7.2 Adjusting the Incoming Beam Energy

If the energy of the injected beam is not correct, it will undergo synchrotron oscillations. Let us assume that the initial relative momentum deviation is δ. The maximum change in the beam energy with respect to the incoming energy occurs after half a synchrotron period, when the relative momentum error is $-\delta$. The orbit difference at a location with dispersion D_x is $\Delta x = -2D_x\delta$. Minimizing the difference orbit measured at this time with respect to the first turn corrects the energy of the injected beam.

If the 'matched' energy around which the injected beam oscillates does not coincide with a centered closed orbit, one may first have to adjust the ring rf frequency and, for protons or ions, the bending field, such that the equilibrium orbit in the ring is centered and corresponds to the correct energy. Afterwards one can then apply the procedure described above in order to adjust the injected beam energy.

If a longitudinal 'phase' monitor is available, another solution, for any type of beam, is to minimize the phase error measured after a quarter synchrotron period.

7.3 Resonant Depolarization

We can express the relative energy error as

$$\frac{\Delta E}{E} = \beta^2 \frac{\Delta p}{p} = \beta^2 \frac{1}{\alpha_c} \frac{\Delta C}{C}, \tag{12.71}$$

where β is the beam velocity in units of the speed of light (and not the beta function), α_c the momentum compaction factor, and C the ring circumference.

7.4 Approximate Expression for the Momentum Compaction Factor

a) We insert the approximate formula for the average dispersion into the definition of the momentum compaction factor, (7.22), and get

$$\begin{aligned} \alpha_c &= \frac{1}{C} \oint \frac{D_x(s)}{\rho(s)}\, \mathrm{d}s \approx \frac{1}{C} \oint \frac{\langle \beta_x \rangle}{Q_x \rho(s)}\, \mathrm{d}s \\ &\approx \frac{1}{C} \frac{\langle \beta_x \rangle}{Q_x} \frac{C}{\rho} = \frac{\langle \beta_x \rangle}{\rho} \frac{1}{Q_x} \approx \frac{1}{Q_x^2}, \end{aligned} \tag{12.72}$$

where in the last step we have used $Q_x = \oint \mathrm{d}s/\beta/(2\pi) \approx \rho/\langle \beta_x \rangle$. As an example, assuming $Q_x = 50$, we estimate that $\alpha_c \approx 1/Q_x^2 \approx 4 \times 10^{-4}$.

b) A corresponding expression for the transition energy γ_t is easily obtained:

$$\gamma_t = \frac{1}{\sqrt{\alpha_c}} = Q_x. \tag{12.73}$$

7.5 Achieving Design Parameters in the Presence of Unknowns

A possible set up is the following. The rf frequency ω_{rf} is known, and first set so as to horizontally center the beam at the beam position monitors in the steady state after a few radiation damping times. The rf frequency determines the revolution time. Since the electron beam moves at the speed of light, also the ring circumference is now determined.

The energy of the injected beam is set to the design value, and the magnetic field of the ring is adjusted until one observes no horizontal orbit variation at dispersive locations, or no longitudinal phase motion, due to synchrotron oscillations after injection. This might also require an adjustment of the rf phase, in order to avoid synchrotron oscillations due to injection phase errors, in addition to those from magnetic field errors. (The two types of errors can also be distinguished from the phase of the oscillation.)

Note that one possibility of measuring the energy of the injected beam is to monitor the orbit in a dispersive region of the injection transfer line, and its dependence on known step changes in the beam energy, e.g., generated by phasing a klystron.

The problem could be simplified, if additional information on the beam energy in the ring is available, e.g., by resonant depolarization or by a reaction with a target for which the cross section is sensitive to the energy.

7.6 Chromatic Phase Advance[1]

a) Keeping only terms up to order δ, with $x = x_\beta + D_x\eta$ and $y = y_\beta + D_y\eta$ with $D_y = 0$, (7.45) gives

$$\begin{aligned} x_\beta'' + kx_\beta &= kx_\beta\delta - mD_x x_\beta\delta - \frac{m}{2}({x_\beta}^2 - {y_\beta}^2)\,, \\ y_\beta'' - ky_\beta &= -ky_\beta\delta + mD_x y_\beta\delta + mx_\beta y_\beta\,. \end{aligned} \tag{12.74}$$

Keeping only the energy-dependent terms,

$$\begin{aligned} x_\beta'' + kx_\beta &= (k - mD_x)x_\beta\delta \\ y_\beta'' - ky_\beta &= -(k - mD_x)y_\beta\delta\,. \end{aligned} \tag{12.75}$$

b) In the present case, the horizontal focusing error for an off-energy particle is

$$\Delta k_x(s) = -(k - mD_x)\delta\,. \tag{12.76}$$

In the vertical plane, the focusing error is of opposite sign, $\Delta k_y(s) = -\Delta k_x(s)$. Inserting this into

$$\Delta Q = \frac{1}{4\pi}\int_s^{s+C} \beta(s)\Delta k(s)\,\mathrm{d}s\,, \tag{12.77}$$

and noting that $\Delta\phi = \Delta Q 2\pi$, we immediately have

$$\begin{aligned} \Delta\phi_x &= -\frac{\delta}{2}\int_s^{s+C} \beta_x(k - mD_x)\,\mathrm{d}s\,, \\ \Delta\phi_y &= \frac{\delta}{2}\int_s^{s+C} \beta_y(k - mD_x)\,\mathrm{d}s\,. \end{aligned} \tag{12.78}$$

[1] adapted from [31]

c) Using the definition of the chromaticity

$$Q'_{x,y} = \frac{\Delta Q}{\delta} = \frac{\Delta\phi_{x,y}}{2\pi\delta} , \tag{12.79}$$

and the expressions for $\Delta\phi_{x,y}$ derived in b), we obtain

$$\begin{aligned} Q'_x &= -\frac{1}{4\pi}\int \beta_x(k - mD_x)\,\mathrm{d}s , \\ Q'_y &= \frac{1}{4\pi}\int \beta_y(k - mD_x)\,\mathrm{d}s . \end{aligned} \tag{12.80}$$

8.1 Phase Tolerances in a Bunch Compressor

From (8.8) we have

$$\frac{\partial\phi_3}{\partial\phi_1} = 1 + R_{56}\frac{\omega_{\mathrm{rf}}}{c}\frac{eV}{E} . \tag{12.81}$$

The error in the final phase $\partial\phi_3$ as a function of error in the injection phase $\partial\phi_1$ therefore depends linearly on the compressor voltage. Minimum sensitivity is achieved for $\frac{\partial\phi_3}{\partial\phi_1} = 0$ or $V = 33.4$ MV.

8.2 Bunch Precompression

a) We assume that the bunch length is small compared to the wavelength of the accelerating rf. The bunch centroid is initially ($t < t_0$) at the center of phase space ($\delta = 0, \phi = 0$) and the phase space trajectories are elliptical centered about the bunch centroid with amplitude given by the voltage V_0. During the time $t_0 < t < t_1 = \frac{\mathcal{T}_{s,l}}{8}$ the cavity voltage is lowered to 0.75 V_0. This introduces a shift in synchronous phase and the bunch centroid executes 1/8 of a synchrotron oscillation centered about the new synchronous phase. After the voltage is restored to V_0, during the time ($t_1 < t < t_2 = \frac{\mathcal{T}_{s,h}}{4}$), the bunch executes oscillations about the original synchronous phase angle. If no other changes were made, then both the first and second moments (mean phase and bunch length) of the particle distribution would subsequently vary in time. Application of a second step change in voltage to 0.75 V_0 for a time $t_2 < t < t_3 = \frac{\mathcal{T}_{s,l}}{8}$ shifts again the rf bucket and the bunch oscillates again around the new synchronous phase for 1/8 of a synchrotron period. With a perfectly linear rf system as assumed, the bunch centroid would return to $\delta = 0$ and $\phi = 0$ as initially, and would remain there indefinitely.

b) With a finite bunch length, the distribution of particles at t_3 is mismatched in the original phase space. The different particles have therefore different trajectories in longitudinal phase space. Since the synchrotron tune is approximately the same for all particles, by waiting an appropriate time, eventually (within a fraction of a synchrotron period) the particles will be aligned vertically in phase space with a significantly smaller bunch length, but with an increase in energy variation between particles. In this example with a two step changes in the applied voltage, the first moment is restored while the second moment varies in time; the bunch rotates in phase space with the centroid position at $\delta = 0$ and $\phi = 0$.

Bunch precompression was used in the SLC damping rings to decrease the particle losses in the extraction line prior to injection into the main linac. By decreasing the bunch length at extraction, the energy variation along the bunch introduced by a downstream compressor was reduced. This in turn translated to a smaller dispersive beam size in the transport line which had a restricted horizontal aperture. A slightly different type of bunch precompression is also used at DESY in the PETRA accelerator, in order to reduce the bunch length at extraction for injection into the HERA accelerator for better capture efficiency. Here, two step changes are applied to the rf phase. The first shifts the rf phase by π, which places the beam distribution next to the unstable fixed point. The ensuing slow motion of particles along the separatrices translates into a mismatch, when the second step change moves the phase back to the original position.

8.3 Harmonic Cavities

a) From $eV(0) = U$, $\mathrm{d}V(t)/\mathrm{d}t|_{t=0} = 0$, and $\mathrm{d}^2V(t)/\mathrm{d}t^2|_{t=0} = 0$, we obtain

$$\sin\phi_1 + k\cos n\phi_n = \frac{U}{e\hat{V}} \tag{12.82}$$

$$\cos\phi_1 + kn\cos n\phi_n = 0 \tag{12.83}$$

$$\sin\phi_1 + kn^2\cos n\phi_n = 0\,. \tag{12.84}$$

Combining the last two equations yields

$$\tan n\phi_n = \frac{\tan\phi_1}{n}\,, \tag{12.85}$$

or

$$\sin n\phi_n = \frac{\tan\phi_1}{\sqrt{n^2 + \tan^2\phi_1}}\,, \tag{12.86}$$

and

$$k = -\frac{\cos\phi_1}{n\cos n\phi_n}\,. \tag{12.87}$$

Inserting these relations into (12.82), we get

$$\sin\phi_1 = \frac{n^2}{n^2-1}\frac{U}{e\hat{V}}\,, \tag{12.88}$$

so that (12.86) becomes

$$\tan n\phi_n = \frac{nU/(e\hat{V})}{\sqrt{(n^2-1)^2-(n^2U/(e\hat{V}))^2}}\,. \tag{12.89}$$

Finally, using this result in (12.87) we obtain for the square of the relative voltage amplitude

$$k^2 = \frac{1}{n^2} - \frac{n^2}{(n^2-1)^2}\left(\frac{U}{e\hat{V}}\right)^2 . \tag{12.90}$$

b) As the ratio of radiative losses to primary rf voltage varies from zero (proton beam limit) to slightly less than one, the phase ϕ_1 changes from 0 to almost $\pi/2$, whereas the optimum phase ϕ_n much more slowly increases from $-\pi$ towards $-\pi/2$.

8.4 Minimum Voltage Required for Beam Storage

a) The total radiated power for 10^{11} particles is 16.9 kW.

b) The energy lost per turn is 44.3 keV. Thus, at 44.3 kV voltage the beam could no longer be captured. The synchronous phase at this voltage is $\pi/2$ measured with respect to the zero crossing.

c) Lowering the cavity voltage limits the maximum beam energy and may also reduce the number of particles that can be captured. Note that a large change in voltage is required in order to significantly vary the bunch length. On the other hand, the use of harmonic cavities does not affect the available capture voltage.

8.5 Phase Shift along a Bunch Train

During the passage of the bunch train, the additional voltage $V_{\mathrm{beam}} = ZI_{\mathrm{b}}$ is applied to the cavity, where Z is the impedance, and $I_{\mathrm{b}} = 2I_{\mathrm{dc}}$ is the beam current at the rf frequency. The cavity response is

$$\Delta V = V_{\mathrm{beam}}\left(1 - e^{-\frac{t_{\mathrm{train}}}{\tau_f}}\right) \approx V_{\mathrm{beam}} t_{\mathrm{train}}/\tau_f\,, \tag{12.91}$$

where the cavity fill time is 2 μs. The beam induced voltage is $V_{\mathrm{beam}} = 5$ MV. The change in the cavity voltage along the bunch train amounts to $\Delta V =$ 775 kV. The synchronous phase shift along the train can be estimated as

$$\Delta\phi \approx \frac{\Delta V}{V} \approx 0.0775 \text{ rad} = 4.4 \text{ deg}\,. \tag{12.92}$$

9.1 Septum Fields for Injection and Extraction

The beam separation at the septum can be written as

$$x_{\mathrm{sep}} = \sqrt{\beta_{\mathrm{kic}}\beta_{\mathrm{sep}}}\sin\mu\;\theta_{\mathrm{kic}}\,, \tag{12.93}$$

where the kick angle is

$$\theta_{\rm kic} = \frac{B_{\rm kic} L_{\rm kic}}{(B\rho)}, \tag{12.94}$$

with $(B\rho)$ the magnetic rigidity, $L_{\rm kic}$ the length of the kicker, and $B_{\rm kic}$ the kicker field. We should fulfill

$$x_{\rm sep} > n_s \sigma_x = n_{\rm sep} \sqrt{\beta_{\rm sep} \frac{\epsilon_{x,N}}{\gamma}} \tag{12.95}$$

Inserting (12.93) and solving for $B_{\rm kic}$ we find

$$B_{\rm kic} = \frac{n_s (B\rho) \sqrt{\epsilon_{x,N}/\gamma}}{\sqrt{\beta_{\rm kic}} L_{\rm kic} \sin\mu}. \tag{12.96}$$

Finally using $(B\rho) \approx 3.356\text{Tm}\ E/\text{GeV}$ and $\gamma \approx E/\text{GeV}$ we estimate

$$B_{\rm kic} \approx 3.356\text{Tm} \frac{n_s \sqrt{\epsilon_{x,N}}}{\sqrt{\beta_{\rm kic}} L_{\rm kic} \sin\mu} \sqrt{\frac{E}{\text{GeV}}}. \tag{12.97}$$

Assuming $n_s = 10$, $\beta_{\rm kic} = 100$ m, $\mu = \pi/2$, $L_{\rm kic} = 5$ m, and $\epsilon_{x,N} = 4$ µm, the magnetic field $B_{\rm kic}$ required at a beam energy of 10 GeV and at 10 TeV is 4.3 mT and 0.134 T, respectively.

9.2 Emittance Dilutions due to Injection Errors

a) We estimate the emittance resulting from an injection error x_0 after filamentation as

$$\epsilon_x \approx \frac{x_0^2}{2\beta_x}. \tag{12.98}$$

For $x_0 = 1$ mm and $\beta_x = 100$ m, this gives $\epsilon_x \approx 5$ nm. The corresponding normalized emittances for various particles and energies are listed in the following table, along with the typical design emittances.

	10 GeV	1 TeV	Design norm. emittance
p	55 nm	5.5 µm	3.75 µm
μ	473 nm	47 µm	50 µm
e	98 µm	9.8 mm	3 nm

Obviously, the emittance dilution becomes more severe at higher energies.

b) The rms beam size at $\beta_x = 100$ m is 224 µm for a 7-TeV proton beam (LHC), 554 nm for a 500-GeV electron beam (NLC), and 514 µm for a 2-TeV muon beam (MC).

9.3 Filamentation

The filamented 'point bunch' occupies a circle in phase space. Using $x = r\cos\phi$ and $\phi = \arccos(x/r)$ the projected density is

$$\frac{dN}{dx} = \frac{dN}{d\phi}\frac{d\phi}{dx}. \tag{12.99}$$

Now $\mathrm{d}N/\mathrm{d}\phi = 2/(2\pi) = 1/\pi$, and $\mathrm{d}\phi/\mathrm{d}x = 1/\sqrt{r^2 - x^2}$, so that the projected density becomes

$$\frac{\mathrm{d}N}{\mathrm{d}x} = \frac{1}{\pi\sqrt{r^2 - x^2}} \, . \tag{12.100}$$

9.4 Particle Impact for Slow Extraction

The change in the action variable over three turns is

$$\Delta I = \frac{\partial H}{\partial \psi} 6\pi = \frac{3}{8}(2I)^{3/2}|\tilde{K}_s|\cos(3\psi + \theta_0)2\pi \, . \tag{12.101}$$

On the resonance near the unstable fixed point, but with an asymptotic angle for large amplitudes of $\psi \approx \pm\pi/6$ instead of 0, the cosine factor is approximately constant, equal to 1.

Considering only every 3rd turn, the amplitude at the septum is

$$x_{\text{sep}} = \sqrt{2\beta_{\text{sep}} I}\cos\psi \approx \sqrt{2\beta_{\text{sep}} I} \, , \tag{12.102}$$

where we have roughly approximated $\cos\pi/6 \approx 1$. The change in amplitude at the septum over three turns becomes

$$\Delta x_{\text{sep}} \approx \sqrt{\frac{\beta_{\text{sep}}}{2I}} \Delta I_{\text{sep}} \, , \tag{12.103}$$

which, after inserting (12.101) and (12.102), becomes

$$\Delta x_{\text{sep}} = \frac{3\pi}{4}(2I)|\tilde{K}_s|\sqrt{\beta_{\text{sep}}} = \frac{3\pi}{4}\frac{x_{\text{sep}}^2}{\beta_{\text{sep}}^{1/2}}|\tilde{K}_s| \, . \tag{12.104}$$

9.5 Crystal Channeling

From (9.23), the critical radius is

$$R_c \approx 0.4\text{m}\; p[\text{TeV}/c] \, . \tag{12.105}$$

This translates into a maximum bending angle over the length l of

$$\theta \leq \frac{l}{R_c} \, , \tag{12.106}$$

which for $l = 3$ cm at 7 TeV/c amounts to $\theta \leq 11$ mrad.

10.1 Electrostatic Lenses and Muon Storage Rings

If $\gamma = \sqrt{1 + \frac{1}{a_\mu}}$, we have $(\gamma^2 - 1) = 1/a_\mu$, and the coefficient multiplying the electric field in the equation for the spin precession is zero:

$$\left(a_\mu\gamma - \frac{\gamma}{\gamma^2 - 1}\right) = (a_\mu\gamma - a_\mu\gamma) = 0 \, . \tag{12.107}$$

10.2 Spinors

a) Letting

$$\begin{aligned} \Psi &= \begin{pmatrix} a \\ b \end{pmatrix} \\ \Psi^\dagger &= (a^*,\ b^*)\,, \end{aligned} \tag{12.108}$$

where a and b are to be determined,

$$\begin{aligned} \Psi^\dagger \sigma_x \Psi &= a^* b + b^* a \\ \Psi^\dagger \sigma_s \Psi &= -\mathrm{i} a^* b + \mathrm{i} b^* a \\ \Psi^\dagger \sigma_y \Psi &= |a|^2 - |b|^2\,. \end{aligned} \tag{12.109}$$

Setting these equal to the spin basis vector of interest and using the normalization $|\Psi|^2 = 1$,

$$\begin{aligned} \Psi_x &= \frac{1}{\sqrt{2}} \begin{pmatrix} 1 \\ 1 \end{pmatrix} \\ \Psi_s &= \frac{1}{\sqrt{2}} \begin{pmatrix} 1 \\ i \end{pmatrix} \\ \Psi_y &= \begin{pmatrix} 1 \\ 0 \end{pmatrix} \quad . \end{aligned} \tag{12.110}$$

b) For example,

$$\sigma_s {\sigma^\dagger}_s = \begin{pmatrix} 0 & -\mathrm{i} \\ \mathrm{i} & 0 \end{pmatrix} \begin{pmatrix} 0 & \mathrm{i} \\ -\mathrm{i} & 0 \end{pmatrix}^t = \begin{pmatrix} 1 & 0 \\ 0 & 1 \end{pmatrix} = I\,, \tag{12.111}$$

where ${\sigma^\dagger}_s = \sigma_s$ is obvious.

c) For example, with equal indices,

$$\sigma_s \sigma_s = \delta_{ss} I + \mathrm{i} \Sigma_m \epsilon_{ssm} \sigma_m = I\,, \tag{12.112}$$

since $\delta_{ss} = 1$ and $\epsilon_{ssm} = 0$ for all m. Squaring σ_s gives the same result.

With unequal indices, for example,

$$\begin{aligned} \sigma_y \sigma_s &= \delta_{ys} I + \mathrm{i} \Sigma_m \epsilon_{ysm} \sigma_m \\ &= \mathrm{i}(\epsilon_{ysx}\sigma_x + \epsilon_{yss}\sigma_s + \epsilon_{ysy}\sigma_y) \\ &= \mathrm{i}\epsilon_{ysx}\sigma_x = -\mathrm{i}\sigma_x\,, \end{aligned} \tag{12.113}$$

and $\epsilon_{ysx} = -1$. Direct multiplication of $\sigma_y \sigma_s$ gives the same result.

10.3 Spin Precession in Solenoidal Fields

a) From (10.2) and (10.7),

$$\Omega = -\frac{e}{\gamma m}(1 + G) B_z\,, \tag{12.114}$$

With $\boldsymbol{B} = B_z\hat{s}$, $\boldsymbol{\beta} = \beta\hat{s}$, and $\boldsymbol{E} = 0$, and a solenoidal field of length l,

$$\phi = \Omega t = -\frac{e}{\gamma mc}\frac{l}{\beta}(1+G)B_z \,. \tag{12.115}$$

b) The Lorentz force acting on the particle is $F_L = e\beta cB$ and the centrifugal force $F_c = m\gamma c^2\beta^2/\rho$. The combination gives

$$B\rho = \frac{\beta E}{ec} = \frac{\beta\gamma mc}{e} \,. \tag{12.116}$$

Direct substitution in (12.115) yields

$$\phi = -\frac{B_z l}{B\rho}(1+G) \,. \tag{12.117}$$

c) With $T = 100$ MeV, we obtain $p = 444$ MeV/c and $B\rho = 444/299.8 = 1.48$ T-m. So the required integrated field strength is $B_z l = 1.66$ T-m.

10.4 Periodic Spin Motion

From (10.24) and the definition of the directional cosines,

$$\begin{aligned} M &= e^{-\mathrm{i}\pi\nu_s(\sigma\cdot\hat{n}_0)} = I_2\cos\pi\nu_s - \mathrm{i}(\sigma\cdot\hat{n}_0)\sin\pi\nu_s \\ &= I_2\cos\pi\nu_s - \mathrm{i}(\sigma_x\cos\alpha_x + \sigma_s\cos\alpha_s + \sigma_y\cos\alpha_y)\sin\pi\nu_s \,. \end{aligned} \tag{12.118}$$

Substituting the Pauli matrices gives the result directly.

10.5 SLC '3-state experiment'

Since the spin transport through the arcs is a pure precession, the polarization vector is simply rotated. In particular, using the inverse transformation we can back-propagate the direction of longitudinal polarization at the collision point to the coordinate system of the incoming beam. Using spherical coordinates, we may express the orientation of the back-propagated polarization vector at the injection point as

$$\boldsymbol{P} = P_{\mathrm{IP}}(\sin\phi\sin\theta, \cos\phi\sin\theta, \cos\theta) \,. \tag{12.119}$$

Now, measuring the longitudinal polarization at the IP for the three initial states

$$\begin{aligned} \boldsymbol{P}_1 &= (1,0,0) \\ \boldsymbol{P}_2 &= (0,1,0) \\ \boldsymbol{P}_3 &= (0,0,1) \end{aligned} \tag{12.120}$$

determines the three components of $\boldsymbol{S}$ in (12.119). Now the magnitude of the polarization vector at the IP is simply the sum in quadrature of the three values measured:

$$||P_{\mathrm{IP}}|| = \sqrt{(\boldsymbol{P}_1\cdot\boldsymbol{P}_{\mathrm{IP}})^2 + (\boldsymbol{P}_2\cdot\boldsymbol{P}_{\mathrm{IP}})^2 + (\boldsymbol{P}_3\cdot\boldsymbol{P}_{\mathrm{IP}})^2} \,. \tag{12.121}$$

At the SLC, this 3-state experiment was applied regularly, e.g., every few months, in order to monitor the IP polarization and to optimize the orientation of the spin vector at the collision point. The initial direction of the polarization was varied by means of solenoidal spin rotators located in the transfer line between the electron damping ring and the SLC linac.

10.6 Type-3 Snakes

a) With regard to the orbit, the sum of the horizontal and vertical deflections is zero. With the beam momentum purely longitudinal, there is no deflection due to the longitudinal fields. The type-3 snake is therefore optically transparent with respect to the angle (though in this example, a net offset is introduced). The spin matrices, however, do not commute. The matrix product is

$$\left(e^{\mathrm{i}\frac{\phi}{4}\sigma_x} e^{-\mathrm{i}\frac{\chi}{4}\sigma_s} e^{-\mathrm{i}\frac{\phi}{4}\sigma_x}\right) e^{\mathrm{i}\frac{\chi}{4}\sigma_s} \left(e^{\mathrm{i}\frac{\phi}{4}\sigma_x} e^{-\mathrm{i}\frac{\chi}{4}\sigma_s} e^{-\mathrm{i}\frac{\phi}{4}\sigma_x}\right), \tag{12.122}$$

where σ_x and σ_y are the Pauli matrices. This product can be evaluated using the expansion (12.118) and matrix multiplication. For the given case, the product of the spin matrices is different from unity, if neither χ nor ϕ equal 0 or 2π.

b) The horizontal orbit and the spin orientations for the case $\phi = \chi = \pi$ are shown in Fig. 12.3.

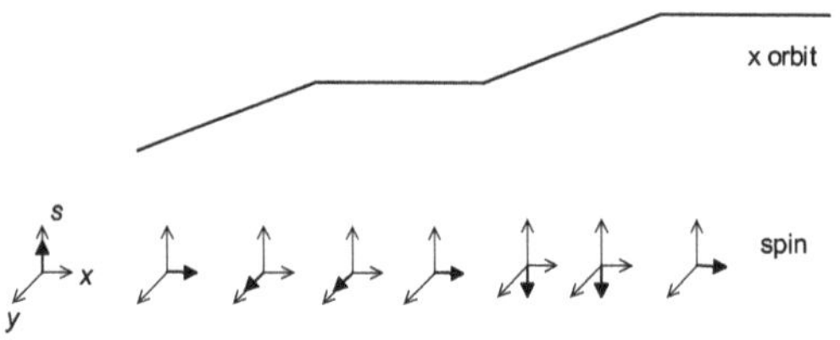

Fig. 12.3. Horizontal orbit and spin orientation along the beam line of Ex. 10.6 with $\phi = \chi = \pi$

This example is derived from experience at the IUCF cooler ring. There, the longitudinal fields were provided by the main cooling solenoid together with the compensating solenoids (1 on each side of the cooling region). The orbital displacements and dipolar deflections arose from the bending fields used to align the proton beam with the cooling electrons.

11.1 Longitudinal Damping Rate with Beam Cooling

The centroid motion is characterized by the centroid momentum $P = (p_1 + p_2)/2$, which fulfills the equation

$$\frac{\mathrm{d}P}{\mathrm{d}t} = \frac{1}{2}\frac{\mathrm{d}(p_1 + p_2)}{\mathrm{d}t} = -\lambda(p_1 + p_2) = -2\lambda P\,. \tag{12.123}$$

From this we obtain the damping rate

$$\frac{1}{\tau_P} = \frac{1}{P}\frac{\mathrm{d}P}{\mathrm{d}t} = -2\lambda\,. \tag{12.124}$$

Since there are only two particles, we can define the momentum spread as the difference between the two particle momenta: $p_{\text{spr}} = p_1 - p_2$. The time derivative of p_{spr} is

$$\frac{\mathrm{d}p_{\text{spr}}}{\mathrm{d}t} = \frac{\mathrm{d}(p_1 - p_2)}{\mathrm{d}t} = -\lambda(p_1 - p_2) = 0\,. \tag{12.125}$$

Hence, the momentum spread is unchanged by the cooling.

11.2 Temperature of a Cooled Beam

a) We have

$$\begin{aligned} T_x &= \frac{\langle p_x^2 \rangle}{k_B m} = \frac{(mc\beta\gamma x')^2}{k_B m} \\ &= \frac{c^2 m \epsilon_{x,N} \beta\gamma}{k_B \beta_x} \end{aligned} \tag{12.126}$$

$$\begin{aligned} T_{||} &= \frac{\langle \Delta p_{||}^2 \rangle}{k_B m} = \frac{mc^2 \left\langle \frac{\Delta p_{||}^2}{(mc)^2} \right\rangle}{k_B} \\ &= \frac{mc^2\beta^2 \left\langle \frac{\Delta p_{\text{lab}}^2}{(\gamma\beta mc)^2} \right\rangle}{k_B} = \frac{mc^2\beta^2 \left\langle \frac{\Delta p_{\text{lab}}^2}{p_{\text{lab}}^2} \right\rangle}{k_B} \\ &= \frac{mc^2\beta^2\sigma_p^2}{k_B}\,, \end{aligned} \tag{12.127}$$

where we have related the momentum deviation in the beam frame and in the laboratory frame via $\Delta p_{\text{lab}} = \gamma \Delta p_{||}$.

b) The velocity and energy of the proton beam follow from $\beta\gamma = 0.7$ or $\beta = 0.57$. The horizontal temperature is $k_B T_x = 32.8$ eV ($T_x = 404$ kK); the longitudinal temperature $k_B T_{||} = 309$ eV ($T_{||} = 3.8$ MK).

The transverse temperature of the electron beam is $T_x = 1230$ K ($k_B T_x = 0.1$ eV). For the longitudinal plane, we use the relation $\Delta p/p = (1/\beta^2)\Delta E/E$; The energy difference ΔE between particles is unchanged by the acceleration, so that $\langle \Delta E \rangle = k_B T^c/2$. Further assuming a Gaussian distributions for the momenta, one can show that $\langle (\Delta E)^2 \rangle = 4\langle \Delta E \rangle^2 = (k_B T^c)^2$. Combining these relations and the definition of $T_{||}$, we find

$$T_{||} = \frac{(k_B T^c)^2}{\beta^2\gamma^2 mc^2}\,. \tag{12.128}$$

For 100 kV accelerating voltage, we have $\gamma \approx 1.2$, $\beta \approx 0.55$, and $T_{||} \approx 4 \times 10^{-8}$ eV. Thus, after acceleration the longitudinal temperature of the electron beam is much smaller than its transverse temperature.

c) For $T_x = 1230$ K, the transverse Debye shielding length (11.22) is $r_D \approx 136$ µm.

d) For a field B of 500 Gauss the electron cyclotron period is $T_{\rm cycl} = 2\pi m_e/(eB) \approx 0.7$ ns. The transverse velocity of the proton beam is

$$v_x^2 \approx \frac{k_B T_x}{m}\,, \tag{12.129}$$

which yields $v_x \approx 2.4 \times 10^6$ m/s. Then the typical impact time is $t_{\rm impact} \sim r_D/u_\perp \sim r_D/v_x \approx 6 \times 10^{-11}$ s. For these parameters, the impact time is about 10 shorter than the cyclotron period. For a field of 5 kG the two times would be equal.

11.3 Recombination of Ion Beams during Electron Cooling

According to (11.27)

$$\frac{1}{\tau} \propto \frac{Z^2}{A}\,. \tag{12.130}$$

Thus the cooling time for fully stripped lead ions is $82^2/207$ times that for protons, or $\tau_{Pb} \approx 308$ μm. The electron capture rate scales as (11.28)

$$\frac{1}{\tau_r} \propto Z^2\,. \tag{12.131}$$

Hence, for the Pb ions it is 15 s.

The fraction of lead ions that would be lost by recombination during one cooling time is

$$\frac{\Delta N}{N} \approx \frac{308\ \mu\text{s}}{15\ \text{s}} \approx 2 \times 10^{-5}\,. \tag{12.132}$$

11.4 Electron-Beam Energy for Electron Cooling

The relativistic Lorentz factor γ should be the same for both beams. Hence, the electron-beam energy required to cool the 7-TeV LHC proton beam is about 3.8 GeV.

11.5 Derivation of the Debye Length

We denote the change in potential experienced by electrons near a single ion charge by $\phi(r)$, where r is the distance from the ion. In thermal equilibrium the electron density is described by

$$n(r) = n_0 e^{-\frac{e\phi(r)}{k_B T}} \approx n_0 \left(1 - \frac{e\phi(r)}{k_B T}\right)\,, \tag{12.133}$$

so that the deviation from the unperturbed density n_0 is

$$\Delta n(r) \approx -n_0 \frac{e\phi(r)}{k_B T}\,. \tag{12.134}$$

Noting that the electron charge is $-e$, the Laplace equation for the perturbed potential, ϕ, in the vicinity of the ion is

$$\frac{\partial^2 \phi(r)}{\partial r^2} = \frac{\Delta n e}{\epsilon_0} = -\left(n_0 \frac{e^2}{k_B T \epsilon_0}\right)\phi\,, \tag{12.135}$$

with the solution

$$\phi = \phi_0 e^{-\sqrt{\frac{n_0 e^2}{k_B T \epsilon_0}} r} = \phi_0 e^{-\frac{r}{r_D}} , \tag{12.136}$$

where the Debye shielding length r_D equals

$$r_D = \left(\frac{k_B T}{4\pi n_0 m c^2 r_e} \right)^{1/2} . \tag{12.137}$$

11.6 Interaction Probability with Electron Cooling

The ion velocity u follows from

$$u^2 = \frac{k_B T}{m} = c^2 \beta\gamma \frac{\epsilon_N}{\beta_x} \approx \left(4 \times 10^5 \text{ m/s} \sqrt{\beta\gamma} \right)^2 , \tag{12.138}$$

where β and $\gamma = 1/(1 - \beta^2)$ are the relativistic factors. Using the minimum impact parameter $\rho_{\min}$ from (11.23), we obtain an interaction time of

$$t_{\text{int}} \approx \frac{\rho_{\min}}{u} = \frac{r_e c^2}{u^3} = (\beta\gamma)^{-3/2} \times 4 \times 10^{-15} \text{ s} . \tag{12.139}$$

Denoting the length of the cooling section by l, the travel time is

$$t_{\text{trav}} \approx \frac{l}{\beta c} = 3 \times 10^{-8} / \beta \text{ s} . \tag{12.140}$$

For the parameters chosen, the two times will only be equal for an extremely small value of β (about 2×10^{-14}), which will not occur in practice. However, the interaction time increases with decreasing ion-beam emittance.

11.7 Beam Temperature with Ion-Beam Laser Cooling

(a) From (11.47) we have

$$\Delta\beta \approx \frac{\Delta\omega'}{\omega'} , \tag{12.141}$$

which, for a laser tuning range of $\Delta f = 20$ GHz and a laser wavelength of $\lambda = 280$ nm, gives $\Delta\beta \approx 1.9 \times 10^{-5}$. The velocity of the ion beam is $\beta \approx (100 \text{ kV}/(\frac{1}{2} m_p A c^2))^{1/2} \approx 3 \times 10^{-3}$, so that $\Delta\beta/\beta \approx 6 \times 10^{-3}$.

(b) The minimum temperature according to (11.49) is 120 μK which corresponds to 10^{-8} eV. The temperature is also limited by the recoil energy acquired in the beam frame when absorbing a single photon, which is $(1/2) m_{\text{ion}} v_r^2$. Inserting v_r from (11.48), this amounts to only 5 μK or 4^{10} eV. The larger of these two limits applies, namely 120 μK.

11.8 Damping Times with Electron-Beam Laser Cooling

From (11.55) we get $n_d \approx 1600$ turns. The equivalent damping time for an average ring radius ρ of 1 m is

$$\tau_d = n_d \left(\frac{c}{2\pi\rho} \right) \approx 34 \text{ μs} . \tag{12.142}$$

This is two orders of magnitude smaller than in conventional storage rings.

11.9 Equilibrium Emittances with Electron-Beam Laser Cooling

The equilibrium emittance is given by (11.61)

$$\epsilon_{x,y,N} = \frac{3}{10}\frac{\lambda_c}{\lambda_L}\beta^*_{x,y} \approx 7 \times 10^{-9}\ \mathrm{m}\,, \tag{12.143}$$

and the relative energy spread by

$$\sigma_\delta = \sqrt{\frac{7}{5}\frac{\lambda_c}{\lambda_L}\gamma} \approx 2.6\%\,. \tag{12.144}$$

11.10 Damping Rates and Equilibrium Emittances with Ionization Cooling

(a) From $(\gamma - 1)m_\mu c^2 = 150$ MeV we deduce $\gamma \approx 2.42$. The first term on the right hand side of (11.113) describes the damping. Its average rate is

$$\lambda = \frac{1}{\epsilon_N}\frac{\mathrm{d}\epsilon_N}{\mathrm{d}s} = -\frac{l_{H_2}}{l_{\mathrm{cool}}}\frac{1}{\beta^2}\frac{\mathrm{d}E_\mu}{\mathrm{d}s}\frac{1}{E_\mu} \approx -0.074\ \mathrm{m}^{-1}\,, \tag{12.145}$$

where l_{H_2} and l_{cool} denote the length of the hydrogen cell and the length of the entire cooling stage, respectively. The emittance reduction in a section of length $l_{\mathrm{cool}} = 10$ m is a factor $\exp(\lambda l_{\mathrm{cool}}) \approx 0.48$.

(b) The total length required for a factor 10 emittance reduction is

$$l_{\mathrm{tot}} = -\frac{1}{\lambda}\ln(10) \approx 31\ \mathrm{m}\,, \tag{12.146}$$

or about three 10-m long stages. The muon lifetime in the laboratory system is $\tau_{\mathrm{lab}} = \gamma\tau_{\mu,0} \approx 5.2$ µs, where $\tau_{\mu,0}$ denotes the muon lifetime at rest. The fraction of muons left after traversing $l = 500$ m at 150 MeV is $\exp(-l/(\beta c\tau_{\mathrm{lab}})) \approx 0.7$, or 70%.

(c) The minimum normalized emittance is reached when the time derivative on the left-hand side of (11.113) approaches zero. We can solve the right-hand side for the final emittance and obtain

$$\epsilon_N = \frac{1}{\beta}\frac{\beta_\perp}{2}\frac{(14\ \mathrm{MeV})^2}{(\mathrm{d}E_\mu/\mathrm{d}s)m_\mu c^2 L_R} \approx 4 \times 10^{-4}\ \mathrm{m}\,. \tag{12.147}$$

(d) We first discuss the damping term. The trajectory slope x' is related to the horizontal and longitudinal momenta p_x and p_0 via

$$x' = \frac{p_x}{p_0}\,, \tag{12.148}$$

and due to the ionization energy loss in the direction of the trajectory and re-acceleration in the longitudinal direction over a distance Δs it changes as

$$\Delta x' = -\frac{p_x}{p_0^2}\frac{\Delta p_0}{\Delta s}\Delta s\,. \tag{12.149}$$

Now, $\Delta p_0/\Delta s = (\mathrm{d}E_\mu/\mathrm{d}s)/(\beta c)$, so that

$$\frac{\Delta x'}{\Delta s} = -x' \frac{1}{\beta^2 \gamma m_\mu c^2} \frac{\mathrm{d}E_\mu}{\mathrm{d}s}\,. \tag{12.150}$$

The analogous equation applies in the vertical plane. From (1.15), the change in transverse (horizontal or vertical) action is

$$\frac{\Delta I_\perp}{\Delta s} = -\beta_\perp {x'}^2 \frac{1}{\beta^2 \gamma m_\mu c^2} \frac{\mathrm{d}E_\mu}{\mathrm{d}s}\,. \tag{12.151}$$

Since $\Delta \epsilon_N = \beta\gamma\langle \Delta I_\perp \rangle$, and $\epsilon_N = \beta\gamma\langle \beta_\perp {x'}^2 \rangle$ the damping term in (11.113) follows.

To derive the heating term we start from (1.15) and compute the change in action due to multiple scattering at an angle θ:

$$\Delta I_\perp = \frac{2\beta_\perp \theta(\alpha_\perp x + \beta_\perp x') + \beta_\perp^2 \theta^2}{2\beta_\perp}\,. \tag{12.152}$$

After averaging over the distribution (assuming random betatron phases) only the term quadratic in θ remains, or

$$\langle \Delta I_\perp \rangle = \frac{\beta_\perp \langle \theta^2 \rangle}{2}\,, \tag{12.153}$$

where $\langle \theta^2 \rangle$ is the squared rms scattering angle after a distance s. Inserting (11.114) for this angle, differentiating with respect to s, and noticing again that $\epsilon_N = \beta\gamma\langle \Delta I_\perp \rangle$, the previous equation is rewritten as

$$\begin{aligned} \frac{d\epsilon_N}{ds} &= \frac{\beta_\perp}{2} \frac{\beta\gamma \,(14\ \mathrm{MeV})^2}{(\beta c p)^2} \frac{1}{L_R} \\ &= \frac{\beta_\perp}{2\beta^3} \frac{(14\ \mathrm{MeV})^2}{E_\mu m_\mu c^2 L_R}\,, \end{aligned} \tag{12.154}$$

which equals the expression for the heating in (11.113).

Correction to: Measurement and Control of Charged Particle Beams

Correction to:
M. G. Minty et al., *Measurement and Control of Charged Particle Beams,*
https://doi.org/10.1007/978-3-662-08581-3

The updated version of this book can be found at https://doi.org/10.1007/978-3-662-08581-3

M. G. Minty et al., *Measurement and Control of Charged Particle Beams,*
https://doi.org/10.1007/978-3-662-08581-3_13

References for Chapter 1

1. S.Y. Lee: *Accelerator Physics* (World Scientific, Singapore 1993)
2. H. Wiedemann: *Particle Accelerator Physics 1: Basic Principles and Linear Beam Dynamics* (Springer Verlag, Berlin 1993)
3. K. Wille: *Physik der Teilchenbeschleuniger und Synchrotronstrahlungsquellen* (B.G. Teubner, Stuttgart 1992)
4. E.D. Courant, H.S. Snyder: Annals Phys. **3**, 1 (1958)
5. S. Myers: The LEP Collider, from Design to Approval and Commissioning. In: *CERN Accelerator School, 6th John Adams Memorial Lecture, CERN-91-08, 1991*
6. The PEP-II Design Group: PEP-II An Asymmetric B Factory. SLAC-418 (1993)
7. The KEKB Design Group: KEKB B-Factory Design Report. KEK Report 95-7 (1995)
8. J.T. Seeman: Ann. Rev. Nucl. Part. Sci. **40**, 389 (1990)
9. P. Emma: The Stanford Linear Collider. In: *Proc. of 1995 IEEE PAC, Dallas, 1995* (IEEE, Piscataway 1995) p. 606
10. Y. Kimura: Status of Tristan. In: *Proc. of the XVth International Conference on High Energy Accelerators, Hamburg, 1992* (World Scientific,1992) p. 72
11. SPEAR Storage Ring Group: SPEAR: Status and Improvement Program. In: *Proc. of IXth Conference International Conference on High Energy Accelerators, Stanford, 1974* p. 37
12. M. Cornacchia: 'Conceptual Design of a 1–2 GeV Synchrotron Radiation Source'. In: *Proc. 13th Int. Conf. High Energy Accelerators, Novosibirsk, Russia, Aug 7-11, 1986* LBL-21390
13. P. Lefèvre: LEAR, Present Status: Future Developments. In: *Proc. 3rd LEAR Workshop, Tignes, France, Jan 19-26, 1985* CERN-PS/85-37
14. F. Hinode, S. Kawabata, H. Matsumoto et al.: Accelerator Test Facility–Design and Study Report, KEK Internal Report 95-4 (1995)
15. P. Schmüser: Nucl. Instr. Meth. A **235**, 201 (1985)
16. M. Berndt, V. Bressler, K. Brown et al.: Final Focus Test Beam: Project Design Report. SLAC-0376 (1991)
17. J. Gareyte: The SPS P Anti-P Collider. In: *Proc. of Antiprotons for Colliding Beam Facilities, CERN, Geneva, 1983* CERN report 84-15
18. C. Adolphsen, K. Bane, T. Higo et al.: Phys. Rev. Lett. **74**, 2475 (1995)
19. TESLA Collaboration: Part. Acc. **60**, 53 (1998)
20. G. Jackson: The Fermilab Recycler Ring Technical Design Report. Rev. 1.2, FERMILAB-TM-1981 (1996) p. 153
21. Conceptual Design of the Relativistic Heavy Ion Collider: RHIC, BNL-52195-mc (microfiche) (1989) p. 283

22. K. Johnsen: Present Status and Future Plans for the ISR. In: *Proc. of IXth Conference International Conference on High Energy Accelerators, Stanford, 1974*, p. 32
23. The LHC Study Group: The Large Hadron Collider – Conceptual Design. CERN/AC/95-05 (1995)
24. The NLC Design Group: Zeroth Order Design Report for the Next Linear Collider. SLAC-Report 474 (1996)
25. R. Brinkmann, G. Materlik, J. Rossbach et al.: Conceptual Design of a 500 GeV e^+e^- Linear Collider with Integrated X-ray Laser Facility. DESY 1997-048, ECFA 1997-182 (1997)
26. Muon Collaboration: $\mu^+\mu^-$ Collider. A Feasibility Study. In: *Proc. APS Summer Study, Snowmass 1996, New Directions for High-Energy Physics* BNL-52503
27. M. Sands: The Physics of Electron Storage Rings. SLAC-121 (1970)
28. K.L. Brown: A First and Second-Order Matrix Theory for the Design of Beam Transport Systems and Charged Particle Spectrometers. SLAC-75 (1982)
29. K.L. Brown, F. Rothacker, D. Carey, C. Iselin: Transport. A Computer Program for Designing Charged Particle Beam Transport Systems. SLAC-91 (1977)

References for Chapter 2

1. F. Zimmermann, P. Krejcik, M. Minty, D. Pritzkau, T. Raubenheimer, M. Ross, M. Woodley: Ion Effects in the SLC Electron Damping Ring under Exceptionally Poor Vacuum Conditions, In: *Proc. of the International Workshop on Multibunch Instabilities in Future Electron and Positron Accelerators (MBI97), Tsukuba, 1997*, KEK Proceedings 97-17
2. T. Ieiri, K. Hirata: 'Observation and Simulation of Nonlinear Behavior of Betatron Oscillations During the Beam-Beam Collision'. In: *Proc. of the 1989 IEEE PAC Chicago* (IEEE, Piscataway 1989) p. 709
3. E. Asseo: Causes et Corrections des Erreurs dans la Mesure de Caractéristiques des Oscillations Bétatroniques Obtenues à Partir d'une Transformation de Fourier: Fréquence, Phase, Amplitude et Facteur d'Amortissement, CERN PS/85-9 (LEA) (1985)
4. R. Bartolini, M. Giovannozzi, W. Scandale, A. Bazzani, E. Todesco: Algorithms for a Precise Determination of the Betatron Tune. In: *Proc. of EPAC 96, Sitges, 1996* (IOP, Bristol 1996) Vol. II, p. 1329
5. F.J. Harris: Proc. IEEE **66**, 1 (1978)
6. J. Laskar: Introduction to Frequency Map Analysis. In: *Proc. of the NATO Advanced Study Institute 3DHAM95, S'Agaro, Spain, 1995*
7. J. Laskar, D. Robin: Part. Acc. 54, **183** (1996)
8. Y. Papaphilippou: Frequency Maps of LHC models. In: *Proc. IEEE PAC 99, New York, 1999* (IEEE, Piscataway 1999)
9. Y. Papaphilippou, F. Zimmermann: PRST-AB **2**, 104001 (1999)
10. D. Robin, C. Steier, A. Robinson: Synchr. Radiat. News **13**(6), 33 (2000)
11. N.R. Lomb: Astrophysics and Space Science **39** (1976) p. 447
12. A.-S. Müller: *Precision Measurements of the LEP Beam Energy for the Determination of the W Boson Mass* (Shaker, Aachen, 2000)
13. J. Byrd: Part. Accel. **57**, 159 (1997)
14. G. Carron, D. Möhl, Y. Orlov, F. Pedersen, A. Poncet, S. van der Meer, D.J. Williams: Observation of Transverse Quadrupole Mode Instabilities in Intense Cooled Antiproton Beams in the AA. CERN/PS/89-18 (1989)
15. H. Schmickler: Diagnostics and Control of the Time Evolution of Beam Parameters. In: *Third European Workshop on Beam Diagnostics and Instrumentation at Particle Accelerators (DIPAC 97), 1997* and CERN-SL-97-68
16. C. Boccard, W. Hofle, H. Jakob et al.: 'Tune Measurements in the SPS as Multicycling Machine'. In: *Proc. of EPAC 96, Sitges, 1996* (IOP, Bristol 1996)
17. J. Wenninger: private communication (2001)
18. K.D. Lohmann, M. Placidi, H. Schmickler: 'Design and Functionality of the LEP Q-meter", *Proc. of EPAC 90, Nice, 1990* (Ed. Frontieres, Gif-sur-Yvette 1990) Vol. 1, p. 774

19. L. Jensen, O.R. Jones, H. Schmickler: First Results on Closed-Loop Tune Control in the CERN-SPS. In: *Proc. DIPAC, Chester, 16–18 May, 1999* and CERN SL-99-047
20. J. Borer, R. Jung: Diagnostics. In: *CERN Accelerator School on Antiprotons for Colliding Beam Facilities, CERN 84-15, 1984*, p. 385
21. J. Borer, C. Bovet, A. Burns, G. Morpurgo: Harmonic Analysis of Coherent Bunch Oscillations at LEP. In: *Proc. of EPAC 92, Berlin, 1992* (Ed. Frontieres, Gif-sur-Yvette 1992) p. 1082
22. A. Chao: *Physics of Collective Beam Instabilities in High Energy Accelerators* (Wiley, New York 1993)
23. D. Brandt, P. Castro, K. Cornelis, A. Hofmann, G. Morpugo, G.L. Sabbi, J. Wenninger, B. Zotter: Measurements of Impedance Distributions and Instability Thresholds in LEP. In: *Proc. of 1995 IEEE PAC, Dallas, 1995* (IEEE, Piscataway 1995) p. 570
24. K. Oide: private communication (1997)
25. E.D. Courant, H.S. Snyder: Annals Phys. **3**, 1 (1958)
26. W. Wan, F. Zimmermann: experiment on March 4, 2000, unpublished
27. P. Castro: Betatron Function Measurement at LEP Using the BOM 1000 Turns Facility'. SL/Note 92-63 (1992)
28. J. Corbett: private communication (1998)
29. N. Akasaka, H. Koiso et al.: private communication (1999)
30. S. Peggs: private communication (2000)
31. C. Bovet: The Measurement of Some Machine Parameters. In *Beam Instrumentation, CERN-PE-ED 001-92 (1992), rev. (1994)* ed. by J. Bosser
32. H. Grote, F.C. Iselin: The MAD Program, Version 8.1. CERN/SL/90-13 (1991)
33. T. Lohse, P. Emma: Linear Fitting of BPM Orbits and Lattice Parameters. SLAC Single Pass Collider Memo CN-371 (1989)
34. Y. Cai, M. Donald, J. Irwin, Y. Yan: LEGO: A Modular Accelerator Design Code. In: *Proc. of IEEE PAC97, Vancouver, 1997* (IEEE, Piscataway 1998) and SLAC-PUB-7642 (1997)
35. M. Lee: The Resolve Program, unpublished
36. J. Safranek: Nucl. Instr. Meth. A **388**, 27 (1997)
37. J. Safranek: In: *Beam Measurements. Proc. US-CERN-Japan-Russian School on Beam Measurement, Montreux, Switzerland, 1998.* ed. by S.I. Kurokawa, S.Y. Lee, E. Perevedentsev, S. Turner (World Scientific, 2000)
38. M. Donald: private communication (1998)
39. S. Kamada: Overview on Experimental Techniques and Observations. In: *Proc. of Workshop on Nonlinear Dynamics in Particle Accelerators: Theory and Experiments', Arcidosso, Italy, 1994*
40. H. Koiso, H. Fukuma, Y. Funakoshi, S. Kamada, S. Matsumoto, K. Oide, N. Yamamoto: Beam-Based Measurement of Strength Errors in Quadrupole Magnets with Orbit Bumps. In: *Proc. of EPAC 96, Sitges, Spain, 1996* (IOP, Bristol 1996)
41. E.-S. Kim, S. Matsumoto: Beam-Based Measurement of Focusing Errors in Quadrupole Magnets in ATF Damping Ring. In: *Proc. of the 2nd ATF International Collaboration Meeting', KEK, Tsukuba, 1997* KEK Proceedings 97-9
42. P. Bambade, R. Erickson, W.A. Koska, W. Kozanecki, N. Phinney, S.R. Wagner: Phys. Rev. Lett. **62**, 2949 (1989)
43. W.A. Koska, P. Bambade, W. Kozanecki, N. Phinney, S.R. Wagner: Nucl. Instr. Meth. A **286**, 32 (1990)
44. P. Raimondi, F.-J. Decker: Flat Beam Spot Sizes Measurement in the SLC Final Focus. In: *Proc. 16th IEEE PAC and International Conference on High Energy Accelerators, Dallas, Texas, 1–5 May, 1995* (IEEE, Piscataway 1995)

45. R. Raimondi, F.-J. Decker, P. Chen: Disruption Effects on the Beam Size Measurement. In: *Proc. 16th IEEE PAC and International Conference on High Energy Accelerators, Dallas, Texas, 1–5 May 1995* (IEEE, Piscataway 1995)
46. P. Emma, L.J. Hendrickson, P. Raimondi, F. Zimmermann: Limitations of Interaction Point Spot Size Tuning at the SLC. In: *Proc. 17th IEEE PAC, Vancouver, 12–16 May, 1997* (IEEE, Piscataway 1998)
47. L. Hendrickson, S. Bes, P. Grossberg et al.: 'Luminosity Optimization Feedback in the SLC'. In: *Proc. International Conference on Accelerators and Large Experimental Physics Control Systems, Beijing, China, 1999*
48. N.J. Walker, J. Irwin, M. Woodley: Global Tuning Knobs for the SLC Final Focus. In: *Proc. of 1993 IEEE PAC, Washington, 1993* (IEEE, Piscataway 1993)
49. R.W. Assmann, F.-J. Decker, L.J. Hendrickson, N. Phinney, R.W. Siemann, K.K. Underwood, M. Woodley: Beam-Based Monitoring of the SLC Linac Optics with a Diagnostic Pulse. In: *Proc. IEEE PAC 97, Vancouver, 1997* (IEEE, Piscataway 1998), and SLAC-PUB-7578
50. J. Irwin, C.X. Wang, Y.T. Yan, K.L.F. Bane, Y. Cai, F.-J. Decker, M.G. Minty, G.V. Stupakov, F. Zimmermann: Phys. Rev. Lett. **82**, 1684 (1999)
51. C.X. Wang: Model Independent Analysis of Beam Centroid Dynamics in Accelerators. Ph.D. thesis, Stanford University (2000)
52. J. Irwin, C.X. Wang, Y.T. Yan, K. Bane, Y. Cai, F.-J. Decker, M.G. Minty, G.V. Stupakov, F. Zimmermann: Model-Independent Analysis with BPM Correlation Matrices. In: *Proc. EPAC 98, Stockholm, Sweden 1998* (IOP, Bristol 1999), and SLAC-PUB-7863
53. J. Irwin, C.X. Wang, Y.T. Yan, K. Bane, Y. Cai, F.-J. Decker, M.G. Minty, G.V. Stupakov, F. Zimmermann: Transverse Wake Field Effect Measurement via Model-Independent Analysis. In: *Proc. 19th Intl. Linear Acc. Conf. (Linac 98)*, Chicago, IL (1998), and SLAC-PUB-7917
54. A.-S. Müller, J. Wenninger: Measurements of Coherent Damping and Tune Shifts with Amplitude at LEP. In: *Proc. EPAC 98, Stockholm, 1998* (IOP, Bristol 1999)
55. R. Bartolini, M. Giovannozzi, W. Scandale, A. Verdier, C. Pellegrini, P. Tran, E. Todesco, J. Corbett, M. Cornacchia: Measurements of the Tune Variations Induced by Non-Linearities in Lepton Machines. In: *Proc. of EPAC 96, Sitges, Spain, 1996* (IOP, Bristol 1996) vol. II, p. 917
56. R.E. Meller, A.W. Chao, J.M. Peterson, S.G. Peggs, M. Furman: Decoherence of Kicked Beams. SSC-N-360 (1987)
57. J. Bengtsson: Spectral Analysis of the Motion at a Single Nonlinear Resonance by Canonical Perturbation Theory. CERN internal report PS/LEA/Note 87-03 (1987)
58. R. Bartolini, F. Schmidt: Part. Accelerators **59**, 93 (1998)
59. C.-X. Wang, J. Irwin: Possibility to Measure the Poincaré Section Map of a Circular Accelerator. SLAC-PUB-7547 (1997)
60. R. Bartolini, L.H.A. Leunissen, Y. Papaphilippou et al.: 'Measurement of Resonance Driving Terms from Turn-by-Turn Data'. In: *Proc. IEEE PAC99 New York, 1999* (IEEE, Piscataway 1999) p. 1557
61. M. Bai, S.Y. Lee, J.W. Glenn et al.: Phys. Rev. E **56**, 6002 (1997)
62. Y. Cai: Lattice Performance of the PEP-II High Energy Ring. In: *Proc. Advanced ICFA Beam Dynamics Workshop on Beam Dynamics Issues for e^+e^- Factories, Frascati, Italy, 1997* and SLAC-PUB-7733
63. G. Ripken, F. Willeke: On the Impact of Linear Coupling on Nonlinear Dynamics. DESY 90-001 (1990)

64. G. Ripken: Untersuchungen zur Strahlführung und Stabilität der Teilchenbewegung in Beschleunigern und Storage-Ringen unter strenger Berücksichtigung einer Kopplung der Betatronschwingungen. Int. Rep. DESY R1-70/4 (1970)
65. D.A. Edwards, L.C. Teng: IEEE Trans. on Nucl. Sc. **20**, 885 (1973)
66. L.C. Teng: Concerning n-Dimensional Coupled Motion. Fermilab internal report FN 229 (1971)
67. F. Willeke, G. Ripken: Methods of Beam Optics. In: *Proc. Cornell Summer School, AIP Proceedings 184* (AIP) and DESY 88-114 (1988)
68. G. Guignard: The General Theory of All Sum and Difference Resonances in a Three-Dimensional Magnetic Field in a Synchrotron. CERN 76-06 (1976)
69. T. Raubenheimer: Part. Acc. **36**, 75 (1991)
70. P. Bryant: A Simple Theory for Weak Betatron Coupling. In: *CERN Accelerator School, CERN 89-05, 1989*
71. F. Zimmermann, M.G. Minty, H. Hayano: Measurements of Betatron Coupling at the ATF Damping Ring. SLAC/AAS-95 and KEK ATF Internal Report ATF 98–22 (1998)
72. G. Bourianoff, S. Hunt, D. Mathieson, F. Pilat, R. Talman, G. Morpugo, Determination of Coupled-Lattice Properties Using Turn-by-Turn Data. In: *Proc. of the Workshop on the Stability of Particle Motion in Storage Rings, Upton, New York 1992, AIP Conference Proceedings 292* (AIP)
73. J. Seeman, U. Wienands, and PEP-II team: unpublished (1997)
74. G. Guignard: Adjustment of Emittance Ratio by Coupling Control in Electron-Positron Storage Rings. In: *Proc. of 11th International Conference on High-Energy Accelerators, Geneva, 1980*, p. 682
75. E. Métral: Simple Theory of Emittance Sharing and Exchange due to Linear Betatron Coupling. In: CERN-PS-066-AE (Dec. 2001) p. 11
76. M. Mitsuhashi: In: *Beam Measurements. Proc. US-CERN-Japan-Russian School on Beam Measurement, Montreux, Switzerland, 1998* ed. by S.I. Kurokawa, S.Y. Lee, E. Perevedentsev, S. Turner (World Scientific, 2000)
77. M. Minty, M. Woodley: Summary of Accelerator Physics Studies at KEK/ATF. ATF-98-12 (1998)
78. F. Zimmermann, K.L.F. Bane, T. Kotseroglou et al.: Vertical Emittance Studies at the ATF Damping Ring. SLAC-AP-113 and KEK ATF Internal Report ATF 98–37 (1998)
79. G. Guignard: Phys. Rev. E **51**(6), 6104 (1995)
80. J.P. Koutchouk: Linear Betatron Coupling Measurement and Compensation in the ISR. In: *Proc. XIth Int. Conf. on High Energy Accelerators, CERN, Geneva, 1980* (Birkhäuser, Basel, 1980) p. 491
81. Y. Derbenev: University of Michigan Report No. UM-HE-98-04 (1998)
82. R. Brinkmann, Y. Derbenev, K. Flöttmann: PRST-AB **4**, 053501 (2001)
83. D. Edwards et al. In: *Proceedings of the XXth Int. Linac Conference, Monterey, 2000*
84. D. Boussard: In: *CERN Accelerator School: Course on Advanced Accelerator Physics*; CERN-95-06

References for Chapter 3

1. J. Turner: private communication (1998)
2. D. Robin, G. Portmann, L. Schachinger: Automated Beam Based Alignment of the ALS Quadrupoles. SLAC internal report, NLC-Note 18 (1995)
3. R. Brinkmann, M. Böge: Beam-Based Alignment and Polarization Optimization in the HERA Electron Ring. In: *Proc. of EPAC 94, London, 1994* (World Scientific, Singapore 1994) p. 938
4. J. Corbett, R.O. Hettel, H.-D. Nuhn: Quadrupole Shunt Experiments at SPEAR. In: *Proc. of 7th Beam Instrumentation Workshop (BIW96), Argonne, Illinois, 1996*
5. P. Emma: Beam-Based Alignment of Sector-1 of the SLAC Linac. In: *Proc. of EPAC 92, Berlin, 1992* (Ed. Frontieres, Gif-sur-Yvette 1992) and SLAC-PUB-5787
6. I. Reichel: Beam Position Measurement by Modulation of Quadrupole Strengths. CERN SL note/95-50 (1995)
7. M. Kikuchi, K. Egawa, H. Fukuma, M. Tejima: Beam-Based Alignment of Sextupoles with the Modulation Method. In: *Proc. of 1995 IEEE PAC, Dallas, 1995* (IEEE, Piscataway 1995) p. 603
8. S. Kamada: Overview on Experimental Techniques and Observations. In: *Proc. of Workshop on Nonlinear Dynamics in Particle Accelerators: Theory and Experiments, Arcidosso, Italy, 1994*
9. N. Yamamoto, S. Kamada, Y. Kobayashi, H. Koiso, S. Matsumoto: Beam-Based Alignment of Sextupole Magnets with a π-Bump Orbit. KEK Preprint 96-74 (1996)
10. S. Herb, G.B. Jaczko, F. Willeke: 'Beam-Based Calibration of Beam Position Monitor Offsets in the HERA Proton Ring using Strong Sextupole Fields'. In: *Proc. of Workshop on Beam Diagnostics and Instrumentation for Particle Accelerators, Travemünde, 1995*, DESY Int. Rep. M-95-07
11. T. Raubenheimer: In: *Zeroth Order Design Report for the Next Linear Collider, SLAC Report 474, 1996*, pp. 227–228
12. P. Emma, J. Irwin, N. Phinney, P. Raimondi, N. Toge, N.J. Walker, V. Ziemann: Beam Based Alignment of the SLC Final Focus Sextupoles. In: *Proc. of 1993 IEEE PAC, Washington, DC, 1993* (IEEE, Piscataway 1993) p. 116
13. P. Emma, J. Irwin, N. Phinney, P. Raimondi, N. Toge, N.J. Walker, V. Ziemann: Beam Based Alignment of the SLC Final Focus Sextupoles. In: *Proc. of IEEE PAC 93, Washington, D.C., 1993* (IEEE, Piscataway 1993)
14. P.G. Tenenbaum: Expanded Studies of Linear Collider Final Focus Systems at the Final Focus Test Beam. Ph.D. thesis, Stanford University, SLAC-R-95-475 (1995)
15. C. Adolphsen, K. Bane, T. Higo et al.: Phys. Rev. Lett. **74**, 2475 (1995)
16. M. Seidel, C. Adolphsen, K.L.F. Bane, R.M. Jones, N.M. Kroll, R.H. Miller, D.H. Whittum: Nucl. Instr. Meth. A **404**, 31 (1998)

17. K. Thompson, T. Himel, S. Moore et al.: 'Operational experience with Model Based Steering in the SLC linac'. In: *Proc. Part. Acc. Conf., Chicago, Illinois 1989*, pp. 1675–1677
18. T. Barklow, P. Emma, P. Krejcik, N. Walker: Review of Lattice Measurement Techniques at the SLC. In: *Proc. 5th ICFA Advanced Beam Dynamics Workshop, Corpus Christi, Texas, 1991*
19. T. Lohse, P. Emma: Linear Fitting of BPM Orbits and Lattice Parameters. SLAC Single Pass Collider Memo CN-371 (1989)
20. T. Himel, K. Thompson: Energy Measurements from Betatron Oscillations. In: *Proc. of the 1989 IEEE Part. Acc. Conf., Chicago, Illinois, 1989*
21. C. Adolphsen, T.L. Lavine, W.B. Atwood et al.: 'Beam Based Alignment Technique for the SLC Linac'. In: *Proc. of the 1989 IEEE Part. Acc. Conf., Chicago, Illinois, 1989*, pp. 977–979
22. V. Ziemann: Corrector Ironing. SLAC internal report Single-Pass Collider Note CN-393 (1992)
23. W.H. Press, B.P. Fannery, S.A. Teukolsky, W.T. Vetterling: *Numerical Recipes: The Art of Scientific Computing*, Section 2.9 (Cambridge University Press, Cambridge 1986)
24. The MathWorks Inc., *http://www.mathworks.com* (1994)
25. W.J. Corbett, F. Fong, M. Lee, V. Ziemann: Optimum Steering of Photon Beam Lines in SPEAR. In: *Proc. of 1993 IEEE PAC Washington, 1993* (IEEE, Piscataway 1993) p. 1483
26. T. Limberg, J. Seeman, W.L. Spence: Effects and Tolerances of Injection Jitter in the SLC and Future Linear Colliders. In: *Proc. of EPAC 90, Nice, France, 1990* (Ed. Frontieres, Gif-sur-Yvette 1990) p. 1506
27. A.W. Chao, B. Richter, C.Y. Yao: Nucl. Instr. Meth. **178**, 1–24 (1980)
28. P. Emma, T.H. Fieguth, T. Lohse: Nucl. Instr. Meth. A **288**, 313–334 (1990)
29. V. Balakin, A. Novokhatsky, V. Smirnov: In: *Proc. 12th International Conference on High Energy Accelerators, Fermilab, 1983* (IEEE, New York, 1983)
30. J.T. Seeman, F.-J. Decker, I. Hsu: The Introduction of Trajectory Oscillations to Reduce Emittance Growth in the SLC Linac. In: *Proc. 15th Intl. Conf. High Energy Acc., Hamburg, Germany, 1992* (World Scientific, 1992) pp. 879–881
31. R. Assmann, C. Adolphsen, K. Bane et al.: LIAR – A Computer Program for the Modeling and Simulation of High Performance Linacs. SLAC/AP-103, 1997
32. R. Assmann: Beam dynamics in the SLC. In: *Proc. of 1997 IEEE Part. Acc. Conf., Vanouver, BC, Canada, 1997*
33. R. Assmann, F.-J. Decker, P. Raimondi: Improvements in Emittance Wakefield Optimization for the SLAC Linear Collider. In: *Proc. EPAC 98, Stockholm, Sweden, 1998* (IOP, Bristol 1999)
34. R.W. Assmann, F.J. Decker, L.J. Hendrickson et al.: 'Beam Based Monitoring of the SLC Linac Optics with a Diagnostic Pulse'. In: *Proc. 17th Part. Acc. Conf., Vancouver, Canada, 1997*, pp. 497–499
35. T.O. Raubenheimer, R.D. Ruth: Nucl. Instr. Meth. **A302**, 191–208 (1991)
36. R.W. Assmann, T. Chen, F.J. Decker et al.: 'Quadrupole Alignment and Trajectory Correction for Future Linear Colliders: SLC Tests of a Dispersion-Free Steering Algorithm'. In: *Proc. 4th Int. Workshop on Acc. Alignment, Tsukuba, Japan, 1995*
37. F.-J. Decker, R. Brown, J.T. Seeman: Beam Size Measurements with Noninterceptive Fff-Axis Screens. In *Proc. of Part. Acc. Conf., Washington, DC, 1993*, pp. 2507–2509
38. R. Assmann, T. Chen, F.-J. Decker, M. Minty, P. Raimondi, T.O. Raubenheimer, R. Siemann: Simultaneous Trajectory and Dispersion Correction in the SLC Linac. Unpublished draft (1996)

39. D. McCormick, M. Ross, T. Himel, N. Spencer: Thermal Stabilization of Low Level RF Distribution Systems at SLAC. In: *Proc. of Part. Acc. Conf., Washington, DC, 1993* (IEEE, Piscataway 1993) pp. 1975–1977
40. F.-J. Decker, R. Akre, M. Byrne, et al.: 'Effects of Temperature Variation on the SLC Linac Rf System'. In: *Proc. of Part. Acc. Conf., Dallas, Texas, 1995* (IEEE, Piscataway 1995) pp. 1821–1823
41. J. Bogart, R.W. Assmann, M. Breidenbach et al.: 'A Fast and Accurate Phasing Algorithm for the RF Accelerating Voltages of the SLAC Linac'. In: *Proc. EPAC 98, Sweden, 1998* (IOP, Bristol 1999) pp. 22–26
42. T. Himel: Ann. Rev. Nucl. Part. Sci. **47**, 157 (1997)
43. M. Minty, C. Adolphsen, L.J. Hendrickson, R. Sass, T. Slaton, M. Woodley: Feedback Performance at the Stanford Linear Collider. In: *Proc. of IEEE PAC 95, Dallas, 1995* (IEEE, Piscataway 1995)
44. M. Bai, S.Y. Lee, J.W. Glenn et al.: Phys. Rev. E **56**, 5, 6002 (1997)
45. M. Bai, L. Ahrens, J. Alessi et al.: Phys. Rev. Lett. **80**, 21, 4673 (1998)
46. O. Berrig, W. Hofle, R. Jones et al.: 'Excitation of Large Transverse Beam Oscillations without Emittance Blow-Up using AC-Dipole Principle'. In: *Proc. 5th European Workshop on Diagnostics and Beam Instrumentation, Grenoble, 2001*, and CERN-SL-2001-019-BI

References for Chapter 4

1. F.-J. Decker, R. Pennacchi, R. Stege, J. Turner: Characterizing Transverse Beam Jitter in the SLC Linac. In: *Proc. 6th European Particle Accelerator Conference (EPAC 98), Stockholm, Sweden, 1998* (IOP, Bristol 1999)
2. T. Shintake, K. Oide, N. Yamamoto et al.: 'Experiments of Nanometer Spot Size Monitor at FFTB Using Laser Interferometry'. In: *Proc. 1995 IEEE PAC, Dallas, Texas, 1995*, p. 2444
3. T. Mitsuhashi: Beam Profile and Size Measurement by SR Interferometers. In: *Proc. Joint US-CERN-Russia-Japan School on Particle Accelerators: Beam Measurement, Montreux, 1998* (World Scientific, Singapore, 1999)
4. M. Hüning: Analysis of Surface Roughness Wake Fields and Longitudinal Phase Space in a Linear Electron Accelerator. Ph.D. thesis, Hamburg University. DESY-THESIS-2002-029 (2002)
5. P. Tenenbaum, T. Shintake: Measurement of Small Electron Beam Spots, Ann. Rev. Nucl. Part. Sci. **49**, 125–162 (1999)
6. M.C. Ross, N. Phinney, G. Quickfall et al.: 'Automated Emittance Measurements in the SLC'. In: *Proc. 1987 IEEE PAC, Washington, D.C. 1987* (IEEE, Washington 1987)
7. F.-J. Decker: private communication (1994)
8. M. Woodley, P. Emma: Measurement and Correction of Cross Plane Coupling in Transport Lines. In: *Proc. 20th Intnl. Linac Conf., Monterey, CA 2000*
9. M. Sands: A Beta Mismatch Parameter. SLAC internal report SLAC-AP-85 (1991)
10. W. Spence: private communication (1996)
11. R. Iverson, M. Minty, M. Woodley: ATF Internal Report ATF 12-29 (1997)
12. H. Wiedemann: *Particle Accelerator Physics* (Springer-Verlag, Berlin, 1993)
13. M. Sands: The Physics of Electron Storage Rings, SLAC Internal Report SLAC-121 (1970)
14. K. Robinson: Phys. Rev. **111**, 373 (1958)
15. J.T. Seeman: Observations of the Beam-Beam Interaction, Joint US/CERN School on Particle Accelerators, Sardinia, Italy (1985)
16. M.G. Minty, R. Brown, F.-J. Decker et al.: 'Using a Fast-Gated Camera for Measurements of Transverse Beam Distributions and Damping Times'. In: *Accelerator Instrumentation Workshop, Berkeley, 1992* ed. by J.A. Hinkson, G. Stover (AIP Conf. Proc. **281**, 1992) p. 158
17. M. Minty, R. Akre, F.J. Decker et al.: 'Emittance Reduction via Dynamic RF Frequency Shift at the SLC Damping Rings'. In: *17th Int. Conf. on High Energy Accelerators, Dubna, Russia, 1998*
18. R. Akre, F.-J. Decker, M.G. Minty: RF Frequency Shift during Beam Storage in the SLC Damping Rings. In: *Proc. 1999 IEEE PAC New York, 1999* (IEEE, Piscataway 1999)

19. R.D. Kohaupt, G.A. Voss: Progress and Problems in Performance of e+/e- Storage Rings, Ann. Rev. Nucl. and Part. Sci. **33**, 67 (1983)
20. G. Arduini, R. Assmann, R. Bailey et al.: 'LEP Operation and Performance with 100 GeV Colliding Beams' In: *Proc. EPAC Vienna, Austria, 2000* (European Phys. Soc., Geneva 2000) p. 265
21. T. Sen: Beam Dynamics with the New Interaction Regions. In: *Proc. DESY Beschleuniger-Betriebsseminar Bad Lauterberg, 1998*; DESY HERA 98-04
22. T.O. Raubenheimer, L.Z. Rivkin, R.D. Ruth: Damping Ring Designs for a TeV Linear Collider. In: *DPF Summer Study Snowmass 1988*, and Internal Report SLAC-PUB-4808 (1988)
23. A.M. Kondratenko, B.W. Montague: Polarized Beams in LEP. CERN Internal Report CERN ISR-TH/80-38 (1980)
24. K. Steffen: An Alternative Interaction Geometry for HERA. Internal Report DESY HERA 81/17 (1981)
25. J.T. Seeman: Observation of High Current Effects in High Energy Linear Colliders. In: *1990 Joint US-CERN School on Particle Accelerators: Frontiers of Particle Beams, Intensity Limitations, Hilton Head Island, SC (1990)*; published in CERN US PAS 255–292 (1990)
26. J.T. Seeman, K.L.F. Bane, T. Himel et al.: 'Observation and Control of Emittance Growth in the SLC Linac'. In: Part. Accel., **30**, 97–104 (1989)
27. F.-J. Decker, R. Brown, J.T. Seeman: Beam Size Measurements with Noninterceptive Off-axis Screens. In: *Proc. 1993 IEEE PAC Washington, DC, 1993* (IEEE, Piscataway 1993)
28. V. Balakin, A. Novokhatsky, V. Smirnov. In *Proc. 12th International Conference High Energy Accelerators, Fermilab, 1983* (IEEE, New York, 1983)
29. J.T. Seeman, F-J. Decker, I. Hsu: The Introduction of Trajectory Oscillations to Reduce Emittance Growth in the SLC Linac. In: *Proc. XV International Conference on High Energy Accelerators, Hamburg, Germany, 1992* (World Scientific, 1992)
30. A. Chao: *Physics of Collective Beam Instabilities in High Energy Accelerators* (Wiley, New York 1993)
31. T.O. Raubenheimer, R.D. Ruth: Nucl. Instr. Meth. A **302**, 191 (1991)
32. T.O. Raubenheimer: Nucl. Instr. Meth. A **306**, 61 (1991)
33. R. Assmann, P. Raimondi, G. Roy, J. Wenninger: Phys. Rev. ST Accel. Beams **3**, 121001 (2000)

References for Chapter 5

1. K.-J. Kim: Nucl. Instr. Meth. A **275**, 201 (1989)
2. J.B. Rosenzweig, E. Colby, G. Jackson, T. Nicol: Design of a High Duty Cycle, Asymmetric Emittance RF Photoinjector for Linear Collider Applications. In: *Proc. IEEE PAC 1993, Washington, DC, 1993* (IEEE, Piscataway 1993)
3. B.E. Carlsten: Part. Acc. **49**, 27 (1995)
4. M.E. Jones, B.E. Carlsten: Space-Charge Induced Emittance Growth in the Transport of High-Brightness Electron Beams. In: *Proc. IEEE Part. Acc. Conf., Washington, D.C., 1987*, p. 1391
5. Y. Derbenev: University of Michigan Report No. UM-HE-98-04 (1998)
6. R. Brinkmann, Ya. Derbenev, K. Flöttmann: A Flat Beam Electron Source for Linear Colliders. Internal Report TESLA 99-09 (1999)
7. R. Brinkmann, Y. Derbenev, K. Flöttmann: A Low Emittance, Flat-Beam Electron Source for Linear Colliders. In: *Proc. EPAC 2000, Vienna, 2000* (European Phys. Soc., Geneva 2000)
8. D. Edwards: Notes on the Production of Flat Beams, unpublished (2000)
9. E.D. Courant, H.S. Snyder: Ann. Phys. **281**, 360 (1958)
10. D. Edwards, H. Edwards, N. Holtkamp et al.: 'The Flat Beam Experiment at the FNAL Photoinjector'. In: *Proc. LINAC 2000, Monterey, 2000*

References for Chapter 6

1. V.I. Telnov: Nucl. Instr. Meth. A **260**, 304 (1987)
2. I. Reichel, F. Zimmermann, T. Raubenheimer, P. Tenenbaum: 'Thermal Photon and Residual Gas Scattering in the NLC Beam Delivery'. In: *Proc. of International Computational Accelerator Physics Conference (ICAP 98), Monterey, 1998*
3. L.P. Keller: Calculation of Muon Background in a 0.5 TeV Linear Collider. In: *DPF Summer Study on High Energy Physics, Snowmass, Colorado, 1990*
4. Y. Tsai: Rev. Mod. Phys. **46**(4), 815 (1974)
5. H.-J. Schreiber: In: *7th Workshop of ECFA/DESY Study on Linear Colliders, Hamburg, 2000*
6. R. Assmann, H. Burkhardt, S. Fartoukh et al.: 'Overview of the CLIC Collimation Design'. In: *Proc. IEEE PAC 2001, Chicago, Illinois, 2001* (IEEE, Piscataway, 2001)
7. C. Adolphsen, R. Aiello, R. Alley et al.: Zeroth Order Design Report for the Next Linear Collider, SLAC Internal SLAC Report 474 (1995)
8. S.S. Hertzbach, T.W. Markiewicz, T. Maruyama, R. Messner: 'Backgrounds at the Next Linear Collider'. In: *Proc. DPF and DPB Summer Study, Snowmass, 1996*
9. H. Burkhardt: Machine Background Common to All Machines. In: *Proc. International Workshop on Linear Colliders, Sitges, Barcelona, 1999*
10. K.L.F. Bane, C. Adolphsen, F.-J. Decker, P. Emma, P. Krejcik, F. Zimmermann: 'Measurement of the Effect of Collimator Generated Wake Fields on the Beams in the SLC'. In: *Proc. 16th IEEE Particle Accelerator Conference (PAC 95) and International Conference on High Energy Accelerators, Dallas, Texas, 1995* (IEEE, Piscataway 1995)
11. L. Merminga, J. Irwin, R. Helm, R.D. Ruth: Part. Acc. **48**, 85–108 (1994)
12. F. Zimmermann: New Final Focus Concepts at 5 TeV and Beyond. In: *Proc. 8th Workshop on Advanced Accelerator Concepts, Baltimore, Maryland, 5 - 11 July* ed. by W. Lawson (AIP Conference Proceedings 472, New York 1998)
13. S. Heifets, T. Raubenheimer: Plasma Possibilties in the NLC. In: *Proc. 12th Adv. ICFA Beam Dynamics Workshop on Nonlinear and Collective Phenomena, Arcidosso, Italy, 1996*
14. P. Emma, R. Helm, Y. Nosochkov, R. Pitthan, T. Raubenheimer, K. Thompson, F. Zimmermann: 'Nonlinear Resonant Collimation for Future Linear Colliders'. In: *Proc. 16th Advanced ICFA Beam Dynamics Workshop on Nonlinear and Collective Phenomena in Beam Physics, Arcidosso, Italy, 1998*
15. M. Seidel: The Proton Collimation System of HERA. DESY Internal Report DESY-94-103 (1994)
16. J.B. Jeanneret: Phys. Rev. ST Accel. Beams **1**, 081001 (1998)

17. I. Reichel: Transverse Beam Tails at LEP. In: *Proc. ICFA Beam Dynamics Workshop on e^+e^- Factories, Frascati, 1997* ed. by L. Palumbo, G. Vignola (Frascati Physics Series, Vol. X, Frascati, 1997)
18. H. Burkhardt, I. Reichel, G. Roy: Phys. Rev. ST Accel. Beams **3**, 091001 (2000)
19. T.O. Raubenheimer, F. Zimmermann: Rev. Mod. Phys. **72**, 95–107 (2000)
20. B. Dehning, A.C. Melissinos, F. Perrone et al.: Phys. Rev. Lett. B **249**, 145 (1990)
21. G. von Holtey, A.H. Ball, E. Brambilla et al.: Nucl. Instr. Meth. A **403**, 205 (1998)
22. V. Baglin, I. R. Collins, O. Grobner: Photoelectron Yield and Photon Reflectivity from Candidate LHC Vacuum Chamber Materials with Implications to the Vacuum Chamber Design. In: *Proc. 6th European Particle Accelerator Conference (EPAC 98), Stockholm, 1998* (IOP, Bristol 1999)

References for Chapter 7

1. M.P. Zorzano, R. Wanzenberg: Intrabeam Scattering and the Coasting Beam in the HERA Proton Ring. CERN-SL-2000-072-AP and DESY-HERA-00-06 (2000)
2. W. Kriens, M. Minty: Longitudinal Schottky Signal Monitoring for Protons in HERA. In: *Proc. of EPAC 2000, Vienna, 2000* (European Phys. Soc., Geneva 2000) p. 1762
3. A. Piwinski: Nucl. Instr. Meth. **72**, 79–81 (1969)
4. T. Suzuki: Orbit Theory for Accelerators. In: *Proc. APAC'02, Beijing, 2001*
5. M. Sands: The Physics of Electron Storage Rings, SLAC-121 (1970)
6. H. Hayano, S. Kamada, K. Kubo et al.: 'Optics Diagnostics and Tuning for Low Emittance Beam in KEK-ATF Damping Ring'. In: *Proc. of 1999 IEEE PAC, Chicago, 1999* (IEEE, Piscataway 1999) p. 3432
7. J. Borer, C. Bovet, A. Burns, G. Morpurgo: Harmonic Analysis of Coherent Bunch Oscillations at LEP. In: *Proc. of EPAC 92, Berlin, 1992* (Ed. Frontieres, Gif-sur-Yvette 1992) p. 1082
8. F. Ruggiero: Effect of Residual Dispersion at the RF Cavities on the Dynamic Measurement of Dispersion in LEP, CERN SL/91-38 (1991)
9. M. Minty: unpublished (1997)
10. K. Kubo, T. Okugi: Dispersion Measurement in ATF DR, ATF Internal Report ATF-97-19 (1997)
11. F. Ruggiero, A. Zholents: Resonant Correction of Residual Dispersion in LEP, CERN SL-MD Note 26 (1992)
12. P. Emma, D. McCormick, M.C. Ross: Beam Dispersion Measurements with Wire Scanners in the SLC Final Focus Systems. In: *Proc. of IEEE PAC 1993, Washington, DC, 1993* (IEEE, Piscataway 1993) p. 2160
13. D.C. Carey, K.L. Brown, F. Rothacker: Third Order Transport: A Computer Program for Designing Charged Particle Beam Transport Systems. SLAC-R-0462 and FERMILAB-PUB-95-069 (1995)
14. P. Emma, W. Spence: In: *Proc. of the SLAC/KEK Linear Collider Workshop on Damping Ring, Tsukuba, 1992* KEK Proceedings 92-6
15. A.-S. Müller: Precision Measurements of the LEP Beam Energy. Ph.D. thesis, U. Mainz, Shaker Verlag, 2000, and A.-S. Müller, J. Wenninger: Synchrotron Tune and Beam Energy at LEP-2. In: *Proc. EPAC 2000, Vienna, 2000* (European Phys. Soc., Geneva 2000), and CERN-SL-2000-062-OP
16. E. Perevedentsev: Chromaticity Measurement and Correction. In: *Beam Measurements. Proc. US-CERN-Japan-Russian School on Beam Measurement, Montreux, Switzerland, 1998.* edited by S.I. Kurokawa, S.Y. Lee, E. Perevedentsev, S. Turner (World Scientific, 2000)
17. U. Wienands: private communication (1998)

18. A. Fisher: Instrumentation and Diagnostics for PEP-II. In: *Proc. of 8th Beam Instrumentation Workshop (BIW98), Stanford, California 1998* AIP Conf. Proc. **451**, 475 (Springer Verlag, 1998)
19. R. Assmann, A. Beuret, A. Blondel et al.: 'The Energy Calibration of LEP in the 1993 Scan'. In: *Zeitschrift für Physik* **C66**, 567–582 (1995), and CERN-SL/95-02 (1995)
20. A. Hofmann: private communication (1998)
21. H. Burkhardt: private communication (1998)
22. D. Cocq, O.R. Jones, H. Schmickler: The Measurement of Chromaticity via a Head-Tail Phase Shift, CERN SL-98-062 BI (1998)
23. R.H. Siemann: Bunched Beam Diagnostics In: Proc. Physics of Particle Accelerators, Batavia, IL 1987, p. 41
24. N.A. Vinokurov, V.N. Korchuganov, G.N. Kulipanov et al.: Preprint INP 76-87, Novosibirsk (1976)
25. B. Holzer (DESY): unpublished (1998)
26. O. Meincke, S. Herb, P. Schmüser: Chromaticity Measurements in the HERA Proton Storage Ring. In: *Proc. of EPAC 92, Berlin, 1992* (Ed. Frontieres, Gif-sur-Yvette 1992) p. 1070
27. A. Piwinski, S. Herb: measurements at DORIS-I and HERA, unpublished
28. For LEP this measurement was proposed by A. Hofmann
29. C. Zhang, W.F. Du, Z.G. Cai: Beam Energy Stabilization and Online Measurement in BEPC. In: *Proc. of the XVth International Conference on High Energy Accelerators, Hamburg, 1992* (World Scientific,1992) p. 409
30. H. Schmickler: Measurement of the Central Frequency of LEP, CERN LEP performance note 89 (1993)
31. H. Wiedemann: *Particle Accelerator Physics 1: Basic Principles and Linear Beam Dynamics* (Springer, Berlin 1993)
32. F. Willeke, G. Ripken: Methods of Beam Optics. In: *Proc. Cornell Summer School, AIP Proceedings 184* (AIP) and DESY 88-114 (1988)

References for Chapter 8

1. R. Erickson (ed.): The SLC Design Handbook, Stanford Linear Accelerator Center (1984)
2. The NLC Study Group: Zeroth-Order Design Report for the Next Linear Collider. LBNL-PUB-5424, SLAC Report 474, UCRL-ID-124161 (1996)
3. N. Akasaka, M. Akemoto, S. Anami et al.: JLC Design Study. KEK-REPORT-97-1 (1997)
4. D.A. Edwards (ed.): 'TESLA Test Facility Linac – Design Report', DESY Report: TESLA 95-01 (1995)
5. C. Pellegrini, J. Rosenzweig, H.D. Nuhn et al.: 'A 2-nm to 4-nm High Power FEL on the SLAC Linac', Nucl. Instr. Meth. **A331**, 223–227 (1993)
6. P. Emma, T. Raubenheimer, F. Zimmermann: A Bunch Compressor for the Next Linear Collider, In: *Proc. IEEE Part. Acc. Conf., Dallas, Texas, 1995* 704–706. SLAC-PUB-6787 (1995)
7. M.G. Minty et al.: Operation and Performance of Bunch Precompression for Increased Current Transmission at the SLC, In: Proc. IEEE Part. Acc. Conf., Vancouverr, Canada, 1997. SLAC-PUB-7477 (1997)
8. W. Kriens: Petra Bunch Rotation. DESY-M-97-10N (1997)
9. G. Wiesenfeldt: Untersuchungen zur longitudinalen Strahlanpassung beim Protonentransfer von PETRA nach HERA. Diplomarbeit, Universitaet Hamburg (1995)
10. I. Kourbanis, G.P. Jackson, X. Lu: Performance and Comparison of the Different Coalescing Schemes Used in the Fermilab Main Ring. In: *Proc. IEEE Part. Acc. Conf., Washington, DC, 1993*, pp. 3799–3801
11. P.S. Martin, D.W. Wildman: Bunch Coalescing and Bunch Rotation in the Fermilab Main Ring: Operational Experience and Comparison with Simulations. In: *Proc. EPAC88, Rome, Italy, 1988* (IOP, 1989) p. 785
12. V. Bharadwaj: private communication (1999)
13. J. Dey, I. Kourbanis, D. Wildman: Improvements in Bunch Coalescing in the Fermilab Main Ring. In: *Proc. Pac95, Dallas, TX, 1995* (IEEE, Piscataway 1995) p. 3312
14. M. Benedikt, A. Blas, J. Borburgh et al.: 'The PS Complex Produces the Nominal LHC Beam'. In: *Proc. EPAC, Vienna, Austria, 2000* (European Phys. Soc., Geneva 2000)
15. R. Garoby: Status of the Nominal Proton Beam for LHC in the PS. In: *Proc. of the Workshop on LEP-SPS Performance, Chamonix IX, 1999*
16. R. Garoby: Multiple Splitting in the PS: Results and Alternative Filling Schemes. In: *Proc. of Chamonix XI* CERN-SL-2001-003 DI
17. See for example A. Blas, R. Cappi, R.Garoby, S. Hancock, K. Schindl, J-L. Vallet: Beams in the CERN PS Complex After the rf Upgrades for LHC. In: *Proc. EPAC98, Stockholm, 1998* (IOP, Bristol 1999)

18. G. Jackson: Phase Space Tomography (PST) Monitor for Adjusting Bunch Rotation during Coalescing. FERMILAB-FN-469 (1987)
19. S. Hancock, P. Knaus, M. Lindroos: Tomographic Measurements of Longitudinal Phase Space Density. CERN/PS 98-030 (RF) (1998)
20. R. Garoby, S. Hancock, J.L. Vallet: Demonstration of Bunch Triple Splitting in the CERN PS. In: *Proc. of Eur. Part. Acc. Conf., Vienna, Austria, 2000* (European Phys. Soc., Geneva 2000)
21. R. Garoby: Status of the Nominal Proton Beam for LHC in the PS. In: *Proc. of the Workshop on LEP-SPS Performance, Chamonix IX, 1999.* CERN/PS 99-013 (RF)
22. H. Huang, M. Ball, B. Brabson et al.: Phys. Rev. E **48**, 4678–4688 (1993)
23. D. Li, M. Ball, B. Brabson, et al.: Phys. Rev. E **48**, 1638–1641 (1993)
24. D. Li, M. Ball, B. Brabson et al.: Nucl. Instr. Meth. A **364**, 205–223 (1995)
25. J.M. Byrd, W.H. Cheng, F. Zimmermann: Phys. Rev. E **57**, 4706–4712 (1998)
26. J.E. Griffin: New Method for Control of Longitudinal Emittance During Transition in Proton Synchrotrons. In: *Proc. of IEEE Part. Acc. Conf., Washington, DC, 1993*, pp. 408–410
27. C.M. Bhat, J. Dey, J. Griffin, I. Kourbanis, J. MacLachlan, M. Martens, K. Meisner, K.Y. Nig, J. Shan, D. Wildman: Operational Experience with Third Harmonic RF Cavity for Improved Beam Acceleation Through Transition in the Fermilab Main Ring. In: *Proc. of IEEE Part. Acc. Conf., Washington, DC, 1993*, pp. 405–407
28. J.M. Kats, W.T. Weng: Effects of the Second Harmonic Cavity on RF Capture and Transition Crossing. In: *15th Int. Conf. on High-Energy Acc., Hamburg, Germany, 1992* vol. 2 (World Scientific, 1992) pp. 1052–1054
29. J.M. Byrd, K. Baptiste, S. De Santis et al.: Nucl. Instr. Meth. A **439**, 15–25 (2000)
30. J.M. Byrd, M. Georgson: PRST-AB **4**, 030701 (2001)
31. S. Bartalucci, M. Migliorati, L. Palumbo, B. Spartaro, M. Zobov: A 3rd Harmonic Cavity for Daphne. In: *Proc. of 4th EPAC, London, England, 1994* (World Scientific, Singapore 1994)
32. M. Migliorati, L. Palumbo, M. Zobov: Nucl. Instr. Meth. A **354**, 215–223 (1995)
33. P. Bramham, A. Hofmann, P.B. Wilson: CERN-LEP-70/25 (1977)
34. P.B. Wilson: Rough Design of a Third Harmonic RF Cavity for LEP. CERN-LEP-70/60 (1978)
35. C. Bernardini, G.F. Corazza, G. Di Giugno et al.: Phys. Rev. Lett. **10**, 407–409 (1963)
36. A. Piwinski: The Touschek Effect in Strong Focusing Storage Rings. DESY 98–179 (1998)
37. A. Piwinski: Touschek Effect and Intrabeam Scattering. In: A. Chao, M. Tigner (eds.): *Handbook of Accelerator Physics and Engineering*, second edition (World Scientific, Singapore, 2002)
38. J.M. Byrd, S. De Santis, M. Georgsson et al.: Nucl. Instr. Meth. A **455**, 271–282 (2000)
39. K. Bane, J. Bowers, A. Chao et al.: 'High Intensity Single Bunch Instability Behavior in the New SLC Damping Ring Vacuum Chamber'. In: *Proc. IEEE Part.'Acc. Conf., Dallas, TX, 1995*, pp. 3109–3111
40. J.T. Seeman: Observations and Cures of Wakefield Effects in the SLC Linac. In: *Proc. 5th ICFA Adv. Beam Dyn. Wkshp on Effects of Errors in Accelerators, Their Diagnosis and Correction, Corpus Christi, TX, 1991*, pp. 339–346
41. J.T. Seeman, N. Merminga: Mutual Compensation of Wakefield and Chromatic Effects of Intense Linac Bunches. In: *Proc. 1990 Linac Conf., Albuquerque, NM, 1990*, pp. 387–389

42. G.A. Loew, J.W. Wang: Minimizing the Energy Spread within a Single Bunch by Shaping its Charge Distribution. SLAC/AP-025 (1984)
43. F.-J. Decker, R. Holtzapple, T. Raubenheimer: Overcompression, a Method to Shape the Longitudinal Bunch Distribution for a Reduced Energy Spread. In: *Proc. 17th Intl. Linear Acc. Conf., Tsukuba, Japan, 1994*, pp. 47–49
44. K.L.F. Bane, F.-J. Decker, J.T. Seeman, F. Zimmermann: Measurement of the Longitudinal Wakefield and the Bunch Shape in the SLAC Linac. In: *Proc. IEEE Part. Acc. Conf., Vancouver, Canada, 1997*
45. K.L.F. Bane, F.-J. Decker, F. Zimmermann: Obtaining the Bunch Shape in a Linac from Beam Spectrum Measurement. In: *Proc. IEEE Part. Acc. Conf., New York, NY, 1999*
46. J.E. Clendenin, R. H. Helm, R. K. Jobe, A. Kulikov, J.C. Sheppard: Energy Matching of 1.2 GeV Positron Beam to the SLC Damping Ring. In: *Proc. XIV Intl. Conf. on High Energy Accelerators, Tsukuba, Japan, 1989*
47. P. Wilson: High Energy Electron Linacs; Application to Storage Ring RF Systems and Linear Colliders, SLAC-PUB-2884 (1982)
48. F.-J. Decker, C. Adolphsen, R. Assmann et al.: 'Long-Range Wakefields and Split-Tune Lattice at the SLC'. In: *Proc. Linac96, Geneva, Switzerland, 1996*
49. C. Adolphsen, K. Bane, R. Jones et al.: 'Wakefield and Beam Centering Measurements of a Damped and Detuned X-Band Accelerator Structure'. In: *Proc. IEEE Part. Acc. Conf., New York, NY, 1999*
50. T. Kageyama, K. Akai, N. Akasaka et al.: 'The ARES Cavity for the KEK B Factory'. In: *Proc. Eur. Part. Acc. Conf., Sitges, Spain, 1996*
51. T. Raubenheimer: In: *Zeroth Order Design Report for the Next Linear Collider.* SLAC Report 474, (1995) pp. 227–228
52. F. Hinode, S. Kawabata, H. Matsumoto et al.: Accelerator Test Facility–Design and Study Report, KEK Internal Report 95-4 (1995)
53. S. Kashiwagi, H. Hayano, F. Hinode et al.: 'Preliminary Test of ±f Energy Compensation System'. In: *Proc. LINAC96, Geneva, 1996*
54. M. Sands: The Physics of Electron Storage Rings. SLAC-121 (1970)
55. R.D. Kohaupt, G.A. Voss: Ann. Rev. Nucl. and Part. Sci. **33**, 67 (1983)
56. I. Reichel: private communication (1997)
57. M.G. Minty, R. Brown, F.-J. Decker et al.: 'Using a Fast-Gated Camera for Measurements of Transverse Beam Distributions and Damping Times'. In: J.A. Hinkson, G. Stover (eds.): *Proc. Accelerator Instrumentation Workshop, Berkeley AIP Conf. Proc.* **281**, *1992*, p. 158
58. K. Oide: SAD Accelerator Modelling Code, unpublished
59. R. Akre, F.J. Decker, M.G. Minty et al.: 'RF Frequency Shift during Beam Storage in the SLC Damping Rings', In: *Proc. IEEE Part. Acc. Conf., New York, NY, 1999*
60. G. Hoffstätter: private communication (1999)
61. M.G. Minty, R. Akre, F.-J. Decker, J. Frisch, S. Kuroda, F. Zimmermann: Emittance Reduction via Dynamic RF Frequency Shift at the SLC Damping Rings. In: *Proc. 17th Intl. Conf. on High-Energy Acc., Dubna, Russia, 1998*
62. D. Schulte: Study of Electromagnetic and Hadronic Background in the Interaction Region of the TESLA Collider. Ph.D. thesis, University of Hamburg (1996)
63. G. Hoffstätter, F. Willeke: Electron Dynamics in the HERA Luminosity Upgrade Lattice of the Year 2000. In: *Proc. IEEE Part. Acc. Conf., New York, NY, 1999*

References for Chapter 9

1. G.H. Rees: Injection. In: *Proc. CERN Accelerator School, Gif-sur-Yvette, Paris, France, CERN 85-19, 1985* p. 331
2. G.H. Rees: 'Extraction'. In: *Proc. CERN Accelerator School, Gif-sur-Yvette, Paris, France, CERN 85-19, 1985* p. 346 (1985)
3. A. Piwinski: IEEE Tr. **NS-24**(3), 1364 (1977)
4. C. Gonzales, M. Morvillo, M. D'yachkov: Impedance Measurements on the LHC Dump Kicker Prototype. LHC Project Note 151 (1998)
5. M. Minty, R.H. Siemann: Nucl. Instr. Meth. A **376**, 301 (1996)
6. The PEP-II Design Group: PEP-II An Asymmetric B Factory. SLAC-418 (1993)
7. Ch. Carli, S. Maury, D. Möhl: Combined Longitudinal and Transverse Multiturn Injection in a Heavy Ion Accumulator. In: *Proc. of 1997 IEEE PAC, Vancouver, 1997* (IEEE, Piscataway 1997)
8. P. Baudrenghien, P. Collier: Double Batch Injection into LEP. In: *Proc. of EPAC 96, Sitges, 1996* (IOP, Bristol 1996) p. 415
9. E. Gianfelice, H. Schönauer: Strategies for Longitudinal Painting in the EHF Booster at Injection. In: *Proc. of EPAC 1988, Rome, 1988* (IOP, 1989)
10. G.H. Rees: Injection and Painting. In: *Proc. of the Summer Study on High Energy Physics in the 1990s, Snowmass, 1988*
11. G.I. Budker, G.I. Dimov. In: *Proc. Int. Conf. on High-Energy Accelerators, Dubna 1963, Conf. 114, 1963*, p. 1372
12. E. Crosbie, A. Gorka, E. Parker, C. Potts, L. Ratner: IEEE Tr. **NS-22**(3), 1056 (1975)
13. C.D. Moak: IEEE Tr. **NS-23**(2), 1126 (1976)
14. R.T. Free, J.S. Fraser, C.D.P. Levy: Laser Stripping of the Triumf H^- Beam. In: *Proc. of 1989 IEEE PAC, Chicago, 1989* (IEEE, Piscataway 1989) p. 414
15. T.K. Khoe, R.J. Lari. In: *Proc. Conf. on High-Energy Accelerators, Geneva, 1971*, p. 98
16. J.T. Seeman: Injection Issues of Electron-Positron Storage Rings. In: *Proc. of B factories: The State of the Art in Acclerators, Detectors and Physics, Stanford, April 6–10, 1992*
17. M.G. Minty, R. Brown, F.-J. Decker et al.: 'Using a Fast-Gated Camera for Measurements of Transverse Beam Distributions and Damping Times'. In: J.A. Hinkson, G. Stover (eds.): *Proc. Accelerator Instrumentation Workshop, Berkeley AIP Conf. Proc.* **281**, *1992*, p. 158
18. M.G. Minty, W.L. Spence: Injection Envelope Matching in Storage Rings. In: *Proc. of 1995 IEEE PAC, Dallas, 1995* (IEEE, Piscataway 1995) p. 536
19. M. Sands: A Beta Mismatch Parameter. SLAC internal report SLAC-AP-85 (1991)
20. T.O. Raubenheimer, L.Z. Rivkin, R.D. Ruth: Damping Ring Designs for a TeV Linear Collider. In: *Proc. of the DFP Summer Study Snowmass '88, 1988*

21. The NLC Design Group: Zeroth Order Design Report for the Next Linear Collider. SLAC-Report 474 (1996)
22. F. Hinode, S. Kawabata, H. Matsumoto et al.: Accelerator Test Facility–Design and Study Report. KEK Internal Report 95-4 (1995)
23. F. Bulos: Detailed Design of PEP II Injection Kickers. SLAC PEP-II/AP Note 5-92 (1992)
24. B.I. Grishanov, F.V. Podgorny, J. Rümmler, V.D. Shiltsev: Nucl. Instr. Meth. A **396** (1997) 28 NIM A 396 (1997) p. 28
25. V.D. Shiltsev: Nucl. Instr. Meth. A **374** (1996) 137
26. A.I. Drozhdin, N.V. Mokhov, M. Harrison: Study of Beam Losses during Fast Extraction of 800 GeV Protons from the Tevatron. Fermilab Internal Report FN-418 (1985)
27. A. Drozhdin, N. Mokhov, C. Johnstone, W. Wan, A. Garren: Scraping Beam Halo in $\mu^+\mu^-$ Colliders. In: *Proc. of the 4th International Conference on Physics Potential and Development of Muon Colliders, San Francisco, 1997*
28. G.E. Fischer: Iron Dominated Magnets. Lecture at the 1985 SLAC Summer Institute. SLAC-PUB-3726 (1985)
29. R. Cappi, M. Giovannozzi: Phys. Rev. Lett. **10**, 104801-1 (2002)
30. R. Cappi, M. Giovannozzi: Adiabatic Capture of Charged Particles in Islands of Phase Space: A New Method for Multi-turn Extraction. In: *Proc. EPAC 02, Paris, 2002* (European Phys. Soc., Geneva 2002) 1250
31. G. Acquistapace, J.L. Baldy, A.E. Ball et al.: The CERN Neutrino Beam to Gran Sasso (NGS): Conceptual Technical Design, ed. by K. Elsener. CERN 98-02 (1998)
32. E. Keil: Beam Separation in a Very Large Lepton Collider. CERN-SL-Note-2001-017 (AP)
33. The LEP2 Team: LEP Design Report – Vol. III: LEP2. CERN-AC/96-01 (LEP2) (1996)
34. A.A. Asseev, M.D. Bavishev, A.N. Vasilev et al.: Nucl. Instr. Meth. A **330**, 39 (1993)
35. K. Elsener, G. Fidecaro, M. Gyr et al.: Nucl. Instr. Meth. B **119**, 215 (1996)
36. K. Elsener, G. Fidecaro, M. Gyr et al.: 'What did we Learn from the Extraction Experiments with Bent Crystals at the CERN SPS'. In: *Proc. of EPAC 98, Stockholm, 1998* (IOP, Bristol 1999)
37. C.T. Murphy, R.A. Carrigan, D. Chen et al.: Nucl. Instr. Meth. B **119**, 231 (1996)
38. S. Pape Møller: Crystal Channeling or How to Build a '1000 TESLA Magnet'. Ninth John Adams Memorial Lecture. CERN 94-05 (1994)

References for Chapter 10

1. E.D. Courant: Bull. Am. Phys. Soc. **7**, 33 (1962); E.D. Courant, R.D. Ruth: The Acceleration of Polarized Protons in Circular Accelerators. BNL 51270 (1980)
2. A.W. Chao: Polarization of a Stored Electron Beam. In: *Proc. 1981 Fermilab Summer School, Batavia, 1981* vol. 1, p. 395
3. A.W. Chao: Nucl. Instr. Meth. A **180**, 29 (1981)
4. A.W. Chao, K. Yokoya: An Alternative Longitudinal Polarization Scheme for Tristan. KEK 81-7 (1981)
5. B.W. Montague: Phys. Rep. **113**, 1 (1984)
6. S.Y. Lee: *Spin Dynamics and Snakes in Synchrotrons* (World Scientific, Singapore, 1997)
7. D.P. Barber: Electron and Proton Spin Polarization in Storage Rings – an Introduction. In: *Proc. Advanced ICFA Beam Dynamics Workshop on Quantum Aspects in Beam Physics* ed. by P. Chen (World Scientific, Singapore, 1999)
8. G.H. Hoffstaetter: A Modern View of High Energy Polarized Proton Beams, Springer Tracts in Modern Physics, to be published
9. M. Vogt: Bounds on the Maximum Attainable Equilibrium Spin Polarization of Protons at High Energy in HERA. Ph.D. thesis, University of Hamburg, DESY-THESIS-2000-054 (2000)
10. L.H. Thomas: Philos. Mag. **3**, 1 (1927)
11. V. Bargman, L. Michel, V.L. Telegdi: Phys. Rev. Lett. **2**, 435 (1959)
12. J.D. Jackson: *Classical Electrodynamics* (Wiley and Sons, New York, 1975)
13. T. Roser: In *Handbook of Accelerator Physics and Engineering* ed. by A.W. Chao, M. Tigner, second edition (World Scientific, Singapore, Singapore, 2002)
14. T. Limberg, P. Emma, R. Rossmanith: The North Arc of the SLC as a Spin Rotator. In: *Proc. IEEE PAC 93, Washington, DC, 1993* (IEEE, Piscataway 1993) p. 429
15. P. Emma, T. Limberg, R. Rossmanith: Depolarization in the SLC Collider Arcs. In: *Proc. EPAC 94, London, 1994* (World Scientific, Singapore 1994)
16. T. Khoe, R.L. Kustom, R.L. Martin et al.: Part. Accel. **6**, 213 (1975)
17. E. Grorud, J.L. Laclare, G. Leleux et al.: 'Crossing of Depolarization Resonances at Saturne (Saclay)'. In: *Proc. High Energy Spin Physics* (Brookhaven, 1982)
18. F.Z. Khiari, P.R. Cameron, G.R. Court et al.: Phys. Rev. D **39**, 45 (1989)
19. H. Sato, D. Arakawa, S. Hiramatsu et al.: Nucl. Instr. Meth. A **272**, 617 (1988)
20. A.D. Krisch, S.R. Mane, R.S. Raymond, et al.: Phys. Rev. Lett. **63**, 1137 (1989)
21. J.E. Goodwin, H.O. Meyer, M.G. Minty et al.: Phys. Rev. Lett. **64**, 2779 (1990)
22. J.E. Goodwin: Ph.D. dissertation, Indiana University (1990)

23. H. Spinka, D. Underwood, L. Ahrens et al.: 'Commissioning and Future Plans for Polarized Protons in RHIC'. In: *Proc. PAC 2001, Chicago, Illinois, 2001* (IEEE, Piscataway 2001)
24. D.P. Barber, M. Boge, H.D. Bremer et al.: Nucl. Instr. Meth. A **338**, 166 (1994)
25. R. Rossmanith, R. Schmidt: Nucl. Instr. Meth. A **236**, 231 (1985)
26. D.P. Barber, G. Ripken: In: A. Chao, M. Tigner (eds.): *Handbook of Accelerator Physics and Engineering*, second edition (World Scientific, Singapore, 2002)
27. M. Böge, R. Brinkmann: Optimization of Electron Spin Polarization by Application of a Beam Based Alignment Technique in the HERA Electron Ring. In: *Proc. 1995 Conf. in Accelerator Alignment, Tsukuba, Japan, 1995*, p. 412
28. R. Brinkmann, M. Boege: Beam Based Alignment and Polarization Optimization in the HERA Electron Ring. In: *Proc. EPAC 94, London, 1994* (World Scientific, Singapore 1994)
29. E. Gianfelice: Measurement of Beam Polarization and a 72°/72° Spin Matched Optics. In: *HERA Accelerator Studies 2000* ed. by G. Hoffstaetter. DESY HERA 00-07 (2000)
30. R. Assmann, A. Blondel, B. Dehning et al.: 'Deterministic Harmonic Spin Matching in LEP'. In: *Proc. EPAC 94, London, 1994* (World Scientific, Singapore 1994) p. 932
31. A.A. Sokolov, I.M. Ternov: Sov. Phys. Dokl. **8**, 1203 (1964)
32. M. Froissart, R. Stora: Nucl. Instr. Meth. **7**, 297 (1960)
33. Y. Cho, R.L. Martin, E.F. Parker et al. In: *Proc. of High Energy Spin Symposium, Argonne, 1976* AIP Conf. Proc. **35**(12), 396 (1976)
34. D.D. Caussyn, Ya.S. Derbenev, T.J.P. Ellison et al.: Phys. Rev. Lett. **73**, 2857–2859 (1994)
35. S. Tepikian, S.Y. Lee, E.D. Courant: Part. Accel. **17**, 1 (1986)
36. Ya.S. Derbenev, A.M. Kondratenko: In: *Proc. of the 10th Intl. Conf. on High Energy Accel., Protvino, 1977* v. 2, p. 70
37. Ya.S. Derbenev: Part. Accel. **8**, 115 (1978)
38. K. Steffen: Note on the Choice of Siberian Snake Configurations in Proton Rings. DESY HERA 85-24 (1985)
39. K. Steffen: Design of an Orthogonal Snake Pair for Proton Rings in the Few Gev Energy Range. DESY HERA 87-11 (1987)
40. K. Steffen: How to Avoid Imperfection Spin Resonances in a Protron Ring with Snakes. DESY 89-024 (1989)
41. E.D. Courant: Hybrid Helical Snakes and Rotators for RHIC. BNL-61920, BNL-AD-133 (1995)
42. E. Courant, W. Fischer, A. Luccio et al.: 'The Use of Helical Dipole Magnets in the RHIC Project'. In: *Proc. Spin 96, Amsterdam, 1996*
43. D.P. Barber, M. Boge, H.D. Bremer et al.: Phys. Lett. **B343**, 436 (1995)
44. B.W. Montague: CERN/ISR-LTD/76-2 (1976)
45. J. Buon: Prescriptions for 90-Percent Spin Rotators in Electron Rings. LAL/RT/81-08 (1981)
46. S.Y. Lee, S. Tepikian: Phys. Rev. Lett. **56**, 1635 (1986)
47. R.A. Phelps, V.A. Anferov, B.B. Blinov et al.: Phys. Rev. Lett. **78**, 2772 (1997)
48. T. Roser. In: *Proc. 1990 Partial Siberian Snake Workshop, Brookhaven, NY, 1990*
49. B. Blinov, C.M. Chu, E.D. Courant et al.: Phys. Rev. Lett. **73**, 1621 (1994)
50. H. Huang, L. Ahrens, J.G. Alessi, et al.: Phys. Rev. Lett. **73**, 2982 (1994)
51. M. Bai, L. Ahrens, J.G. Alessi et al.: Phys. Rev. Lett. **80**, 21, 4673 (1998)
52. D.A. Crandell, V.A. Anferov, B.B. Blinov et al.: Phys. Rev. Lett. **77**, 1763, 4673 (1996)
53. S.Y. Lee: Prospects for Polarization at RHIC and SSC. In: AIP Conf. Proc. **224**, (1991) p. 35

References for Chapter 11

1. D. Möhl: Physica Scripta **T22**, 21 (1998)
2. D.V. Pestrikov: Beam Cooling. In: *Beam Measurements. Proc. US-CERN-Japan-Russian School on Beam Measurement, Montreux, Switzerland, 1998*, ed. by S.I. Kurokawa, S.Y. Lee, E. Perevedentsev, S. Turner (World Scientific, 2000)
3. G.I. Budker: Atomnaja Energia **22**, 346 (1967)
4. J. Bosser: Electron Cooling. In: *Proc. 4th Advanced Accelerator Physics Course, CERN Accelerator School, CERN 92-01, 1992*
5. S.P. Møller: Cooling Techniques. In: *Proc. CERN Acc. School, Jyvaeskylae 1992, CERN-94-01, 1994*
6. Ya.S. Derbenev, A.N. Skrinsky: In: *Proc. of the 10th Internat. Conf. on High Energy Accel., Serpukhov, 1977* v. 1, p. 516
7. A. Chao: *Physics of Collective Beam Instabilities in High Energy Accelerators* (Wiley, New York 1993)
8. J. Bosser, C. Carli, M. Chanel, L. Marie, D. Möhl, G. Tranquille: Nucl. Instr. Meth. A **441**, 60 (2000)
9. J. Bosser, C. Carli, M. Chanel, C. Hill, A. Lombardi, R. Maccaferri, S. Maury, D. Möhl, G. Molinari, S. Rossi, E. Tanke, G. Tranquille, M. Vretenar: Part. Acc. **63**, 171 (1999)
10. G. Tranquille: Optimum Parameters for Electron Cooling. Presented at *Beam Cooling and Related Topics Workshop, 13–18 May, 2001, Bad Honnef.* CERN/PS 2001-056 (BD) (2001)
11. C. Rubbia: In: *Proc. of the Workshop on Producing High Luminosity High Energy Proton-Antiproton Collisions, Berkeley, LBL-7574, 1978*, p. 98
12. S.Y. Lee, P. Colestock, K.Y. Ng: Electron Cooling in High Energy Colliders. FERMILAB-FN-657 (1997)
13. I. Ben-Zvi, J. Kewisch, J. Murphy, S. Peggs: Nucl. Instr. Meth. A **463**, 94 (2001)
14. I. Koop, V. Parkhomchuk, V. Reva et al.: In: *Proc. IEEE PAC 2001, Chicago, 2001* (IEEE, Piscataway 2001)
15. D. Möhl: Nucl. Instr. Meth. A **391**, 164 (1997)
16. D. Möhl: Cooling of Particle Beams. In: *Advances of Accelerator Physics and Technologies*, Adv. Ser. Direct. High Energy Phys. **12**, ed. by H. Schopper (World Scientific, Singapore, 1993) p. 359
17. F. Sacherer: Stochastic Cooling Theory. CERN-IST-TH/78-11. Talk given at the ISR seminar on 24th April 1978
18. H. Herr, D. Möhl: Bunched Beam Stochastic Cooling. In: *Proc. Workshop on Cooling High Energy Beams, Madison 1978*, CERN-PS-DI Note 79-3 and CERN-EP Note 79-34
19. A.A. Mikhailichenko, M.S. Zolotorev: Phys. Rev. Lett. **71**, 4146 (1993)
20. M.S. Zolotorev, A.A. Zholents: Phys. Rev. E **50**, 3087 (1994)

21. D. Boussard: Evaluation of Transverse Emittance Growth from Damper Noise in the Collider. CERN Internal Report SL/Note 92-79 (RFS)
22. D. Möhl: Stochastic Cooling. In: *Proc. CERN Accelerator School, Oxford, CERN 87-03, 1987*
23. P.J. Channel: J. Appl. Phys. **52**, 3791 (1981)
24. S. Schröder, R. Klein, N. Boos et al.: Phys. Rev. Lett. **64**, 2901 (1990)
25. J.S. Hangst, J.S. Nielsen, O. Poulsen et al.: Phys. Rev. Lett. **67**, 1238 (1995)
26. J.S. Hangst, J.S. Nielsen, O. Poulsen, P. Shi, J.P. Schiffer: Phys. Rev. Lett. **74**, 4432 (1995)
27. T. Kihara, H. Okamoto, Y. Iwashita, K. Oide, G. Lamanna, J. Wei: Study of Three-Dimensional Laser Cooling Method Based on Resonant Linear Coupling. KEK Preprint 98-158 (1998)
28. H. Okamoto, A.M. Sessler, D. Möhl: Phys. Rev. Lett. **72**, 3977 (1994)
29. H. Okamoto: Phys. Rev. E **50**, 4982 (1994)
30. V. Telnov: Laser Cooling of Electron Beams for Linear Colliders. SLAC-PUB-7337 (1996)
31. Z. Huang, R.D. Ruth: Phys. Rev. Lett. **80**, 976 (1998)
32. K.-J. Kim, S. Chattopadhyay, C.V. Shank: Nucl. Instr. Meth. in Phys. Res. A **341**, 351 (1994)
33. A. Hofmann: SSRL ACD-Note 38 (1986)
34. J. Urakawa: private communication (2001)
35. A. Tsunemi, A. Endo, I. Pogorelsky et al.: 'Ultra-Bright X-Ray Generation using Inverse Compton Scattering of Picosecond CO_2 Laser Pulses'. In: *Proc. IEEE PAC 99, New York, 1999* (IEEE, Piscataway 1999) p. 2552
36. E.N. Dementev, N.S. Dikanski, A.S. Medvedko, V.V. Parkhomchuk, D.V. Pestrikov: Sov. Phys. Tech. Phys. **25**, 1001 (1980)
37. N.S. Dikanski, D.V. Pestrikov. In: *Proc. of Workshop on Electron Cooling and Related Applications, Karlsruhe 1984*, ed. by H. Poth, KfK 3846 (1984)
38. A. Rahman, J.P. Schiffer: Phys. Rev. Lett. **57**(9), 1133 (1986)
39. A. Rahman, J.P. Schiffer: Z. Phys. A **331**, 71 (1988)
40. J. Wei, X.-P. Li, A.M. Sessler: Phys. Rev. Lett. **73**, 3089 (1994)
41. J. Wei, H. Okamoto, A.M. Sessler: Phys. Rev. Lett. **80**, 2660 (1998)
42. X.-P. Li, A.M. Sessler, J. Wei: Crystalline Beam in a Storage Ring: How Long Can it Last? In: *Proc. EPAC 94, London, 1994* (World Scientific, Singapore 1994) p. 1379
43. R.W. Hasse, M. Steck: Ordered Ion Beams. In: *Proc. EPAC 2000, Vienna, Austria, 1999* (European Phys. Soc., Geneva 2000) p. 274
44. G. Stupakov: Echo Effect in Hadron Colliders. SSCL-579 (1992)
45. This picture arose in a discussion with K.-J. Kim during the 1999 US accelerator school in Argonne
46. L.K. Spentzouris, P. Colestock: Coherent Nonlinear Longitudinal Phenomena in Unbunched Sycnhrotron Beams. In: *Proc. IEEE PAC 97, Vancouver, 1997* (IEEE, Piscataway 1998)
47. O.S. Brüning: On the Possibility of Measuring Longitudinal Echos in the SPS. CERN SL/95-83 (1995)
48. I.V. Agapov, G.H. Hoffstaetter, E. Vogel: Bunched Beam Echoes in the HERA Proton Ring. In: *Proc. EPAC 2002, Paris, 2002* (European Phys. Soc., Geneva 2002)
49. G. Arduini, F. Ruggiero, F. Zimmermann. M-P. Zorzano: Transverse Beam Echo Measurements on a Single Proton Bunch at the SPS. CERN SL-Note-2000-048 MD (2000)
50. C. Ankenbrandt, M. Atac, B. Autin et al.: Phys. Rev. S.T. Accel. Beams **2**, 081001 (1999)

51. R. Palmer, A. Tollestrup, A. Sessler et al. (The $\mu^+\mu^-$ Collider Collaboration): '$\mu^+\mu^-$ Collider. A Feasibility Study'. Submitted to the APS Summer Study, Snowmass 1996, on 'New Directions for High-Energy Physics'. BNL-52503 (1996)
52. The MUCOOL Collaboration (C. Ankenbrandt et al.): Ionization Cooling Research and Development Program for a High Luminosity Muon Collider. FERMILAB-P-0904 (1998)
53. D. Neuffer: Part. Acc. **14**, 75 (1983)
54. R.C. Fernow, J.C. Gallardo: Phys. Rev. E **52**, 1039 (1995)
55. M.J. Syphers, D.A. Edwards: *An Introduction to the Physics of High Energy Accelerators* (Wiley, New York 1993)

References for Chapter 12

1. R. Assmann, A. Chao: Dispersion in the Presence of Strong Transverse Wakefields. In: *Proc. PAC 97, Vancouver, 1997* (IEEE, Piscataway 1998) p. 1523

Index

www.ingramcontent.com/pod-product-compliance
Lightning Source LLC
LaVergne TN
LVHW082114200926
848484LV00008B/1947
9781013270499